Linear Differential Equations in the Complex Domain: Problems of Analytic Continuation

Translations of
MATHEMATICAL MONOGRAPHS

Volume 82

Linear Differential Equations in the Complex Domain: Problems of Analytic Continuation

Yasutaka Sibuya

American Mathematical Society
Providence, Rhode Island

複素領域における線型常微分方程式

解析接続の問題

Translated from the Japanese by Yasutaka Sibuya

2000 *Mathematics Subject Classification*. Primary 34–XX, 32D20, 32E10, 34E05, 30E15.

ABSTRACT. The main part of this book is an almost faithful translation of the book originally written in Japanese in 1976. In this book, focusing attention on intrinsic aspects as much as possible, we shall explain some problems of linear ordinary differential equations in complex domains. Since the original book was published, some new ideas have been introduced and rapidly developed. By adding five appendices to the translation, those new ideas are briefly explained in this book. Over 100 literatures are added to the original references.

Bibliography: 149 titles. 10 figures.

Library of Congress Cataloging-in-Publication Data

Sibuya, Yasutaka, 1930–

Linear differential equations in the complex domain: problems of analytic continuation / Yasutaka Sibuya.

p. cm. — (Translations of mathematical monographs; v. 82)

Includes bibliographical references (p.) and index.

ISBN 0-8218-4535-7 (alk. paper)

1. Differential equations. 2. Functions of several complex variables. 3. Analytic continuation. I. Title. II. Series.

QA372.S473 1990 90-825

515′.354—dc20 CIP

AMS softcover ISBN: 978-0-8218-4676-6

10 9 8 7 6 5 4 3 2 1 13 12 11 10 09 08

To My Mother

Table of Contents

Translator's Preface

Charles Darwin claims that the ape and the human being came from the same origin. The reader may or may not believe him. However, the author and the translator certainly came from the same origin. But they are not exactly the same person. They live in two different environments. Since this book was published, many research papers and books concerning similar subjects have appeared. Thus a new environment was created. So, the translator has tried to provide the reader a guided tour through this new environment by adding §§6.8–6.11 and five appendices. This is like sightseeing in the usual sense. If the reader wishes to discover more, he must go back there himself and look around more extensively. The translation is as faithful as the translator could achieve. Also, wherever necessary and suitable, he inserted remarks in [ ] to help the reader.

The main part of this translation was done at the University of Southern California in the academic year 1988/89 during the translator's sabbatical leave from the University of Minnesota. Also the translator was supported partially by a grant from the National Science Foundation. The translator wishes to express his appreciation to Professor William A. Harris, Jr. (U.S.C.), Mr. Hiroshi Mizuno (Kinokuniya), Ms. Mary C. Lane (A.M.S.), and Ms. Susan Connell (A.M.S.) for their kind help in the publication of this translation.

February, 1990, at Roseville Yasutaka Sibuya

Preface to the Japanese Edition

Research on differential equations is usually oriented toward obtaining explicit results, as it is motivated by applications. In this way many clever ideas have been found one after another. However, when a really difficult problem arises, we must dig out something intrinsic. As the study of topology, algebraic geometry, and functions of several complex variables has advanced, many methods which are useful in such fields have been introduced into the territory of differential equations. In this book, focusing attention on intrinsic aspects as much as possible, we shall explain some problems of linear ordinary differential equations in complex domains. The Riemann problem was solved quite naturally as the study of functions of several complex variables advanced. This was the most remarkable achievement since the beginning of this century. In this book, we shall first explain this problem. Also, we shall try to treat the generalized Riemann problem (or the Birkhoff problem) in a similar manner. Further, we shall explain the characterization of regular singular points due to W. Jurkat and D. A. Lutz. This result is recently attracting attention, because it is closely related to singularities of hypersurfaces.

The author of this book has never given a lecture of this kind. Professor Tosihusa Kimura of the University of Tokyo visited the University of Minnesota during the academic year 1971/72. He gave a series of lectures on algebraic differential equations, hypergeometric functions of several complex variables, and the Riemann problem. Also, during his stay, the two of us discussed various problems of mathematics, sometimes dangling fishing lines from a boat on a lake in Minnesota. Since then it has been four years. The author started writing this book as a term paper for the course given by Professor Kimura. The author tried his best. But this is the first draft without any extensive revisions. So he expects that the reader will find some pages hard to read and other pages containing mistakes. In particular, wherever the author's understanding of the material is shallow and uncertain, the reader will sense it immediately by detecting his somewhat vague way of writing. It is left to the reader to determine what grade Professor Kimura will give to this term paper. In the near future, the author will publish the contents of the last chapter [Chapter 6] as a research paper, because it contains new results [cf. Y. Sibuya [134]].

The author would like to express his appreciation to Professor Tosihusa Kimura (Univ. of Tokyo), Professor Seizo Ito (Univ. of Tokyo), Mr. Kenichi Uzuoka (Kinokuniya), and Mr. Kazumasa Yokota (Kinokuniya) for their kind help in the publication of this book. It should also be clearly noted that this work was partially supported by grant NSF-GP 38955 from the National Science Foundation.

August, 1976, at Roseville Yasutaka Sibuya

CHAPTER I

Structures of Analytic Continuation

§1.0. Preliminaries

The main purpose of this chapter is to explain the method of analytic continuation in a way suitable to the study of linear ordinary differential equations. We shall explain everything in detail with proofs. [The translator omitted some proofs and left them to the reader as exercises.] The reader will find more general materials in a book [10] of N. Steenrod (§13: Bundles of a totally disconnected group, pp. 59–67). In §1.10, we shall introduce the concept of a system of connection matrices defined by a system of fundamental solutions of a differential equation. Systems of connection matrices are the main concern of the connection problem of linear ordinary differential equations. Much research concentrates effort on finding connection matrices for given differential equations. In this book, we shall treat more basic problems rather than technical problems.

As usual, a topology is introduced in the space $GL(n, \mathbf{C})$ of $n \times n$ invertible matrices with complex entries through some norm. However, in §§1.4–1.7, we shall utilize the discrete topology on $GL(n, \mathbf{C})$.

§1.1. Holomorphic functions

Let us denote by x and y the real and imaginary parts of a complex number z respectively, i.e., $z = x + iy$, where $i = \sqrt{-1}$. For an open set $\mathscr{D}$, we denote by $\mathscr{O}(\mathscr{D})$ the set of all complex-valued functions which are holomorphic on $\mathscr{D}$. In order that a complex-valued function $f(z)$ of z belong to $\mathscr{O}(\mathscr{D})$, the two conditions below are necessary and sufficient:

(I) $f(z)$ is continuously differentiable in x and y on $\mathscr{D}$,

(II) $f(z)$ satisfies the Cauchy–Riemann equation:

$$\frac{\partial f}{\partial \bar{z}} = 0 \qquad \left(\frac{\partial}{\partial \bar{z}} = \frac{1}{2}\frac{\partial}{\partial x} + \frac{i}{2}\frac{\partial}{\partial y}\right)$$

on $\mathscr{D}$.

We denote by $\Delta(z_0, r)$ the open disk $\{z; |z - z_0| < r\}$. Since conditions (I) and (II) are local, in order that $f(z)$ belong to $\mathscr{O}(\mathscr{D})$, the two conditions below are necessary and sufficient:

(A) $f(z)$ is defined at every point of $\mathscr{D}$,

(B) for each fixed point z_0 of $\mathscr{D}$, there exists a positive number $r(z_0)$ such that $f(z)$ belongs to $\mathscr{O}(\Delta(z_0, r(z_0)))$.

Let $\mathscr{D}_0$ be a bounded open set such that its closure $\overline{\mathscr{D}}_0$ is contained in an open set $\mathscr{D}$ and that the boundary $\partial\mathscr{D}_0$ of $\mathscr{D}_0$ consists of a finite number of smooth Jordan curves. Then, if $f(z)$ belongs to $\mathscr{O}(\mathscr{D})$, the Cauchy integral representation

$$f(z) = \frac{1}{2\pi i}\int_{\partial\mathscr{D}_0} \frac{f(\zeta)}{\zeta - z} d\zeta \tag{1.1.1}$$

holds at each point z of $\mathscr{D}_0$, where $\partial\mathscr{D}_0$ is oriented counterclockwise so that $\mathscr{D}_0$ is on the left side of $\partial\mathscr{D}_0$.

Fundamental properties of $\mathscr{O}(\mathscr{D})$ are derived from (I), (II) (or (A),(B)) and integral representation (1.1.1). For example:

(i) $f + g$ and fg belong to $\mathscr{O}(\mathscr{D})$ if f and g belong to $\mathscr{O}(\mathscr{D})$,

(ii) $\dfrac{df}{dz}$ belongs to $\mathscr{O}(\mathscr{D})$ if f belongs to $\mathscr{O}(\mathscr{D})$,

(iii) $\lim_{n\to+\infty} f_n$ belongs to $\mathscr{O}(\mathscr{D})$ if the f_n belongs to $\mathscr{O}(\mathscr{D})$ and converges uniformly on each compact subset of $\mathscr{D}$; furthermore, the $\dfrac{df_n}{dz}$ also converges to $\dfrac{d(\lim_{n\to+\infty} f_n)}{dz}$ uniformly on each compact subset of $\mathscr{D}$,

(iv) $\mathbf{C} \subset \mathbf{C}[z] \subset \mathscr{O}(\mathscr{D})$, where $\mathbf{C}$ is the field of complex numbers and $\mathbf{C}[z]$ is the ring of all polynomials in z with complex coefficients.

The following formulas are derived from integral representation (1.1.1):

$$f^{(n)}(z) = \frac{n!}{2\pi i}\int_{\partial\mathscr{D}_0} \frac{f(\zeta)}{(\zeta - z)^{n+1}} d\zeta \qquad (n = 0, 1, \cdots), \tag{1.1.2}$$

where $z \in \mathscr{D}_0$ and $f^{(n)} = d^n f/dx^n$.

§1.2. Power series

For two formal power series in $z - z_0$ with complex coefficients

$$p = \sum_{n\geq 0} a_n(z - z_0)^n \quad \text{and} \quad q = \sum_{n\geq 0} b_n(z - z_0)^n$$

we define $p + q$ and pq by

$$p+q = \sum_{n\geq 0} (a_n + b_n)(z - z_0)^n \quad \text{and} \quad pq = \sum_{n\geq 0}\left(\sum_{n\geq k\geq 0} a_{n-k}b_k\right)(z - z_0)^n$$

respectively. We also write $p = q$ when $a_n = b_n$ for all nonnegative integers n. The derivative $\frac{dp}{dz}$ of p with respect to z is defined by

$$\frac{dp}{dz} = \sum_{n \geq 0} (n+1) a_{n+1} (z - z_0)^n.$$

We denote by $\mathbf{C}[[z - z_0]]$ the set of all formal power series in $z - z_0$ with complex coefficients.

There are three possibilities for each p in $\mathbf{C}[[z - z_0]]$, i.e.,

(1) p is uniformly and absolutely convergent on every compact subset of the z-plane,
(2) there exists a positive number r such that p is uniformly and absolutely convergent on every compact subset of $\Delta(z_0, r)$ and that p is divergent for $|z - z_0| > r$,
(3) p is divergent at every point z except for z_0.

The radius of convergence $r(p)$ of p is defined to be ∞, r, and 0 according to these three cases.

We denote by $\mathbf{C}\{z - z_0\}$ the set of all convergent power series in $z - z_0$ with complex coefficients, i.e.,

$$\mathbf{C}\{z - z_0\} = \{p \in \mathbf{C}[[z - z_0]]; \quad r(p) > 0\}.$$

Note that $\mathbf{C}\{z - z_0\}$ has the following three properties:

(i) $p + q$ and pq belong to $\mathbf{C}\{z - z_0\}$ if p and q belong to $\mathbf{C}\{z - z_0\}$,
(ii) $\dfrac{dp}{dz}$ belongs to $\mathbf{C}\{z - z_0\}$ if p belongs to $\mathbf{C}\{z - z_0\}$,
(iii) $\mathbf{C} \subset \mathbf{C}[z] \subset \mathbf{C}\{z - z_0\}$.

We denote by $S(p)$ the sum of p in $\mathbf{C}\{z - z_0\}$. The sum $S(p)$ has the following properties:

(a) $S(p) \in \mathcal{O}(\Delta(z_0, r(p)))$,
(b) $S(p + q) = S(p) + S(q)$ and $S(pq) = S(p)S(q)$,
(c)

$$\frac{dS(p)}{dz} = S\left(\frac{dp}{dz}\right),$$

(d) $S(p) = S(q)$ if and only if $p = q$.

If a power series $p = \sum_{n \geq 0} a_n (z - z_0)^n$ belongs to $\mathbf{C}\{z - z_0\}$, then

$$\frac{d^n}{dz^n}(S(p))|_{z=z_0} = (n!) \; a_n \tag{1.2.1}$$

for all nonnegative integers n.

§1.3. Taylor series

The Taylor series of f in $\mathscr{O}(\mathscr{D})$ at a point z_0 in $\mathscr{D}$ is defined by

$$T(f, z_0) = \sum_{n\geq 0} \frac{1}{n!} f^{(n)}(z_0)(z - z_0)^n. \tag{1.3.1}$$

The Taylor series have the properties below:

(a′) $T(f, z_0) \in \mathbf{C}[[z - z_0]]$,

(b′) $T(f+g, z_0) = T(f, z_0)+T(g, z_0)$ and $T(fg, z_0) = T(f, z_0)T(g, z_0)$,

(c′)

$$\frac{d}{dz}T(f, z_0) = T\left(\frac{df}{dz}, z_0\right).$$

If $p \in \mathbf{C}\{z - z_0\}$ and $\mathscr{D} = \Delta(z_0, r(p))$, then (1.2.1) implies that

$$T(S(p), z_0) = p. \tag{1.3.2}$$

Let $d(z_0, \partial\mathscr{D})$ be the distance from z_0 to the boundary $\partial\mathscr{D}$ of $\mathscr{D}$. If $f \in \mathscr{O}(\mathscr{D})$ and $z_0 \in \mathscr{D}$, then we can derive the following three results from (1.1.2):

(α) $T(f, z_0) \in \mathbf{C}\{z - z_0\}$,

(β) $r(T(f, z_0)) \geq d(z_0, \partial\mathscr{D})$,

(γ) $S(T(f, z_0)) = f$ on $\Delta(z_0, d(z_0, \partial\mathscr{D}))$.

In the case when $\mathscr{D}$ and $\tilde{\mathscr{D}}$ are two open sets in the z-plane, if $f \in \mathscr{O}(\mathscr{D})$ and $g \in \mathscr{O}(\tilde{\mathscr{D}})$ and if $f = g$ on an open disk $\Delta(z_0, r)$ contained in $\mathscr{D} \cap \tilde{\mathscr{D}}$, then $f = g$ on the connected component containing z_0 of $\mathscr{D} \cup \tilde{\mathscr{D}}$. This result can be proved by the method of analytic continuation.

The content of §§1.1, 1.2, and 1.3 can be extended to the case of several complex variables as well as functions defined on a complex analytic manifold. Leaving this to the reader, we shall advance to the next subject.

§1.4. Structures of analytic continuation

Fix an open set $\mathscr{D}$ in the z-plane. A sequence $\{\Delta(z_j, r_j); j = 1, 2, \cdots\}$ of open disks is called a covering of $\mathscr{D}$ if

(a) $\Delta(z_j, r_j) \subset \mathscr{D}$,

(b) $\mathscr{D} = \bigcup_{j\geq 1} \Delta(z_j, r_j)$.

Condition (a) is equivalent to

(a′) $z_j \in \mathscr{D}$ and $0 < r_j \leq d(z_j, \partial\mathscr{D})$.

DEFINITION 1.4.1. For an open set $\mathscr{D}$ in the z-plane, **a system of connection matrices** on $\mathscr{D}$ is given by $\{\Delta(z_j, r_j), C_{jk}\}$ where $\{\Delta(z_j, r_j)\}$ is a covering of $\mathscr{D}$ and $\{C_{jk}\}$ is a set of $n \times n$ matrices such that

(I) $C_{jk} \in GL(n, \mathbf{C})$ is given whenever $\Delta(z_j, r_j) \cap \Delta(z_k, r_k) \neq \varnothing$, where $GL(n, \mathbf{C})$ is the set of all invertible $n \times n$ matrices with complex entries,

(II) $C_{jk}C_{kh} = C_{jh}$ if $\Delta(z_j, r_j) \cap \Delta(z_k, r_k) \cap \Delta(z_h, r_h) \neq \varnothing$.

EXAMPLE 1.4.2. Selecting $C_1, C_2, \cdots$ from $GL(n, \mathbf{C})$, set $C_{jk} = C_j^{-1}C_k$ if $\Delta(z_j, r_j) \cap \Delta(z_k, r_k) \neq \varnothing$. Then $\{\Delta(z_j, r_j), C_{jk}\}$ is a system of connection matrices on $\mathscr{D}$.

If we set $j = k = h$ in condition (II) of Definition 1.4.1, we have $C_{jj}C_{jj} = C_{jj}$. Hence

$$C_{jj} = I, \tag{1.4.1}$$

where I is the $n \times n$ identity matrix. Instead if we set $j = h$, we have $C_{jk}C_{kj} = C_{jj} = I$. Hence

$$C_{jk}^{-1} = C_{kj}. \tag{1.4.2}$$

DEFINITION 1.4.3. A system $\{\Delta(z_j, r_j), C_{jk}\}$ of connection matrices on $\mathscr{D}$ is said to be equivalent to another system $\{\Delta(\zeta_j, \rho_j), \Gamma_{jk}\}$ if there exists a system $\{G_{jk}\}$ of matrices in $GL(n, \mathbf{C})$ such that G_{jk} is given whenever $\Delta(z_j, r_j) \cap \Delta(\zeta_k, \rho_k) \neq \varnothing$ and that

(α) $C_{jk}G_{kh} = G_{jh}$ if $\Delta(z_j, r_j) \cap \Delta(z_k, r_k) \cap \Delta(\zeta_h, \rho_h) \neq \varnothing$,

(β) $G_{jk}\Gamma_{kh} = G_{jh}$ if $\Delta(z_j, r_j) \cap \Delta(\zeta_k, \rho_k) \cap \Delta(\zeta_h, \rho_h) \neq \varnothing$.

We denote by

$$\{\Delta(z_j, r_j), C_{jk}\} \sim \{\Delta(\zeta_j, \rho_j), \Gamma_{jk}\}$$

that $\{\Delta(z_j, r_j), C_{jk}\}$ is equivalent to $\{\Delta(\zeta_j, \rho_j), \Gamma_{jk}\}$.

EXAMPLE 1.4.4. For any two systems $\{C_j\}$ and $\{\Gamma_j\}$ of matrices selected from $GL(n, \mathbf{C})$, we have

$$\{\Delta(z_j, r_j), C_j^{-1}C_k\} \quad \sim \quad \{\Delta(\zeta_j, \rho_j), \Gamma_j^{-1}\Gamma_k\}.$$

In fact, $G_{jk} = C_j^{-1}\Gamma_k$ satisfy conditions (α) and (β) of Definition 1.4.3.

THEOREM 1.4.5. *Let A, B, and C be systems $\{\Delta(z_j, r_j), C_{jk}\}$, $\{\Delta(\zeta_j, \rho_j), \Gamma_{jk}\}$, and $\{\Delta(w_j, s_j), D_{jk}\}$ of connection matrices on $\mathscr{D}$, respectively. Then*

(i) $A \sim A$,

(ii) $B \sim A$ *if* $A \sim B$,

(iii) $A \sim C$ *if* $A \sim B$ *and* $B \sim C$

i.e., $\sim$ *is an equivalence relation.*

Proof. To prove (i), setting $\Gamma_{jk} = G_{jk} = C_{jk}$, compare condition (II) of Definition 1.4.1 and conditions (α) and (β) of Definition 1.4.3.

To prove (ii), assuming that $A \sim B$, we utilize the system $\{G_{jk}\}$ satisfying conditions (α) and (β) of Definition 1.4.3. In fact, if we set $\hat{G}_{jk} = G_{kj}^{-1}$, we can derive $\hat{G}_{hj}C_{jk} = \hat{G}_{hk}$ from (α) and $\Gamma_{kh}\hat{G}_{hj} = \hat{G}_{kj}$ from (β). Hence $B \sim A$.

It takes a little longer to prove (iii). Since we assume that $A \sim B$ and $B \sim C$, we can select, in addition to the system $\{G_{jk}\}$ of Definition 1.4.3, another system $\{H_{jk}\}$ of matrices from $GL(n, \mathbf{C})$ in such a way that H_{jk} is defined whenever $\Delta(\zeta_j, \rho_j) \cap \Delta(w_j, s_j) \neq \varnothing$ and that

$$\Gamma_{jk}H_{kh} = H_{jh} \quad \text{if } \Delta(\zeta_j, \rho_j) \cap \Delta(\zeta_k, \rho_k) \cap \Delta(w_h, s_h) \neq \varnothing \tag{1.4.3}$$

and

$$H_{jk}D_{kh} = H_{jh} \quad \text{if } \Delta(\zeta_j, \rho_j) \cap \Delta(w_k, s_k) \cap \Delta(w_h, s_h) \neq \varnothing. \tag{1.4.4}$$

Let us assume that $\Delta(z_j, r_j) \cap \Delta(w_k, s_k) \neq \varnothing$. Since $\{\Delta(\zeta_j, \rho_j)\}$ is a covering of $\mathscr{D}$, we can choose $\Delta(\zeta_p, \rho_p)$ so that

$$\Delta(z_j, r_j) \cap \Delta(\zeta_p, \rho_p) \cap \Delta(w_k, s_k) \neq \varnothing. \tag{1.4.5}$$

We claim that $G_{jp}H_{pk}$ is independent of the choice of such a p. To verify this, we assume further that

$$\Delta(z_j, r_j) \cap \Delta(\zeta_q, \rho_q) \cap \Delta(w_k, s_k) \neq \varnothing \tag{1.4.6}$$

for some q. We shall show that

$$G_{jq}H_{qk} = G_{jp}H_{pk}. \tag{1.4.7}$$

Let us consider first the case when

$$\Delta(z_j, r_j) \cap \Delta(\zeta_p, \rho_p) \cap \Delta(\zeta_q, \rho_q) \cap \Delta(w_k, s_k) \neq \varnothing. \tag{1.4.8}$$

It follows from condition (β) of Definition 1.4.3 and (1.4.3) that

$$G_{jp}\Gamma_{pq} = G_{jq}, \quad \Gamma_{qp}H_{pk} = H_{qk}.$$

Since $\Gamma_{pq} = \Gamma_{qp}^{-1}$, we have $G_{jp}^{-1}G_{jq} = H_{pk}H_{qk}^{-1}$, and hence (1.4.7) is verified. If condition (1.4.8) is not satisfied, let us fix a point z_0 in set (1.4.5) and a point $\tilde{z}_0$ in set (1.4.6), and join these two points by a line segment ℓ. Being a compact subset of $\Delta(z_j, r_j) \cap \Delta(w_k, s_k)$, the segment ℓ is covered by a finite number of disks $\Delta(\zeta_h, \rho_h)$. Since ℓ is connected, we can prove (1.4.7) by repeating the process given above for a finite number of times.

Now, set $K_{jk} = G_{jp}H_{pk}$ for (j,k) such that $\Delta(z_j,r_j) \cap \Delta(w_k,s_k) \neq \emptyset$, where p is selected by condition (1.4.5). We shall derive $A \sim C$ by using the system $\{K_{jk}\}$. Let us assume first that

$$\Delta(z_j,r_j) \cap \Delta(z_k,r_k) \cap \Delta(w_h,s_h) \neq \emptyset.$$

Then, if we choose p so that

$$\Delta(z_j,r_j) \cap \Delta(z_k,r_k) \cap \Delta(w_h,s_h) \cap \Delta(\zeta_p,\rho_p) \neq \emptyset,$$

we have

$$C_{jk}K_{kh} = C_{jk}G_{kp}H_{ph} = G_{jp}H_{ph} = K_{jh}.$$

Similarly we have

$$K_{jk}D_{kh} = G_{jp}H_{pk}D_{kh} = G_{jp}H_{ph} = K_{jh}$$

if $\Delta(z_j,r_j) \cap \Delta(w_k,s_k) \cap \Delta(w_h,s_h) \neq \emptyset$. □

DEFINITION 1.4.6. Classifying all systems of connection matrices on $\mathscr{D}$ by the equivalence relation $\sim$, we call each equivalence class **a structure of analytic continuation** on $\mathscr{D}$.

In the theory of fibre bundles, a system of connection matrices is called *a system of coordinate transformations on* $\mathscr{D}$ *with values in* $GL(n,\mathbf{C})$ (cf. N. Steenrod [10], § 3.1, p. 14).

THEOREM 1.4.7. *In order that two systems* $\{\Delta(z_j,r_j), C_{jk}\}$ *and* $\{\Delta(z_j,r_j), \Gamma_{jk}\}$ *of connection matrices be equivalent, it is necessary and sufficient that we choose a system* $\{C_j\}$ *of matrices from* $GL(n,\mathbf{C})$ *so that*

$$C_{jk}C_k = C_j\Gamma_{jk} \quad \textit{if} \quad \Delta(z_j,r_j) \cap \Delta(z_k,r_k) \neq \emptyset. \tag{1.4.9}$$

PROOF. Let us first assume that

$$\{\Delta(z_j,r_j), C_{jk}\} \sim \{\Delta(z_j,r_j), \Gamma_{jk}\}. \tag{1.4.10}$$

According to Definition 1.4.3, we can choose a matrix G_{jk} from $GL(n,\mathbf{C})$ for each (j,k) satisfying $\Delta(z_j,r_j) \cap \Delta(z_k,r_k) \neq \emptyset$ so that

$$C_{jk}G_{kh} = G_{jh}, \quad G_{jk}\Gamma_{kh} = G_{jh}$$

if

$$\Delta(z_j,r_j) \cap \Delta(z_k,r_k) \cap \Delta(z_h,r_h) \neq \emptyset. \tag{1.4.11}$$

In particular, we have $C_{jk}C_k = C_j\Gamma_{jk} = G_{jk}$ if we set $C_j = G_{jj}$. Thus we obtain (1.4.9).

Next, let us assume (1.4.9). Then, if we set $G_{jk} = C_{jk}C_k = C_j\Gamma_{jk}$, we have

$$C_{jk}G_{kh} = C_{jk}C_{kh}C_h = C_{jh}C_h = G_{jh}$$

and

$$G_{jk}\Gamma_{kh} = C_j\Gamma_{jk}\Gamma_{kh} = C_j\Gamma_{jh} = G_{jh}.$$

Hence (1.4.10) is derived. □

The purpose of the materials of this section and §§ 1.5–1.7 will be made clear in §§ 1.9–1.11. The concept of systems of connection matrices can be defined not only on an open set $\mathscr{D}$ in the z-plane, but also on a higher dimensional space as well as on a general topological space.

§1.5. Connection matrices along a curve

So far matrices C_{jk} have been given only for those (j,k) such that $\Delta(z_j,r_j)\cap\Delta(z_k,r_k)\neq\varnothing$. In this section we shall explain how to remove this restriction. We consider the case when a continuous curve

$$\ell : z = z(t) \in \mathscr{D} \qquad (0 \leq t \leq 1) \tag{1.5.1}$$

is given in $\mathscr{D}$ by a continuous function $z(t)$ of a real variable t which is defined on the interval $0 \leq t \leq 1$ and whose values are contained in $\mathscr{D}$.

DEFINITION 1.5.1. An ordered set $\{U_1, \cdots, U_N\}$ of a finite number of disks is called a covering of ℓ selected (in the order) from a covering $\{\Delta(z_j,r_j)\}$ of $\mathscr{D}$ if the conditions below are satisfied:

(a) $U_\nu = \Delta(z_{j_\nu}, r_{j_\nu}) \in \{\Delta(z_j,r_j)\} \qquad (\nu = 1,\cdots,N)$,

(b) we can select N t-intervals $I_\nu = \{t;\ t_\nu \leq t \leq s_\nu\}$ $(\nu = 1,\cdots,N)$ in the following way:

$$t_1 = 0, \quad s_N = 1, \tag{1.5.2}$$

$$t_\nu < t_{\nu+1}, \quad s_\nu < s_{\nu+1} \qquad (\nu = 1,\cdots,N-1), \tag{1.5.3}$$

$$t_\nu < s_{\nu-1} < t_{\nu+1} < s_\nu \qquad (\nu = 2,\cdots,N-1), \tag{1.5.4}$$

$$z(t) \in U_\nu \quad (t \in I_\nu) \qquad (\nu = 1,\cdots,N). \tag{1.5.5}$$

Let us look at condition (b) more closely. In case $N = 1$,

$$I_1 = J = \{t;\ 0 \leq t \leq 1\}, \quad \ell \subset U_1,$$

and hence we do not need (1.5.3) and (1.5.4). If $N = 2$, condition (1.5.4) becomes $t_2 < s_1$, and (1.5.3) and (1.5.4) can be combined into a condition $t_1 < t_2 < s_1 < s_2$. Two intervals I_1, I_2 and two disks U_1, U_2 are related in the following way:

$$J = I_1 \cup I_2, \quad \ell \subset U_1 \cup U_2,$$
$$z(t) \in U_\nu \ \text{ if } \ t \in I_\nu,$$
$$I_1 \cap I_2 \neq \varnothing, \quad U_1 \cap U_2 \neq \varnothing.$$

In case $N \geq 3$, we have

(i) $J = I_1 \cup I_2 \cup \cdots \cup I_N$,
(ii) I_ν does not overlap with any intervals I_j other than $I_{\nu-1}$ and $I_{\nu+1}$,
(iii) $z(t) \in U_\nu$ if $t \in I_\nu$; in particular, $z(0) \in U_1$, $z(1) \in U_N$
(iv) $U_\nu \cap U_{\nu+1} \neq \varnothing$ $(\nu = 1, \cdots, N-1)$,
(v) $\ell \subset U_1 \cup U_2 \cup \cdots \cup U_N$.

For a system $\{\Delta(z_j, r_j), C_{jk}\}$ of connection matrices on $\mathscr{D}$ and a covering $\{U_1, \cdots, U_N\}$ of ℓ selected from the covering $\{\Delta(z_j, r_j)\}$ of $\mathscr{D}$, we set

$$c(\ell; U_1, \cdots, U_N) = \begin{cases} I & \text{if } N = 1, \\ C_{j_1 j_2} & \text{if } N = 2, \\ C_{j_1 j_2} C_{j_2 j_3} \cdots C_{j_{N-1} j_N} & \text{if } N \geq 3, \end{cases} \tag{1.5.6}$$

where $U_\nu = \Delta(z_{j_\nu}, r_{j_\nu})$ and I is the $n \times n$ identity matrix.

LEMMA 1.5.2. *Let $\{U_1, \cdots, U_N\}$ and $\{V_1, \cdots, V_M\}$ be two coverings of ℓ selected from a covering $\{\Delta(z_j, r_j)\}$, and let $U_\nu = \Delta(z_{j_\nu}, r_{j_\nu})$ and $V_\mu = \Delta(z_{h_\mu}, r_{h_\mu})$. Then, for a system $\{\Delta(z_j, r_j), C_{jk}\}$ of connection matrices on $\mathscr{D}$, we have*

$$c(\ell;\ U_1, \cdots, U_N) C_{j_N h_M} = C_{j_1 h_1}\ c(\ell;\ V_1, \cdots, V_M). \tag{1.5.7}$$

PROOF. This lemma is fundamental and its proof is not simple. Theorem 1.5.3 will show clearly the significance of this lemma.

There exist N intervals $I_1, \cdots, I_N$ satisfying condition (b) of Definition 1.5.1, since $\{U_1, \cdots, U_N\}$ is a covering of ℓ selected from $\{\Delta(z_j, r_j)\}$. For each real number τ of the interval $J = \{t; 0 \leq t \leq 1\}$, we define a continuous curve $\ell(\tau)$ in $\mathscr{D}$ by

$$\ell(\tau): \quad z = z(t) \qquad (0 \leq t \leq \tau). \tag{1.5.8}$$

If $\tau \in I_\nu$, then $\{U_1, \cdots, U_\nu\}$ is a covering of $\ell(\tau)$ selected from $\{\Delta(z_j, r_j)\}$. Hence $c(\ell(\tau); U_1, \cdots, U_\nu)$ is defined and independent of τ as long as $\tau \in I_\nu$. On the other hand, since $\{V_1, \cdots, V_M\}$ is also a covering of ℓ selected from $\{\Delta(z_j, r_j)\}$, we can choose an integer $\mu(\tau)$ and a positive number $\varepsilon(\tau)$ for $\tau \in J$ so that

(1) $1 \leq \mu(\tau) \leq M$, $\mu(1) = M$, and $\mu(\tau_1) \leq \mu(\tau_2)$ if $\tau_1 < \tau_2$,
(2) $\mu(\sigma) = \mu(\tau)$ if $\tau \leq \sigma \leq \tau + \varepsilon(\tau)$,
(3) $\mu(\sigma) \geq \mu(\tau) - 1$ if $\tau - \varepsilon(\tau) \leq \sigma \leq \tau$,
(4) $\{V_1, \cdots, V_{\mu(\tau)}\}$ is a covering of $\ell(\sigma)$ selected from $\{\Delta(z_j, r_j)\}$ if $|\sigma - \tau| \leq \varepsilon(\tau)$.

We leave the precise choice of $\mu(\tau)$ and $\varepsilon(\tau)$ to the reader. Anyway, if we choose $\mu(\tau)$ and $\varepsilon(\tau)$ in this manner, the matrix $c(\ell(\sigma); V_1, \cdots, V_{\mu(\tau)})$ is well defined for every real number τ in the interval J and independent of σ as long as the inequality $|\sigma - \tau| \leq \varepsilon(\tau)$ is satisfied.

We shall now prove that the relation

$$c(\ell(\tau); U_1, \cdots, U_\nu)C_{j_\nu h_{\mu(\tau)}} = C_{j_1 h_1} c(\ell(\tau); V_1, \cdots, V_{\mu(\tau)}) \tag{1.5.9}$$

holds for every real τ in the interval I_ν. In particular, if we let $\tau = 1$ and $\nu = N$, relation (1.5.9) becomes (1.5.7), and hence Lemma 1.5.2 will be proved.

Let us look at the case $\tau \in I_1$. In this case $\ell(\tau)$ being contained in U_1, the relation

$$c(\ell(\tau); U_1)C_{j_1 h_{\mu(\tau)}} = C_{j_1 h_{\mu(\tau)}}$$

holds. On the other hand, we have

$$U_1 \cap V_\mu \cap V_{\mu+1} \neq \varnothing \quad \text{for} \quad 1 \leq \mu \leq \mu(\tau) - 1,$$

since $\{V_1, \cdots, V_{\mu(\tau)}\}$ is a covering of $\ell(\tau)$ selected from $\{\Delta(z_j, r_j)\}$. Hence

$$C_{h_\mu h_{\mu+1}} = C_{h_\mu j_1} C_{j_1 h_{\mu+1}} = C_{j_1 h_\mu}^{-1} C_{j_1 h_{\mu+1}}$$

for such a μ. It follows from this result that

$$c(\ell(\tau); V_1, \cdots, V_{\mu(\tau)}) = C_{j_1 h_1}^{-1} C_{j_1 h_{\mu(\tau)}}.$$

Therefore

$$C_{j_1 h_1}\ c(\ell(\tau); V_1, \cdots, V_{\mu(\tau)}) = c(\ell(\tau); U_1)\ C_{j_1 h_{\mu(\tau)}}.$$

This is exactly (1.5.9) for $\nu = 1$.

Now assuming (1.5.9) for $\tau \in I_\nu$, we shall derive

$$c(\ell(\tau); U_1, \cdots, U_{\nu+1})C_{j_{\nu+1} h_{\mu(\tau)}} = C_{j_1 h_1}\ c(\ell(\tau); V_1, \cdots, V_{\mu(\tau)}) \tag{1.5.9$'$}$$

for $\tau \in I_{\nu+1}$. We first look at the case when τ is in $I_\nu \cap I_{\nu+1}$. Then $z(\tau)$ is contained in $U_\nu \cap U_{\nu+1}$. On the other hand, $z(\tau)$ is also contained in $V_{\mu(\tau)}$, since $\{V_1, \cdots, V_{\mu(\tau)}\}$ is a covering of $\ell(\tau)$ selected from $\{\Delta(z_j, r_j)\}$. Consequently, $U_\nu \cap U_{\nu+1} \cap V_{\mu(\tau)} \neq \varnothing$, and hence

$$\begin{aligned} &c(\ell(\tau); U_1, \cdots, U_{\nu+1})C_{j_{\nu+1} h_{\mu(\tau)}} \\ &= c(\ell(\tau); U_1, \cdots, U_\nu)C_{j_\nu j_{\nu+1}} C_{j_{\nu+1} h_{\mu(\tau)}} \\ &= c(\ell(\tau); U_1, \cdots, U_\nu)C_{j_\nu h_{\mu(\tau)}} \\ &= C_{j_1 h_1}\ c(\ell(\tau); V_1, \cdots, V_{\mu(\tau)}). \end{aligned}$$

Therefore (1.5.9$'$) is verified on the interval $t_{\nu+1} \leq \tau \leq s_\nu$.

If $t_{\nu+1} \leq \tau \leq \sigma \leq \tau + \varepsilon(\tau)$, we have $\mu(\sigma) = \mu(\tau)$. Consequently

$$c(\ell(\sigma); V_1, \cdots, V_{\mu(\sigma)}) = c(\ell(\sigma); V_1, \cdots, V_{\mu(\tau)}) = c(\ell(\tau); V_1, \cdots, V_{\mu(\tau)}).$$

If we have also $\sigma \in I_{\nu+1}$, then

$$c(\ell(\sigma); U_1, \cdots, U_{\nu+1}) = c(\ell(\tau); U_1, \cdots, U_{\nu+1}).$$

This means that (1.5.9′) holds for $t_{\nu+1} \leq \tau \leq \tau_0 + \delta$ if (1.5.9′) holds for $t_{\nu+1} \leq \tau \leq \tau_0$ $(< s_{\nu+1})$, provided a positive number δ satisfies two inequalities: $\delta \leq \varepsilon(\tau_0)$ and $\tau_0 + \delta \leq s_{\nu+1}$.

Let us assume that (1.5.9′) holds on the interval $t_{\nu+1} \leq \tau < \tau_0$ $(\leq s_{\nu+1})$. Then

$$\mu(\tau) = \mu(\tau_0) \quad \text{or} \quad \mu(\tau_0) - 1 \quad \text{for} \quad \tau_0 - \varepsilon(\tau_0) \leq \tau < \tau_0.$$

If $\mu(\tau) = \mu(\tau_0)$, then

$$\begin{aligned} &C_{j_1 h_1}\ c(\ell(\tau_0); V_1, \cdots, V_{\mu(\tau_0)}) \\ &= C_{j_1 h_1}\ c(\ell(\tau); V_1, \cdots, V_{\mu(\tau)}) \\ &= c(\ell(\tau); U_1, \cdots, U_{\nu+1}) C_{j_{\nu+1} h_{\mu(\tau)}} \\ &= c(\ell(\tau_0); U_1, \cdots, U_{\nu+1}) C_{j_{\nu+1} h_{\mu(\tau_0)}}. \end{aligned}$$

Therefore (1.5.9′) holds also at $\tau = \tau_0$. Now assume that

$$t_{\nu+1} \leq \tau < \tau_0, \qquad \tau_0 - \varepsilon(\tau_0) \leq \tau, \qquad \mu(\tau) = \mu(\tau_0) - 1.$$

In this case $z(\tau)$ is contained not only in $U_{\nu+1}$ and $V_{\mu(\tau)}$ but also in $V_{\mu(\tau_0)}$. This is due to the properties of $\{V_1, \cdots, V_{\mu(\tau_0)}\}$ as a covering of $\ell(\tau)$ selected from $\{\Delta(z_j, r_j)\}$ (cf. condition (4) given above). Hence $V_{\mu(\tau)} \cap V_{\mu(\tau_0)} \cap U_{\nu+1} \neq \varnothing$. Therefore

$$\begin{aligned} &C_{j_1 h_1}\ c(\ell(\tau_0); V_1, \cdots, V_{\mu(\tau_0)}) \\ &= C_{j_1 h_1}\ c(\ell(\tau); V_1, \cdots, V_{\mu(\tau)}) C_{h_{\mu(\tau)} h_{\mu(\tau_0)}} \\ &= c(\ell(\tau); U_1, \cdots, U_{\nu+1}) C_{j_{\nu+1} h_{\mu(\tau)}} C_{h_{\mu(\tau)} h_{\mu(\tau_0)}} \\ &= c(\ell(\tau_0); U_1, \cdots, U_{\nu+1}) C_{j_{\nu+1} h_{\mu(\tau_0)}}. \end{aligned}$$

This means that (1.5.9′) holds also at $\tau = \tau_0$.

Let $\mathscr{I}$ be the set of all τ_0 such that (1.5.9′) holds on the interval $t_{\nu+1} \leq \tau \leq \tau_0$. Then $\mathscr{I}$ is an open and closed subset of $I_{\nu+1}$ and $\mathscr{I}$ is not an empty set. Since $I_{\nu+1}$ is connected, we have $\mathscr{I} = I_{\nu+1}$. □

THEOREM 1.5.3. *If* $\{U_1, \cdots, U_N\}$ *is a covering of* ℓ *selected from* $\{\Delta(z_j, r_j)\}$ *and if* $\{\Delta(z_j, r_j), C_{jk}\}$ *is a system of connection matrices on* $\mathscr{D}$, *then*

(i) $c(\ell; U_1, \cdots, U_N) \in GL(n, \mathbf{C})$,

(ii) $c(\ell; U_1, \cdots, U_N)$ *is determined by* ℓ, j_1 *and* j_N, *where* $U_\nu = \Delta(z_{j_\nu}, r_{j_\nu})$.

[Proof was omitted by the translator and left to the reader as an exercise.]

Suppose that a point z_0 of $\Delta(z_j, r_j)$ and a point $\tilde{z}_0$ of $\Delta(z_k, r_k)$ are in the same connected component of $\mathscr{D}$. Then there exists a continuous curve ℓ joining z_0 and $\tilde{z}_0$ in $\mathscr{D}$. Let us select a covering $\{U_1, \cdots, U_N\}$ of ℓ from $\{\Delta(z_j, r_j)\}$ so that $U_1 = \Delta(z_j, r_j)$ and $U_N = \Delta(z_k, r_k)$. The matrix $c(\ell; U_1, \cdots, U_N)$ being determined by ℓ, j, k, we denote it by $C(\ell; j, k)$.

DEFINITION 1.5.4. The matrix $C(\ell; j, k)$ is called the connection matrix from $\Delta(z_j, r_j)$ to $\Delta(z_k, r_k)$ along ℓ defined by the system $\{\Delta(z_j, r_j), C_{jk}\}$ of connection matrices on $\mathscr{D}$.

THEOREM 1.5.5. *The matrix $C(\ell; j, k)$ has the following properties*:

$$C(\ell; j, k) \in GL(n, \mathbf{C}), \tag{1.5.10}$$

and

$$C(\ell; j, k)C_{kq} = C_{jp}C(\ell; p, q) \tag{1.5.11}$$

if both sides of (1.5.11) *are well defined.*

A more general theorem can be derived from Theorem 1.5.5.

THEOREM 1.5.6. *Assume that the following conditions are satisfied*:

(i) $\{\Delta(z_j, r_j), C_{jk}\}$ *and* $\{\Delta(\zeta_j, \rho_j), \Gamma_{jk}\}$ *are two systems of connection matrices on* $\mathscr{D}$,

(ii) $\{\Delta(z_j, r_j), C_{jk}\}$ *and* $\{\Delta(\zeta_j, \rho_j), \Gamma_{jk}\}$ *are equivalent and a system* $\{G_{jk}\}$ *of matrices satisfies conditions* (α) *and* (β) *of Definition* 1.4.3,

(iii) $z_0 \in \Delta(z_j, r_j) \cap \Delta(\zeta_p, \rho_p)$ *and* $\tilde{z}_0 \in \Delta(z_k, r_k) \cap \Delta(\zeta_q, \rho_q)$,

(iv) z_0 *and* $\tilde{z}_0$ *are joined by a continuous curve* ℓ *in* $\mathscr{D}$,

(v) *the matrix* $C(\ell; j, k)$ *is the connection matrix from* $\Delta(z_j, r_j)$ *to* $\Delta(z_k, r_k)$ *along* ℓ *defined by* $\{\Delta(z_j, r_j), C_{jk}\}$,

(vi) *the matrix* $\Gamma(\ell; p, q)$ *is the connection matrix from* $\Delta(\zeta_p, \rho_p)$ *to* $\Delta(\zeta_q, \rho_q)$ *along* ℓ *defined by* $\{\Delta(\zeta_j, \rho_j), \Gamma_{jk}\}$.

Then

$$C(\ell; j, k)G_{kq} = G_{jp}\Gamma(\ell; p, q). \tag{1.5.12}$$

[Proof was omitted by the translator and left to the reader as an exercise.]

In the theory of fibre bundles, the content of this section is explained as "translating a fibre along a curve" (cf. N. Steenrod [10], §§13.1 and 13.2, pp. 59–60).

§1.6. Structures of monodromy

Let $\mathscr{D}$ be an open set in the z-plane, and let J be the interval $0 \leq t \leq 1$ of a variable t. Assume that two continuous curves in $\mathscr{D}$:

$$\ell_0: \quad z = z(t_0) \in \mathscr{D} \qquad (0 \leq t \leq 1) \tag{1.6.1}$$

$$\ell_1 : \quad z = z(t_1) \in \mathscr{D} \qquad (0 \le t \le 1) \tag{1.6.2}$$

are given in terms of two continuous functions $z_0(t)$ and $z_1(t)$ which are defined on J and take values in $\mathscr{D}$.

DEFINITION 1.6.1. Two curves ℓ_0 and ℓ_1 are said to be **mutually homotopic** if

(i) $z_0(0) = z_1(0), \qquad z_0(1) = z_1(1),$

(ii) we can choose a suitable continuous map $f(t,s){:}J \times J \to \mathscr{D}$ so that

$$f(0,s) = z_0(0) = z_1(0), \qquad f(1,s) = z_0(1) = z_1(1) \qquad \text{for } 0 \le s \le 1$$

and

$$f(t,0) = z_0(t), \qquad f(t,1) = z_1(t) \qquad \text{for } 0 \le t \le 1.$$

We denote it by $\ell_0 \approx \ell_1$ that ℓ_0 and ℓ_1 are mutually homotopic.

The relation $\approx$ is an equivalence relation, and hence we can classify all continuous curves in $\mathscr{D}$ by this relation. Each of such equivalence classes is called a homotopy class, and we denote by $[\ell]$ the homotopy class containing ℓ.

In the case when $z_0(1) = z_1(0)$, we define product $\ell_0\ell_1$ of two continuous curves ℓ_0 and ℓ_1 by

$$\ell_0\ell_1 : \quad z = \begin{cases} z_0(2t) & \left(0 \le t \le \dfrac{1}{2}\right) \\ z_1(2t-1) & \left(\dfrac{1}{2} \le t \le 1\right). \end{cases} \tag{1.6.3}$$

We also define inverse ℓ^{-1} of a continuous curve $\ell : \ z = z(t) \ (0 \le t \le 1)$ by

$$\ell^{-1} : \quad z = z(1-t) \qquad (0 \le t \le 1). \tag{1.6.4}$$

For four curves $\ell_0, \ell_1, \tilde{\ell}_0$, and $\tilde{\ell}_1$, we have

$$\ell_0\ell_1 \approx \tilde{\ell}_0\tilde{\ell}_1 \tag{1.6.5}$$

if $\ell_0 \approx \tilde{\ell}_0$ and $\ell_1 \approx \tilde{\ell}_1$ provided that $\ell_0\ell_1$ and $\tilde{\ell}_0\tilde{\ell}_1$ are well defined. Also

$$\ell_0^{-1} \approx \ell_1^{-1} \tag{1.6.6}$$

if $\ell_0 \approx \ell_1$. Consequently we can define the product of two homotopy classes and the inverse of a homotopy class by

$$[\ell_0][\ell_1] = [\ell_0\ell_1] \tag{1.6.7}$$

and

$$[\ell]^{-1} = [\ell^{-1}] \tag{1.6.8}$$

respectively. Note that (1.6.7) applies only to the case when $\ell_0\ell_1$ is well defined.

For a fixed point z_0 of $\mathscr{D}$, we denote by $\Omega(z_0)$ the set of all continuous closed curves in $\mathscr{D}$ which start from and terminate at z_0. If ℓ_0 and ℓ_1 belong to $\Omega(z_0)$, then $\ell_0\ell_1$ is well defined and belong to $\Omega(z_0)$. Also $\ell^{-1} \in \Omega(z_0)$ if $\ell \in \Omega(z_0)$. Therefore if we set

$$\pi_1(\mathscr{D}, z_0) = \{\ [\ell];\quad \ell \in \Omega(z_0)\ \}, \tag{1.6.9}$$

$\pi_1(\mathscr{D}, z_0)$ is a group because of the facts explained above. The unit element of this group is the homotopy class containing "the constant curve", i.e.,

$$1_{z_0}: \quad z(t) = z_0 \qquad (0 \le t \le 1). \tag{1.6.10}$$

DEFINITION 1.6.2. The group $\pi_1(\mathscr{D}, z_0)$ is called **the fundamental group of $\mathscr{D}$ with the base point at z_0.**

If two points z_0 and z_1 are in the same connected component of $\mathscr{D}$, then $\pi_1(\mathscr{D}, z_0)$ and $\pi_1(\mathscr{D}, z_1)$ are isomorphic. If ℓ is a curve joining z_0 and z_1 in $\mathscr{D}$, then

$$\phi(\sigma;\ell) = [\ell^{-1}]\sigma[\ell] \qquad (\sigma \in \pi_1(\mathscr{D}, z_0)) \tag{1.6.11}$$

gives an isomorphism: $\pi_1(\mathscr{D}, z_0) \to \pi_1(\mathscr{D}, z_1)$.

If a map $\rho : \pi_1(\mathscr{D}, z_0) \to GL(n, \mathbf{C})$ satisfies the condition

$$\rho(\sigma\tau) = \rho(\sigma)\rho(\tau) \qquad (\sigma, \tau \in \pi_1(\mathscr{D}, z_0)), \tag{1.6.12}$$

then ρ is called a representation of $\pi_1(\mathscr{D}, z_0)$ in $\mathbf{C}^n$. Two representations ρ_1 and ρ_2 of $\pi_1(\mathscr{D}, z_0)$ in $\mathbf{C}^n$ are said to be equivalent if there exists a matrix G in $GL(n, \mathbf{C})$ such that

$$\rho_2(\sigma) = G^{-1}\rho_1(\sigma)G \qquad (\sigma \in \pi_1(\mathscr{D}, z_0)). \tag{1.6.13}$$

This is an equivalence relation, and we can classify all representations of $\pi_1(\mathscr{D}, z_0)$ in $\mathbf{C}^n$ by this relation. We denote by $[\rho, z_0]$ the equivalence class containing the representation ρ.

DEFINITION 1.6.3. An equivalence class $[\rho, z_0]$ is called **a structure of monodromy on $\mathscr{D}$ with the base point at z_0.**

If two points z_0 and z_1 are contained in the same connected component of $\mathscr{D}$, an isomorphism : $\pi_1(\mathscr{D}, z_0) \to \pi_1(\mathscr{D}, z_1)$ is given by (1.6.11) where ℓ is a curve joining z_0 to z_1 in $\mathscr{D}$. This isomorphism has the properties below:

(a) $\phi(\cdot;\ell^{-1}) = \phi(\cdot;\ell)^{-1}$ (i.e., the inverse of $\phi(\cdot;\ell)$),

(b) $\phi(\cdot;\ell_1\ell_2) = \phi(\cdot;\ell_2) \circ \phi(\cdot;\ell_1)$,

(c) if ρ is a representation of $\pi_1(\mathscr{D}, z_1)$ in $\mathbf{C}^n$, then $\rho \circ \phi(\cdot;\ell)$ is a representation of $\pi_1(\mathscr{D}, z_0)$ in $\mathbf{C}^n$; we denote this representation by $\phi^*(\rho;\ell)$, i.e.,

$$\phi^*(\rho;\ell) = \rho \circ \phi(\cdot;\ell). \tag{1.6.14}$$

LEMMA 1.6.4. *Let ρ be a representation of $\pi_1(\mathscr{D}, z_0)$ in $\mathbf{C}^n$ and $\ell \in \Omega(z_0)$, then*

$$[\phi^*(\rho;\ell), z_0] = [\rho, z_0] \tag{1.6.15}$$

PROOF. If we set $G = \rho([\ell])$, then

$$G^{-1}\rho(\sigma)G = \rho([\ell]^{-1}\sigma[\ell]) = \rho(\phi(\sigma;\ell)). \quad \square$$

LEMMA 1.6.5. *If ℓ_1 and ℓ_2 are two continuous curves joining z_0 and z_1 in $\mathscr{D}$, then*

$$[\phi^*(\rho;\ell_1), z_0] = [\phi^*(\rho;\ell_2), z_0] \tag{1.6.16}$$

for every representation ρ of $\pi_1(\mathscr{D}, z_1)$ in $\mathbf{C}^n$.

PROOF. Since $\ell_2\ell_1^{-1}$ belongs to $\Omega(z_0)$, we have

$$[\phi^*(\phi^*(\rho;\ell_1);\ell_2\ell_1^{-1}), z_0] = [\phi^*(\rho;\ell_1), z_0]$$

because of Lemma 1.6.4. Moreover

$$\begin{aligned} \phi^*(\phi^*(\rho;\ell_1);\ell_2\ell_1^{-1}) &= \phi^*(\rho;\ell_1) \circ \phi(\cdot;\ell_2\ell_1^{-1}) \\ &= \rho \circ \phi(\cdot;\ell_1) \circ \phi(\cdot;\ell_2\ell_1^{-1}) \\ &= \rho \circ \phi(\cdot;\ell_2) = \phi^*(\rho;\ell_2). \quad \square \end{aligned}$$

LEMMA 1.6.6. *If two representations ρ_1 and ρ_2 of $\pi_1(\mathscr{D}, z_1)$ in $\mathbf{C}^n$ satisfy the condition $[\rho_1, z_1] = [\rho_2, z_1]$, then*

$$[\phi^*(\rho_1;\ell), z_0] = [\phi^*(\rho_2;\ell), z_0]. \tag{1.6.17}$$

[Proof was omitted by the translator and left to the reader as an exercise.]

Having these preparations done, let us put

$$\phi^*_{z_0z_1}([\rho, z_1]) = [\phi^*(\rho;\ell), z_0], \tag{1.6.18}$$

where ρ is a representation of $\pi_1(\mathscr{D}, z_1)$ in $\mathbf{C}^n$ and ℓ is a continuous curve joining z_0 to z_1 in $\mathscr{D}$. The right-hand side of (1.6.18) is independent of ℓ because of Lemma 1.6.5. Furthermore Lemma 1.6.6 states that the right-hand side of (1.6.18) is determined by $[\rho, z_1]$. The map $\phi^*_{z_0z_1}$ is a bijection from the set of all structures of monodromy on $\mathscr{D}$ with the base point at z_1 to the set of all structures of monodromy on $\mathscr{D}$ with the base point at z_0. The inverse of $\phi^*_{z_0z_1}$ is $\phi^*_{z_1z_0}$.

Definition 1.6.7. Let $\mathscr{D}$ be a connected open set in the z-plane. For each point z of $\mathscr{D}$ let $\mathscr{M}_z$ be a structure of monodromy on $\mathscr{D}$ with the base point at z. Then $\{\mathscr{M}_z, z \in \mathscr{D}\}$ is called **a structure of monodromy on** $\mathscr{D}$ if

$$\mathscr{M}_{\tilde{z}} = \phi^*_{\tilde{z}z}(\mathscr{M}_z) \tag{1.6.19}$$

for all z and $\tilde{z}$ in $\mathscr{D}$.

Lemma 1.6.8. *Let $\mathscr{D}$ be a connected open set in the z-plane and let $\{\mathscr{M}_z, z \in \mathscr{D}\}$ and $\{\tilde{\mathscr{M}}_z, z \in \mathscr{D}\}$ be two structures of monodromy on $\mathscr{D}$. Then $\{\mathscr{M}_z, z \in \mathscr{D}\} = \{\tilde{\mathscr{M}}_z, z \in \mathscr{D}\}$ if and only if $\mathscr{M}_z = \tilde{\mathscr{M}}_z$ at a point z in $\mathscr{D}$.*

This lemma follows immediately from Definition 1.6.7.

§1.7. One-to-one correspondence between structures of analytic continuation and structures of monodromy

In this section, assuming that $\mathscr{D}$ is a connected open set in the z-plane, we shall explain that

(I) a structure $\mathscr{C}$ of analytic continuation in $\mathscr{D}$ determines a structure $\mathscr{M}(\mathscr{C})$ of monodromy on $\mathscr{D}$,

(II) a structure $\mathscr{M}$ of monodromy on $\mathscr{D}$ determines a structure $\mathscr{C}(\mathscr{M})$ of analytic continuation on $\mathscr{D}$,

(III) $\mathscr{M}(\mathscr{C}(\mathscr{M})) = \mathscr{M}$, $\mathscr{C}(\mathscr{M}(\mathscr{C})) = \mathscr{C}$.

In the theory of fibre bundles, these situations are explained as a classification of all fibre bundles in terms of "characteristic classes" (cf. N. Steenrod [10], §§ 13.3–13.9, pp. 60–66).

From (1.5.6), Theorem 1.5.3, and Definition 1.5.4 of § 1.5 we derive the lemma below.

Lemma 1.7.1. *Let $\mathscr{D}$ be an open set in the z-plane and let $\{\Delta(z_j, r_j), C_{jk}\}$ be a system of connection matrices on $\mathscr{D}$. Then*

(I)

$$C(\ell; j, k)^{-1} = C(\ell^{-1}; k, j) \tag{1.7.1}$$

if $z_0 \in \Delta(z_j, r_j)$, $\tilde{z}_0 \in \Delta(z_k, r_k)$, and ℓ is a continuous curve joining z_0 to $\tilde{z}_0$ in $\mathscr{D}$,

(II)

$$C(\ell_1\ell_2; j, h) = C(\ell_1; j, k)C(\ell_2; k, h) \tag{1.7.2}$$

if $z_0 \in \Delta(z_j, r_j)$, $z_0' \in \Delta(z_k, r_k)$, z_0''in$\Delta(z_h, r_h)$, and if ℓ_1 is a continuous curve joining z_0 to z_0' in $\mathscr{D}$ and ℓ_2 is a continuous curve joining z_0' to z_0'' in $\mathscr{D}$,

(III)

$$C(\ell_1; j, k) = C(\ell_2; j, k) \tag{1.7.3}$$

if $z_0 \in \Delta(z_j, r_j)$, $\tilde{z}_0 \in \Delta(z_k, r_k)$ *and if* ℓ_1 *and* ℓ_2 *are continuous curves joining* z_0 *and* $\tilde{z}_0$ *in* $\mathscr{D}$ *such that* $\ell_1 \approx \ell_2$.

We shall leave the proof of this lemma to the reader.

THEOREM 1.7.2. *Assume that*

(i) $\mathscr{D}$ *is an open set in the z-plane and* $\mathscr{C}$ *is a structure of analytic continuation on* $\mathscr{D}$,
(ii) z_0 *is a fixed point of* $\mathscr{D}$, *and* $\Omega(z_0)$ *is the set of all continuous closed curves in* $\mathscr{D}$ *which start from and terminate at* z_0,
(iii) $\{\Delta(z_j, r_j), C_{jk}\}$ *is a system of connection matrices on* $\mathscr{D}$ *which belongs to* $\mathscr{C}$,
(iv) $z_0 \in \Delta(z_h, r_h)$.

Set

$$\rho_h([\ell]) = C(\ell; h, h) \tag{1.7.4}$$

for every closed curve ℓ *in* $\Omega(z_0)$. *Then*

(a) ρ_h *is a representation of* $\pi_1(\mathscr{D}, z_0)$ *in* $\mathbf{C}^n$,
(b) *if* $z_0 \in \Delta(z_h, r_h) \cap \Delta(z_k, r_k)$, *then*

$$[\rho_h, z_0] = [\rho_k, z_0], \tag{1.7.5}$$

(c) $[\rho_h, z_0]$ *is determined by* $\mathscr{C}$ *and independent of the choice of* $\{\Delta(z_j, r_j), C_{jk}\}$.

PROOF. Because of (III) of Lemma 1.7.1, the right-hand side of (1.7.4) is determined by $[\ell]$. Moreover, because of (II) of the same lemma, ρ_h is a representation of $\pi_1(\mathscr{D}, z_0)$ in $\mathbf{C}^n$. Therefore (a) is proved. Now, for every closed curve in $\Omega(z_0)$ a relation

$$C(\ell; h, h)C_{hk} = C_{hk}C(\ell; k, k)$$

holds (cf. Theorem 1.5.5). Hence, if we put $G = C_{hk}$, we have

$$\rho_k(\sigma) = G^{-1}\rho_h(\sigma)G.$$

Therefore (b) is proved. To prove (c) we consider another system $\{\Delta(w_j, s_j), D_{jk}\}$ of connection matrices on $\mathscr{D}$ which belongs to $\mathscr{C}$. Assuming $z_0 \in \Delta(w_k, s_k)$, we set

$$\tilde{\rho}_k([\ell]) = D(\ell; k, k)$$

for every closed curve ℓ in $\Omega(z_0)$, where $D(\ell; j,k)$ is the connection matrix from $\Delta(w_j, s_j)$ to $\Delta(w_k, s_k)$ along ℓ defined by $\{\Delta(w_j, s_j), D_{jk}\}$ (cf. Definition 1.5.4). To prove (c) we must show that

$$[\rho_h, z_0] = [\tilde{\rho}_k, z_0]. \tag{1.7.6}$$

Since two systems $\{\Delta(z_j, r_j), C_{jk}\}$ and $\{\Delta(w_j, s_j), D_{jk}\}$ of connection matrices belong to the same structure $\mathscr{C}$ of analytic continuation, we can derive (1.7.6) by utilizing Theorem 1.5.6. □

The equivalence class $[\rho_h, z_0]$ of Theorem 1.7.2 is determined by the structure $\mathscr{C}$ of analytic continuation. Furthermore every representation in the class $[\rho_h, z_0]$ is written in form (1.7.4) of Theorem 1.7.2 if we choose a suitable system of connection matrices from $\mathscr{C}$. In fact, assume that $\rho \in [\rho_h, z_0]$, i.e., $\rho(\sigma) = G^{-1}\rho_h(\sigma)G$, where G is a suitable matrix in $GL(n, \mathbf{C})$. The system $\{\Delta(z_j, r_j), G^{-1}C_{jk}G\}$ of connection matrices belongs to $\mathscr{C}$, since this system is equivalent to $\{\Delta(z_j, r_j), C_{jk}\}$ (cf. Theorem 1.4.7). If we use this system of connection matrices, we can write ρ in form (1.7.4).

THEOREM 1.7.3. *Let $\mathscr{D}$ be a connected open set in the z-plane and assume that $[\rho_h, z_0]$ is defined by Theorem 1.7.2. In this case, if we set $\mathscr{M}_{z_0}(\mathscr{C}) = [\rho_h, z_0]$, then $\{\mathscr{M}_z(\mathscr{C}), z \in \mathscr{D}\}$ is a structure of monodromy on $\mathscr{D}$.*

PROOF. Fixing a system $\{\Delta(z_j, r_j), C_{jk}\}$ of connection matrices in the structure $\mathscr{C}$ of analytic continuation, we set

$$\rho_h([\ell]) = C(\ell; h, h), \quad \tilde{\rho}_k([\tilde{\ell}]) = C(\tilde{\ell}; k, k),$$

where we assume that $z_0 \in \Delta(z_h, r_h)$, $\tilde{z}_0 \in \Delta(z_k, r_k)$, and that $\ell \in \Omega(z_0)$, $\tilde{\ell} \in \Omega(\tilde{z}_0)$. If a curve ℓ_0 joins z_0 to $\tilde{z}_0$ in $\mathscr{D}$, we can define an isomorphism $\phi(\sigma; \ell_0)\colon \pi_1(\mathscr{D}, z_0) \to \pi_1(\mathscr{D}, \tilde{z}_0)$ by means of (1.6.11). Furthermore

$$\begin{aligned}\tilde{\rho}_k(\phi([\ell]; \ell_0)) = \tilde{\rho}_k([\ell_0^{-1}\ell\ell_0]) &= C(\ell_0^{-1}\ell\ell_0; k, k)\\ &= C(\ell_0^{-1}; k, h)C(\ell; h, h)C(\ell_0; h, k)\end{aligned}$$

for every $\ell \in \Omega(z_0)$. Hence, if we set $G = C(\ell_0; h, k)$, we have

$$\tilde{\rho}_k(\phi(\sigma; \ell_0)) = G^{-1}\rho_h(\sigma)G \tag{1.7.7}$$

for every σ in $\pi_1(\mathscr{D}, z_0)$, i.e.,

$$[\phi^*(\tilde{\rho}_k; \ell_0), z_0] = [\rho_h, z_0]. \tag{1.7.8}$$

We can write (1.7.8) in the form

$$\phi^*_{z_0\tilde{z}_0}(\mathscr{M}_{\tilde{z}_0}(\mathscr{C})) = \mathscr{M}_{z_0}(\mathscr{C}), \tag{1.7.9}$$

where the notation $\phi^*_{z_0\tilde{z}_0}$ is defined by (1.6.18), and $\mathscr{M}_{z_0}(\mathscr{C}) = [\rho_h, z_0]$, $\mathscr{M}_{\tilde{z}_0}(\mathscr{C}) = [\tilde{\rho}_k, \tilde{z}_0]$. Therefore $\{\mathscr{M}_z(\mathscr{C}), z \in \mathscr{D}\}$ is a structure of monodromy on $\mathscr{D}$. □

THEOREM 1.7.4. *Assume that*

(i) $\mathscr{D}$ *is a connected open set in the z-plane,*
(ii) $\mathscr{M} = \{\mathscr{M}_z, z \in \mathscr{D}\}$ *is a structure of monodromy on* $\mathscr{D}$,
(iii) $\{\Delta(z_j, r_j)\}$ *is a covering of* $\mathscr{D}$,
(iv) *we denote by* ℓ_{jk} *a curve joining* z_j *to* z_k *in* $\Delta(z_j, r_j) \cap \Delta(z_k, r_k)$ *if* $\Delta(z_j, r_j) \cap \Delta(z_k, r_k) \neq \varnothing$,
(v) *fixing a point* z_0 *in* $\mathscr{D}$, *we denote by* ℓ_j *a curve joining* z_0 *to* z_j *in* $\mathscr{D}$,
(vi) ρ *is a representation of* $\pi_1(\mathscr{D}, z_0)$ *belonging to* $\mathscr{M}_{z_0}$.

Set

$$C_{jk} = \rho([\ell_j \ell_{jk} \ell_k^{-1}] \tag{1.7.10}$$

whenever $\Delta(z_j, r_j) \cap \Delta(z_k, r_k). \neq \varnothing$. *Then*

(a) $\{\Delta(z_j, r_j), C_{jk}\}$ *is a system of connection matrices on* $\mathscr{D}$,
(b) *the system* $\mathscr{C}$ *of analytic continuation to which* $\{\Delta(z_j, r_j), C_{jk}\}$ *belongs is determined by* $\mathscr{M}$ *and independent of the choice of* $\{\Delta(z_j, r_j)\}$, ℓ_{jk}, z_0,ℓ_j, *and* ρ.

PROOF. [Proof of (a) was omitted by the translator and left to the reader.] The proof of (b) is complicated. If $\tilde{\rho}$ is a representation belonging to $\mathscr{M}_{z_0}$, there exists a matrix $G \in GL(n, \mathbf{C})$ such that $\tilde{\rho}(\sigma) = G^{-1}\rho(\sigma)G$ for every σ in $\pi_1(\mathscr{D}, z_0)$. Set $\tilde{C}_{jk} = \tilde{\rho}([\ell_j \ell_{jk} \ell_k^{-1}])$ whenever $\Delta(z_j, r_j) \cap \Delta(z_k, r_k) \neq \varnothing$. Then $C_{jk}G = G\tilde{C}_{jk}$. This means that two systems $\{\Delta(z_j, r_j), C_{jk}\}$ and $\{\Delta(z_j, r_j), \tilde{C}_{jk}\}$ of connection matrices are equivalent (i.e., they belong to the same structure of analytic continuation on $\mathscr{D}$).

Let $\tilde{\ell}_j$ be another curve joining z_0 to z_j in $\mathscr{D}$, and set $\tilde{C}_{jk} = \rho([\tilde{\ell}_j \ell_{jk} \tilde{\ell}_k^{-1}])$ whenever $\Delta(z_j, r_j) \cap \Delta(z_k, r_k) \neq \varnothing$. Set also $G_j = \rho([\ell_j \tilde{\ell}_j^{-1}])$. Then $C_{jk}G_k = G_j\tilde{C}_{jk}$. This means that $\{\Delta(z_j, r_j), C_{jk}\}$ and $\{\Delta(z_j, r_j), \tilde{C}_{jk}\}$ belong to the same structure of analytic continuation on $\mathscr{D}$.

Let $\tilde{z}_0$ be another point in $\mathscr{D}$ and let $\tilde{\ell}_j$ be a curve joining $\tilde{z}_0$ to z_j in $\mathscr{D}$. Choosing a representation $\tilde{\rho}$ from $\mathscr{M}_{\tilde{z}_0}$, set $\tilde{C}_{jk} = \tilde{\rho}([\tilde{\ell}_j \ell_{jk} \tilde{\ell}_k^{-1}]$ whenever $\Delta(z_j, r_j) \cap \Delta(z_k, r_k) \neq \varnothing$. Since $\mathscr{M} = \{\mathscr{M}_z, z \in \mathscr{D}\}$ is a structure of monodromy in $\mathscr{D}$, we have

$$\phi^*_{z_0 \tilde{z}_0}(\mathscr{M}_{\tilde{z}_0}) = \mathscr{M}_{z_0}, \tag{1.7.11}$$

i.e., the relation

$$[\phi^*(\tilde{\rho}; \ell), z_0] = [\rho, z_0] \tag{1.7.12}$$

holds for every curve ℓ joining z_0 to $\tilde{z}_0$ in $\mathscr{D}$. Therefore there exists a matrix $G \in GL(n, \mathbf{C})$ such that

$$\tilde{\rho}(\phi(\sigma; \ell)) = G^{-1}\rho(\sigma)G \tag{1.7.13}$$

holds for every σ in $\pi_1(\mathscr{D}, z_0)$. In particular

$$G^{-1}C_{jk}G = G^{-1}\rho([\ell_j\ell_{jk}\ell_k^{-1}])G = \tilde{\rho}(\phi([\ell_j\ell_{jk}\ell_k^{-1}];\ell))$$
$$= \tilde{\rho}([\ell^{-1}\ell_j\ell_{jk}\ell_k^{-1}\ell]).$$

The curve $\ell^{-1}\ell_j$ joins $\tilde{z}_0$ to z_j in $\mathscr{D}$. Hence if we set $H_j = \tilde{\rho}([\ell^{-1}\ell_j\tilde{\ell}_j^{-1}])$, we have

$$G^{-1}C_{jk}G = H_j\tilde{C}_{jk}H_k^{-1},$$

i.e., the relation $C_{jk}GH_k = GH_j\tilde{C}_{jk}$ holds whenever $\Delta(z_j, r_j) \cap \Delta(z_k, r_k) \neq \emptyset$. This means that $\{\Delta(z_j, r_j), C_{jk}\}$ and $\{\Delta(z_j, r_j), \tilde{C}_{jk}\}$ belong to the same structure of analytic continuation on $\mathscr{D}$.

In the case when $\Delta(z_j, r_j) \cap \Delta(z_k, r_k) \neq \emptyset$, let $\tilde{\ell}_{jk}$ be another curve joining z_j to z_k in $\Delta(z_j, r_j) \cap \Delta(z_k, r_k)$. Then $\tilde{\ell}_{jk} \approx \ell_{jk}$. Hence C_{jk} is independent of the choice of ℓ_{jk}.

Let $\{\Delta(w_j, s_j)\}$ be another covering of $\mathscr{D}$ and let $\tilde{\ell}_{jk}$ be a curve joining w_j to w_k in $\Delta(w_j, s_j) \cap \Delta(w_k, s_k)$ ($\neq \emptyset$). Also let $\tilde{\ell}_j$ be a curve joining z_0 to w_j in $\mathscr{D}$. If we set

$$D_{jk} = \rho([\tilde{\ell}_j\tilde{\ell}_{jk}\tilde{\ell}_k^{-1}])$$

whenever $\Delta(w_j, s_j) \cap \Delta(w_k, s_k) \neq \emptyset$, then $\{\Delta(w_j, s_j), D_{jk}\}$ is a system of connection matrices on $\mathscr{D}$. If we apply the same procedure to the covering $\{\Delta(z_j, r_j)\} \cup \{\Delta(w_j, s_j)\}$ of $\mathscr{D}$, we can construct a system of connection matrices on $\mathscr{D}$ which contains $\{\Delta(z_j, r_j), C_{jk}\}$ and $\{\Delta(w_j, s_j), D_{jk}\}$. Consequently

$$\{\Delta(z_j, r_j), C_{jk}\} \sim \{\Delta(w_j, s_j), D_{jk}\}. \tag{1.7.14}$$

□

Thus we have constructed a structure of monodromy by means of a given structure $\mathscr{C}$ of analytic continuation through Theorems 1.7.2 and 1.7.3. This structure of monodromy is denoted by $\mathscr{M}(\mathscr{C})$. Also we have constructed a structure of analytic continuation by means of a given structure $\mathscr{M}$ of monodromy through Theorem 1.7.4. This structure of analytic continuation is denoted by $\mathscr{C}(\mathscr{M})$. In either case we assumed that $\mathscr{D}$ was a connected open set in the z-plane.

THEOREM 1.7.5. *In the case when $\mathscr{D}$ is a connected open set in the z-plane, a one-to-one correspondence between the set of all structures of analytic continuations on $\mathscr{D}$ and the set of all structures of monodromy on $\mathscr{D}$ is given by $\mathscr{M}(\mathscr{C})$ and $\mathscr{C}(\mathscr{M})$, i.e.,*

$$\mathscr{M}(\mathscr{C}(\mathscr{M})) = \mathscr{M} \tag{1.7.15}$$

for every structure $\mathscr{M}$ of monodromy on $\mathscr{D}$ and

$$\mathscr{C}(\mathscr{M}(\mathscr{C})) = \mathscr{C} \tag{1.7.16}$$

for every structure $\mathscr{C}$ of analytic continuation on $\mathscr{D}$.

PROOF. In this proof we use the notations given in Theorems 1.7.2, 1.7.3, and 1.7.4. To verify (1.7.15) we define a matrix C_{jk} by (1.7.10). Then $\{\Delta(z_j, r_j), C_{jk}\}$ is a system of connection matrices belonging to $\mathscr{C}(\mathscr{M})$. In terms of this system of connection matrices, we define a representation ρ_h by (1.7.4). This representation belongs to $\mathscr{M}_{z_0}(\mathscr{C}(\mathscr{M}))$. By virtue of Lemma 1.6.8, in order to prove (1.7.15), it suffices to show

$$\mathscr{M}_{z_0}(\mathscr{C}(\mathscr{M})) = \mathscr{M}_{z_0}. \tag{1.7.17}$$

In fact we can show that

$$\rho_h = \rho, \tag{1.7.18}$$

where ρ is the representation utilized in (1.7.10). Evidently (1.7.18) implies (1.7.17). [Proof of (1.7.18) was omitted by the translator and left to the reader.]

To prove (1.7.16), we define ρ_h by (1.7.4). Then ρ_h belongs to $\mathscr{M}_{z_0}(\mathscr{C})$. In order to apply Theorem 1.7.4 to $\mathscr{M}(\mathscr{C})$, we use ρ_h as the representation ρ there. Accordingly we set

$$\tilde{C}_{jk} = \rho_h([\ell_j \ell_{jk} \ell_k^{-1}]). \tag{1.7.10$'$}$$

Then we must show that

$$\{\Delta(z_j, r_j), \tilde{C}_{jk}\} \sim \{\Delta(z_j, r_j), C_{jk}\} \tag{1.7.19}$$

to complete the proof of (1.7.16). Since ℓ_{jk} is contained in $\Delta(z_j, r_j) \cap \Delta(z_k, r_k)$, definition (1.5.6) implies that $C(\ell_{jk}; j, k) = C_{jk}$. Therefore

$$\begin{aligned}\tilde{C}_{jk} &= \rho_h([\ell_j \ell_{jk} \ell_k^{-1}]) = C(\ell_j \ell_{jk} \ell_k^{-1}; h, h) \\ &= C(\ell_j; h, j) C(\ell_{jk}; j, k) C(\ell_k^{-1}; k, h) \\ &= G_j^{-1} C_{jk} G_k,\end{aligned}$$

where $G_j = C(\ell_j^{-1}; j, h)$. □

By examining closely how we proved $\mathscr{C}(\mathscr{M}(\mathscr{C})) = \mathscr{C}$, we see that, for a given system $\{\Delta(z_j, r_j), C_{jk}\}$ of connection matrices and a given covering $\{\Delta(\zeta_j, \rho_j)\}$ of $\mathscr{D}$, there exists a system $\{\Gamma_{jk}\}$ of matrices such that

$$\{\Delta(z_j, r_j), C_{jk}\} \sim \{\Delta(\zeta_j, \rho_j), \Gamma_{jk}\}.$$

If the matrices C_{jk} of Theorem 1.7.2 belong to a subgroup of $GL(n, \mathbf{C})$, then $\rho_h([\ell])$ defined by (1.7.4) also belongs to the same subgroup. Also, if the matrices $\rho([\ell])$ of Theorem 1.7.4 belong to a subgroup of $GL(n, \mathbf{C})$, then C_{jk} defined by (1.7.10) also belongs to the same subgroup.

§1.8. Differential equations in the ring of power series

We shall denote by $C[[z - z_0]]$ the set of all formal power series in $z - z_0$ with complex coefficients (cf. § 1.2), and by $gl(n, C[[z - z_0]])$ the set of all $n \times n$ matrices with entries in $C[[z - z_0]]$.

Let

$$a_{jk} = \sum_{m \geq 0} a_{jkm} (z - z_0)^m$$

be the (j, k)-entry of a matrix A in $gl(n, C[[z - z_0]])$. For every m, let A_m be the matrix whose entries are the a_{jkm}. Then we can write A in the form

$$A = \sum_{m \geq 0} A_m (z - z_0)^m.$$

If we denote by $gl(n, \mathbf{C})$ the set of all $n \times n$ matrices with complex entries, then $A_m \in gl(n, \mathbf{C})$ for every m. If dA/dz be the matrix whose entries are the da_{jk}/dz, then

$$\frac{dA}{dz} = \sum_{m \geq 0} (m + 1) A_{m+1} (z - z_0)^m.$$

Furthermore, if

$$B = \sum_{m \geq 0} B_m (z - z_0)^m \qquad (B_m \in gl(n, \mathbf{C}))$$

is another matrix in $gl(n, C[[z - z_0]])$, then

$$A + B = \sum_{m \geq 0} (A_m + B_m) (z - z_0)^m,$$

$$AB = \sum_{m \geq 0} \left(\sum_{h+k=m} A_h B_k \right) (z - z_0)^m.$$

Selecting a matrix A from $gl(n, C[[z - z_0]])$, we consider a differential equation:

$$\frac{dX}{dz} = AX, \tag{1.8.1}$$

where X is an unknown in $gl(n, C[[z - z_0]])$. If we set

$$X = \sum_{m \geq 0} X_m (z - z_0)^m,$$

then

$$\frac{dX}{dz} = \sum_{m\geq 0} (m+1)X_{m+1}\,(z-z_0)^m,$$

$$AX = \sum_{m\geq 0}\left(\sum_{h+k=m} A_h X_k\right)(z-z_0)^m.$$

Hence differential equation (1.8.1) is equivalent to a system of infinitely many equations:

$$(m+1)X_{m+1} = \sum_{h+k=m} A_h X_k \qquad (m = 0, 1, \cdots) \tag{1.8.2}$$

in $gl(n, \mathbf{C})$. If an $X_0 \in gl(n, \mathbf{C})$ is arbitrarily given, system (1.8.2) can be solved successively. Hence a solution to (1.8.1) is uniquely determined for a given matrix X_0. We regard X_0 as an initial value of X at $z = z_0$. Then we can say that equation (1.8.1) admits a unique solution for a given initial value X_0 at $z = z_0$. Similarly the differential equation

$$\frac{dY}{dz} = -YA$$

also admits a unique solution for a given initial value Y_0 at $z = z_0$. Since

$$\frac{d}{dz}(YX) = \frac{dY}{dz}X + Y\frac{dX}{dz} = -YAX + YAX = 0,$$

we have $YX = Y_0X_0$, where $=$ denotes the equality as formal power series in $z - z_0$. In particular, if $X_0 \in GL(n, \mathbf{C})$ and $Y_0 = X_0^{-1}$, we have $YX = I$, i.e., X admits an inverse in $gl(n, C[[z-z_0]])$. We denote by $GL(n, C[[z-z_0]])$ the set of all matrices in $gl(n, C[[z-z_0]])$ which are invertible in $gl(n, C[[z-z_0]])$.

THEOREM 1.8.1. *Differential equation* (1.8.1) *in* $gl(n, C[[z-z_0]])$ *admits a unique solution for a given initial value* X_0 *at* $z = z_0$. *Furthermore, if* $X_0 \in GL(n, \mathbf{C})$, *the unique solution is in* $GL(n, C[[z-z_0]])$.

We denote by $C\{z-z_0\}$ the set of all convergent power series in $z - z_0$ with complex coefficients (cf. §1.2), and by $gl(n, C\{z-z_0\})$ the set of all $n \times n$ matrices whose entries belong to $C\{z-z_0\}$. The set $gl(n, C\{z-z_0\})$ is contained in $gl(n, C[[z-z_0]])$. Furthermore the inverse of a matrix in $gl(n, C\{z-z_0\})$ also belongs to $gl(n, C\{z-z_0\})$. We denote by $GL(n, C\{z-z_0\})$ the set of all invertible matrices in $gl(n, C\{z-z_0\})$. Then

$$GL(n, C\{z-z_0\}) = gl(n, C\{z-z_0\}) \cap GL(n, C[[z-z_0]]).$$

If the matrix A of differential equation (1.8.1) is in $gl(n, C\{z - z_0\})$, we can show that $X \in gl(n, C\{z - z_0\})$ by utilizing (1.8.2). We state this result as a theorem.

THEOREM 1.8.2. *Differential equation* (1.8.1) *in* $gl(n, C\{z - z_0\})$ *admits a unique solution for a given initial value* X_0 *at* $z = z_0$. *Furthermore, if* $X_0 \in GL(n, \mathbf{C})$, *the unique solution is in* $GL(n, C\{z - z_0\})$.

In the case when the radius of convergence of every entry of A is greater than a positive number r, the inequalities

$$|A_m| \leq \frac{M}{r^m} \qquad (m = 0, 1, \cdots) \tag{1.8.3}$$

hold, where $|A_m|$ denotes the sum of absolute values of all entries, and M is a suitable positive number. Fixing a positive number ρ so that $0 < \rho < r$, set

$$x_m = \rho^m |X_m| \qquad (m = 0, 1, \cdots). \tag{1.8.4}$$

Then (1.8.2) implies that

$$(m + 1)x_{m+1} \leq \rho M \sum_{h+k=m} \left(\frac{\rho}{r}\right)^h x_k \qquad (m = 0, 1, \cdots). \tag{1.8.5}$$

Consequently, if we set $p_m = \max\{x_0, x_1, \cdots, x_m\}$, we have

$$x_{m+1} \leq \frac{1}{m+1} \frac{\rho M}{1 - \frac{\rho}{r}} p_m \qquad (m = 0, 1, \cdots). \tag{1.8.6}$$

Hence $x_{m+1} \leq p_m$ if m is sufficiently large. Therefore

$$|X_m| \leq \frac{M(\rho)}{\rho^m} \qquad (m = 0, 1, \cdots)$$

for a suitable positive number $M(\rho)$. This means that the radius of convergence of every entry of X is not less than r. We summarize this result in the theorem below.

THEOREM 1.8.3. *If the radius of convergence of every entry of the matrix* A *of differential equation* (1.8.1) *is not less than a positive number* r, *then the radius of convergence of every entry of the solution* X *is also not less than* r.

§1.9. A system of fundamental solutions of a linear ordinary differential equation

Letting $\mathscr{D}$ be a connected open set in the z-plane, we denote by $\mathscr{O}(\mathscr{D})$ the set of all complex-valued functions which are holomorphic on $\mathscr{D}$ (cf.

§1.1). The set of all $n \times n$ matrices whose entries are in $\mathcal{O}(\mathcal{D})$ is denoted by $gl(n, \mathcal{O}(\mathcal{D}))$. The set of all matrices in $gl(n, \mathcal{O}(\mathcal{D}))$ whose inverses are also in $gl(n, \mathcal{O}(\mathcal{D}))$ is denoted by $GL(n, \mathcal{O}(\mathcal{D}))$. An $n \times n$ matrix $A(z)$ in $gl(n, \mathcal{O}(\mathcal{D}))$ belongs to $GL(n, \mathcal{O}(\mathcal{D}))$ if and only if $A(z)$ belongs to $GL(n, \mathbf{C})$ at every point z of $\mathcal{D}$.

For a matrix $A(z)$ in $gl(n, \mathcal{O}(\mathcal{D}))$, we denote by $T(A, z_0)$ the matrix whose entries are the Taylor series of the entries of A at a point z_0 of $\mathcal{D}$ (cf. §1.3). The matrix $T(A, z_0)$ belongs to $gl(n, C\{z - z_0\})$, and the radius of convergence of every entry is not less than $d(z_0, \partial\mathcal{D})$, where $d(z_0, \partial\mathcal{D})$ is the distance from z_0 to the boundary $\partial\mathcal{D}$ of $\mathcal{D}$.

For a matrix B in $gl(n, C\{z - z_0\})$, we denote by $S(B)$ the matrix whose entries are the sums of the corresponding entries of B (cf. §1.2). If the radius of convergence of every entry of B is not less than a positive number r, then $S(B)$ belongs to $gl(n, \mathcal{O}(\Delta(z_0, r))$ and

$$\frac{d}{dz}S(B) = S\left(\frac{dB}{dz}\right) \tag{1.9.1}$$

holds, where the left-hand side of (1.9.1) is the matrix whose entries are the derivatives of the entries of $S(B)$. For a matrix $A(z)$ in $gl(n, \mathcal{O}(\mathcal{D}))$ we have

$$S(T(A, z_0)) = A \tag{1.9.2}$$

on $\Delta(z_0, d(z_0, \partial\mathcal{D}))$.

Selecting a matrix $A(z)$ from $gl(n, \mathcal{O}(\mathcal{D}))$, we consider a differential equation

$$\frac{dW}{dz} = A(z)W, \tag{1.9.3}$$

where W is an $n \times n$ matrix whose entries are *unknowns*. The meaning of differential equation (1.9.3) in this form is not clear. To avoid this ambiguity, we consider the differential equation

$$\frac{dX}{dz} = T(A, z_0)X \tag{1.9.4}$$

in $gl(, n, C\{z - z_0\})$. If Theorem 1.8.2 is applied to (1.9.4), this equation has a unique solution for a given initial value X_0 at $z = z_0$. Furthermore, by virtue of Theorem 1.8.3, the radius of convergence of every entry of this unique solution is not less than $d(z_0, \partial\mathcal{D})$. If we denote by X the unique solution for the initial value X_0, then $S(X)$ belongs to $gl(n, \mathcal{O}(\Delta(z_0, d(z_0, \partial\mathcal{D}))))$, and

$$\frac{d}{dz}S(X) = AS(X)$$

holds in $\Delta(z_0, d(z_0, \partial\mathscr{D}))$ because of (1.9.1) and (1.9.2). In other words, if we regard (1.9.3) as a differential equation in $gl(n, \mathscr{O}(\Delta(z_0, d(z_0, \partial\mathscr{D}))))$, then (1.9.3) admits the solution $W = S(X)$ which takes the initial value X_0 at $z = z_0$. Similarly, the differential equation

$$\frac{dU}{dz} = -UA(z) \tag{1.9.5}$$

admits a unique solution in $gl(n, \mathscr{O}(\Delta(z_0, d(z_0, \partial\mathscr{D}))))$ for a given initial value Y_0 at $z = z_0$. Moreover

$$\frac{d}{dz}(UW) = -UAW + UAW = 0.$$

Hence UW is identically equal to the constant matrix $Y_0 X_0$ on $\Delta(z_0, d(z_0, \partial\mathscr{D}))$. In particular if $X_0 \in GL(n, \mathbf{C})$ and $Y_0 = X_0^{-1}$ then $UW = I$ identically in $\Delta(z_0, d(z_0, \partial\mathscr{D}))$. This means that

$$W \in GL(n, \mathscr{O}(\Delta(z_0, d(z_0, \partial\mathscr{D})))).$$

We summarize this result in the following theorem.

THEOREM 1.9.1. *If the matrix $A(z)$ is in $gl(n, \mathscr{O}(\mathscr{D}))$, differential equation (1.9.3) admits a solution in $gl(n, \mathscr{O}(\Delta(z_0, d(z_0, \partial\mathscr{D}))))$ for a given initial value X_0 at $z = z_0$ in $\mathscr{D}$. Furthermore, if $X_0 \in GL(n, \mathbf{C})$, this solution belongs to $GL(n, \mathscr{O}(\Delta(z_0, d(z_0, \partial\mathscr{D}))))$.*

Utilizing the method given above, we can also prove the uniqueness of the solution for a given initial value X_0. More generally, if W_1 and W_2 are two solutions of (1.9.3) belonging to $gl(n, \mathscr{O}(\Delta(z_0, d(z_0, \partial\mathscr{D}))))$ and if $W_1 \in GL(n, \mathscr{O}(\Delta(z_0, d(z_0, \partial\mathscr{D}))))$, then W_1^{-1} is a solution of (1.9.5) and $W_1^{-1} W_2$ is identically equal to a constant matrix on $\Delta(z_0, d(z_0, \partial\mathscr{D}))$. We leave the details to the reader.

In the case when $\{\Delta(z_j, r_j)\}$ is a covering of $\mathscr{D}$ (cf. §1.4), differential equation (1.9.3) gives a differential equation on each disk $\Delta(z_j, r_j)$. By means of the global property of $A(z)$ (i.e., $A(z) \in gl(n, \mathscr{O}(\mathscr{D}))$), those local differential equations are connected together. If we look at differential equation (1.9.3) in this way, the meaning of (1.9.3) becomes much clearer. By virtue of Theorem 1.9.1, differential equation (1.9.3) admits a solution belonging to $GL(n, \mathscr{O}(\Delta(z_j, r_j)))$ on each disk $\Delta(z_j, r_j)$. Such a solution is called **a fundamental solution** of (1.9.3) on $\Delta(z_j, r_j)$.

DEFINITION 1.9.2. Let $\{\Delta(z_j, r_j)\}$ be a covering of $\mathscr{D}$ and let $W_j(z)$ be a fundamental solution of (1.9.3) on $\Delta(z_j, r_j)$. Then $\{\Delta(z_j, r_j), W_j(z)\}$ is called **a system of fundamental solutions of differential equation (1.9.3) on the domain $\mathscr{D}$.**

For more discussions on the meaning of differential equation (1.9.3), see P. Deligne [1].

§1.10. Structures of analytic continuation defined by linear differential equations

Let $\mathscr{D}$ be a connected open set in the z-plane and let $A(z)$ be a matrix in $gl(n,\mathscr{O}(\mathscr{D}))$. Assume that $\{\Delta(z_j,r_j),W_j(z)\}$ is a system of fundamental solutions of the differential equation

$$\frac{dW}{dz}=A(z) \tag{1.10.1}$$

(cf. Definition 1.9.2). Set

$$C_{jk}=W_j(z)^{-1}W_k(z) \tag{1.10.2}$$

whenever $\Delta(z_j,r_j)\cap\Delta(z_k,r_k)\neq\varnothing$. Then the C_{jk} are constant matrices in $GL(n,\mathbf{C})$. Moreover if $\Delta(z_j,r_j)\cap\Delta(z_k,r_k)\cap\Delta(z_h,r_h)\neq\varnothing$, we have

$$C_{jk}C_{kh}=W_j^{-1}W_kW_k^{-1}W_h=W_j^{-1}W_h=C_{jh}.$$

Hence $\{\Delta(z_j,r_j),C_{jk}\}$ is a system of connection matrices on $\mathscr{D}$.

DEFINITION 1.10.1. The system $\{\Delta(z_j,r_j),C_{jk}\}$ is called the system of connection matrices on $\mathscr{D}$ defined by the system $\{\Delta(z_j,r_j),W_j(z)\}$ of fundamental solutions of (1.10.1),

THEOREM 1.10.2. *If two systems $\{\Delta(z_j,r_j),C_{jk}\}$ and $\{\Delta(\zeta_j,\rho_j),\Gamma_{jk}\}$ of connection matrices on $\mathscr{D}$ are defined by two systems $\{\Delta(z_j,r_j),W_j(z)\}$ and $\{\Delta(\zeta_j,\rho_j),V_j(z)\}$ of fundamental solutions of differential equation* (1.10.1) *respectively, then these two systems of connection matrices are equivalent.*

[Proof was omitted by the translator and left to the reader as an exercise.]

Thus, by virtue of Theorem 1.10.2, a unique structure $\mathscr{C}$ of analytic continuation on $\mathscr{D}$ is defined by differential equation (1.10.1).

DEFINITION 1.10.3. The structure $\mathscr{C}$ of analytic continuation on $\mathscr{D}$ defined as explained above is called the structure of analytic continuation defined by differential equation (1.10.1).

THEOREM 1.10.4. *If a system $\{\Delta(z_j,r_j),D_{jk}\}$ of connection matrices belongs to the structure of analytic continuation defined by differential equation* (1.10.1), *then there exists a system $\{\Delta(z_j,r_j),V_j(z)\}$ of fundamental solutions of* (1.10.1) *that defines the system $\{\Delta(z_j,r_j),D_{jk}\}$, i.e., we have*

$$D_{jk}=V_j(z)^{-1}V_k(z)$$

whenever $\Delta(z_j,r_j)\cap\Delta(z_k,r_k)\neq\varnothing$.

PROOF. Let $\{\Delta(z_j, r_j), W_j(z)\}$ be a system of fundamental solutions of (1.10.1) and define matrices C_{jk} by (1.10.2). Then $\{\Delta(z_j, r_j), C_{jk}\}$ is a system of connection matrices such that

$$\{\Delta(z_j, r_j), D_{jk}\} \sim \{\Delta(z_j, r_j), C_{jk}\}.$$

Consequently, by virtue of Theorem 1.4.7, there exists a system $\{C_j\}$ of matrices in $GL(n, \mathbf{C})$ such that $C_{jk}C_k = C_jD_{jk}$. Set $V_j = W_j(z)C_j$, then $\{\Delta(z_j, r_j), V_j(z)\}$ is a system of fundamental solutions of (1.10.1) and we have

$$V_j(z)^{-1}V_k(z) = C_j^{-1}W_j(z)^{-1}W_k(z)C_k = C_j^{-1}C_{jk}C_k = D_{jk}. \quad \square$$

DEFINITION 1.10.5. Let $\mathscr{C}$ be the structure of analytic continuation defined by differential equation (1.10.1). Then the structure $\mathscr{M}(\mathscr{C})$ of monodromy (cf. §1.7) is called the structure of monodromy defined by (1.10.1).

§1.11. Two linear ordinary differential equations which define the same structure of analytic continuation

Let $\mathscr{D}$ be a connected open set in the z-plane and assume that two matrices $A_1(z)$ and $A_2(z)$ belong to $gl(n, \mathscr{O}(\mathscr{D}))$. Let $\mathscr{C}_1$ and $\mathscr{C}_2$ be two structures of analytic continuation on $\mathscr{D}$ which are defined by the two differential equations

$$\frac{dW}{dz} = A_1(z)W \tag{1.11.1}$$

and

$$\frac{dV}{dz} = A_2(z)V \tag{1.11.2}$$

respectively.

THEOREM 1.11.1. *In order that the two structures $\mathscr{C}_1$ and $\mathscr{C}_2$ defined above be identical, it is necessary and sufficient that there exists a matrix $P(z)$ in $GL(n, \mathscr{O}(\mathscr{D}))$ such that differential equation* (1.11.1) *is changed to differential equation* (1.11.2) *by the transformation*

$$W = P(z)V. \tag{1.11.3}$$

PROOF. Assume first that $\mathscr{C}_1 = \mathscr{C}_2 = \mathscr{C}$ and that $\{\Delta(z_j, r_j), C_{jk}\}$ is a system of connection matrices in $\mathscr{C}$. Then, by virtue of Theorem 1.10.4, there exist systems $\{\Delta(z_j, r_j), W_j(z)\}$ and $\{\Delta(z_j, r_j), V_j(z)\}$ of fundamental solutions of (1.11.1) and (1.11.2) respectively such that

$$C_{jk} = W_j(z)^{-1}W_k(z) = V_j(z)^{-1}V_k(z)$$

whenever $\Delta(z_j, r_j) \cap \Delta(z_k, r_k) \neq \emptyset$. Hence if we set

$$P(z) = W_j(z)^{-1} V_j(z) \quad \text{on} \quad \Delta(z_j, r_j)$$

for every j, the matrix $P(z)$ belongs to $GL(n, \mathcal{O}(\mathcal{D}))$. Since $W_j(z) = P(z)V_j(z)$ for every j, transformation (1.11.3) changes (1.11.1) to (1.11.2). Thus the necessity of the condition of Theorem 1.11.1 is proved. We leave the proof of the sufficiency to the reader. □

If $\mathcal{D}$ is connected and simply connected, then the fundamental group $\pi_1(\mathcal{D}, z_0)$ of $\mathcal{D}$ consists of the unit element only. Hence there is only one structure of monodromy on $\mathcal{D}$. Therefore there is also only one structure of analytic continuation on $\mathcal{D}$. Because of this situation, we can apply Theorem 1.11.1 to any two differential equations (1.11.1) and (1.11.2). In particular, if we choose the zero matrix as $A_2(z)$, we can derive the theorem below.

THEOREM 1.11.2. *Assume that $\mathcal{D}$ is connected, simply connected and open in the z-plane and that a matrix $A(z)$ belongs to $gl(n, \mathcal{O}(\mathcal{D}))$. Then there exists a matrix $P(z)$ in $GL(n, \mathcal{O}(\mathcal{D}))$ such that the differential equation*

$$\frac{dW}{dz} = A(z)W$$

is changed to

$$\frac{dV}{dz} = 0 \tag{1.11.4}$$

by the transformation $W = P(z)V$.

In the case when $\mathcal{D}$ is connected and simply connected, we can regard differential equation (1.11.4) as an equation of standard form on $\mathcal{D}$. If $\mathcal{D}$ is merely connected, we can classify all differential equations on $\mathcal{D}$ by structures of analytic continuation (or structures of monodromy) defined by these differential equations. Selecting an equation of standard form from each equivalence class of differential equations we can derive a result similar to Theorem 1.11.2. To investigate such a situation further, the following existence theorem of Grauert is useful.

THEOREM 1.11.3 (Grauert). *Let $\mathcal{D}$ be a connected open set in the z-plane and let $\mathcal{C}$ be a structure of analytic continuation on $\mathcal{D}$. Then there exists a matrix $A(z)$ in $gl(n, \mathcal{O}(\mathcal{D}))$ such that the differential equation*

$$\frac{dW}{dz} = A(z)W \tag{1.11.5}$$

defines the structure $\mathcal{C}$.

By claiming that differential equation (1.11.5) defines the structure $\mathscr{C}$ we mean that the system $\{\Delta(z_j, r_j), C_{jk}\}$ of connection matrices defined by a system $\{\Delta(z_j, r_j), W_j(z)\}$ of fundamental solutions of (1.11.5) belongs to $\mathscr{C}$. The matrices C_{jk} are given by $C_{jk} = W_j(z)^{-1}W_k(z)$ whenever $\Delta(z_j, r_j) \cap \Delta(z_k, r_k) \neq \varnothing$. If we look into this situation more carefully, we can derive the lemma below.

LEMMA 1.11.4. *Let $\mathscr{D}$ be a connected open set in z-plane and let $\{\Delta(z_j, r_j)\}$ be a covering of $\mathscr{D}$. Assume that, for every j, a matrix $W_j(z)$ is given in $GL(n, \mathscr{O}(\Delta(z_j, r_j)))$. Then if the matrix*

$$C_{jk} = W_j(z)^{-1}W_k(z) \tag{1.11.6}$$

is a constant matrix independent of z for every (j, k) such that $\Delta(z_j, r_j) \cap \Delta(z_k, r_k) \neq \varnothing$, the matrix

$$A(z) = \frac{dW_j(z)}{dz}W_j(z)^{-1} \qquad (z \in \Delta(z_j, r_j)) \tag{1.11.7}$$

belongs to $gl(n, \mathscr{O}(\mathscr{D}))$ and $\{\Delta(z_j, r_j), W_j(z)\}$ is a system of fundamental solutions of differential equation (1.11.5).

Utilizing this lemma, we can restate Theorem 1.11.3 as follows.

THEOREM 1.11.3′ (Grauert). *Let $\mathscr{D}$ be a connected open set in the z-plane and let $\{\Delta(z_j, r_j), C_{jk}\}$ be a system of connection matrices on $\mathscr{D}$. Then, for every j, there exists a matrix $W_j(z)$ in $GL(n, \mathscr{O}(\Delta(z_j, r_j)))$ such that $C_{jk} = W_j(z)^{-1}W_k(z)$ whenever $\Delta(z_j, r_j) \cap \Delta(z_k, r_k) \neq \varnothing$.*

In Chapter 2 we shall explain Theorem 1.11.3′ from a more general standpoint. In particular, we shall examine the case when the matrices C_{jk} belong to a subgroup of $GL(n, \mathbf{C})$. We leave the proof of Lemma 1.11.4 to the reader.

CHAPTER II

Existence Theorem of Grauert

§2.0. Preliminaries

To prove the theorem of Grauert given at the end of Chapter 1 (cf. Theorems 1.11.3 and 1.11.3′), we shall explain a more general result.

THEOREM 2.0.1 (Grauert). *Assume that*

(I) *M is an open Riemann surface,*
(II) *G is a connected complex Lie group,*
(III) *$\{\Omega_j\}$ is a covering of M by open sets Ω_j,*
(IV) *a holomorphic map $F_{jk}(z) : \Omega_j \cap \Omega_k \to G$ is given whenever $\Omega_j \cap \Omega_k \neq \varnothing$,*
(V)
$$F_{jk}(z)F_{kh}(z) = F_{jh}(z) \tag{2.0.1}$$
holds at all points of $\Omega_j \cap \Omega_k \cap \Omega_h$ whenever $\Omega_j \cap \Omega_k \cap \Omega_h \neq \varnothing$.

Then, for every j, there exists a holomorphic map $F_j(z) : \Omega_j \to G$ such that
$$F_{jk}(z) = F_j(z)^{-1}F_k(z) \tag{2.0.2}$$
holds for $z \in \Omega_j \cap \Omega_k$ ($\neq \varnothing$).

We shall explain a proof of this theorem, focusing our attention mostly on the facts below.

(i) an open Riemann surface is a Stein manifold (cf. §2.10),
(ii) a Stein manifold is approximated by Oka–Weil domains (cf. §2.9),
(iii) an Oka–Weil domain is holomorphically embedded into a polycylinder in a higher dimensional space (cf. §2.9),
(iv) a decomposition theorem of Cartan holds on a polycylinder (cf. §2.2),
(v) for an open Riemann surface, there exists a continuous map $S_j(z) : \Omega_j \to G$ such that $F_{jk}(z) = S_j(z)^{-1}S_k(z)$ on $\Omega_j \cap \Omega_k$ ($\neq \varnothing$) (cf. §2.10),

(vi) a continuous map can be deformed to a holomorphic map continuously (cf. §2.5),

(vii) if a holomorphic map is continuously deformed to the constant map whose values are identically equal to the unit element of G, then such a deformation can be also achieved holomorphically (cf. §2.5).

Since it is impossible for the author to prove all of these facts in this book, literatures such as Gunning–Rossi [2], N. Steenrod [10] and others will be used as references. We also restrict G to a subgroup of $GL(n, C)$ (cf. §2.1). In this chapter we introduce a topology into $GL(n, C)$ by means of a usual norm.

We borrowed the terminology concerning Lie groups and manifolds mostly from N. Iwahori [6] and the Iwanami Dictionary of Mathematics.

§2.1. Linear groups

If we denote by $gl(n, C)$ the set of all $n \times n$ matrices with complex entries, then $gl(n, C)$ is an n^2-dimensional vector space over C. If we define a commutator $[A, B]$ of two matrices A and B by $[A, B] = AB - BA$, then $gl(n, C)$ becomes a Lie algebra. This Lie algebra is the Lie algebra of the connected complex Lie group $GL(n, C)$. There is a one-to-one correspondence between connected complex Lie subgroups of $GL(n, C)$ and subalgebras of the Lie algebra $gl(n, C)$. Connected complex Lie subgroups of the Lie group $GL(n, C)$ are called **linear groups**. We can characterize the linear groups by the following two properties:

(1) they are subgroups of $GL(n, C)$ as abstract groups

(2) they are connected submanifolds of the complex analytic manifold $GL(n, C)$.

If g is the Lie algebra of a linear group G, then the dimension of the vector space g over C is equal to the dimension of the manifold G. For every element E of the Lie algebra g, $\exp E = \sum_{m \geq 0} \frac{1}{m!} E^m$ is an element of G. Moreover, if the absolute value of every entry of E tends to 0, then $\exp E$ tends to the identity matrix I within G. On the other hand, if we choose a neighborhood U of the identity matrix I suitably, $\log A = -\sum_{m \geq 1} \frac{1}{m}(I - A)^m$ is an element of g for every A in U. Let $\{E_1, E_2, \cdots, E_N\}$ be a basis for the vector space g over C. Denote by $(\alpha_1, \alpha_2, \cdots, \alpha_N)$ an N-tuple of complex numbers. Then for a given small positive number ε, a neighborhood of the identity matrix I in G is given by

$$U(\varepsilon) = \left\{ \exp\,[\sum_{j=1}^{N} \alpha_j E_j\,]; \quad |\alpha_j| < \varepsilon \quad (j = 1, \cdots, N) \right\}. \tag{2.1.1}$$

In the neighborhood $U(\varepsilon)$, we can use $(\alpha_1, \cdots, \alpha_N)$ as a system of coordinates of the manifold G. This system of coordinates is called **the system of standard coordinates** (of the first kind) determined by the basis $\{E_1, \cdots .E_N\}$. Since G is connected, G is generated by elements in a neighborhood $U(\varepsilon)$ of the identity matrix I.

In the case when G is not closed as a subset of the topological space $GL(n, C)$, the relative topology induced on G by $GL(n, C)$ differs from the inner topology of G as a manifold. However, in a neighborhood $U(\varepsilon)$ of the identity matrix I of G, the relative topology and the inner topology are identical. Since the manifold G is locally compact, every Cauchy sequence of elements of G converges to an element of G.

The following six lemmas will be utilized in the proofs of a decomposition theorem of Cartan (cf. §§2.2 and 2.3) and its generalizations (cf. §§2.4–2.7). Fix a small positive number ε_0 so that we can use a system of standard coordinates in the neighborhood $U(\varepsilon_0)$ of the identity matrix I. Further, fix another small positive number ε_1 so that the product ABC is in $U(\varepsilon_0)$ if three matrices A, B, and C are in $U(\varepsilon_1)$. We denote by $|A|$ the sum of absolute values of all entries of a matrix A.

LEMMA 2.1.1. *If we choose a suitable positive number k_0, every matrix A in $U(\varepsilon)$ satisfies the inequality*

$$|I - A| \leq k_0\,\varepsilon, \tag{2.1.2}$$

where $0 < \varepsilon \leq \varepsilon_0$ and k_0 is independent of ε.

LEMMA 2.1.2. *If a matrix A in $U(\varepsilon_0)$ satisfies an inequality $|I - A| \leq \delta$, then A is contained in $U(k_1\delta)$, where k_1 is a positive number independent of δ.*

LEMMA 2.1.3. *If two matrices A and B satisfy the condition*:

$$A \in U(\varepsilon_0), \quad B \in U(\varepsilon), \quad AB \in U(\varepsilon_0), \tag{2.1.3}$$

then the inequality

$$|\log(AB) - \log\,A| \leq k_2\,\varepsilon \tag{2.1.4}$$

holds, where $0 < \varepsilon \leq \varepsilon_0$ and k_2 is a positive number independent of ε.

The proofs of these three lemmas are left to the reader.

LEMMA 2.1.4. *If we choose a small positive number p less than one in a suitable way, the product AB of two matrices A and B in $U(p\varepsilon)$ is in $U(\varepsilon)$, where $0 < \varepsilon \leq \varepsilon_0$ and p does not depend on ε.*

PROOF. If p is a positive number less than one, the inequality $0 < p\,\varepsilon < \varepsilon \leq \varepsilon_1$ holds. Consequently AB is in $U(\varepsilon_0)$ if A and B are in $U(p\,\varepsilon)$. From the inequality

$$|I - AB| \leq |I - A| + |A||I - B|$$

and Lemma 2.1.1 it follows that $|I - AB| \leq (2 + k_0\,\varepsilon_1)k_0 p\,\varepsilon$. By virtue of Lemma 2.1.2, AB is contained in $U((2 + k_0\,\varepsilon_1)k_0 k_1 p\,\varepsilon)$. Hence it suffices to determine p by $(2 + k_0\,\varepsilon_1)k_0 k_1 p\,\varepsilon \leq 1$. □

LEMMA 2.1.5. *Assume that a positive number p satisfies the conditions given in Lemma 2.1.4. If the matrices A_k are in $U(p^k\varepsilon)$ for all positive integers k, then the products $A_1 A_2 \cdots A_m$ converge as $m \to \infty$, and the limit is contained in the closure of $U(\varepsilon)$ in G provided $0 < \varepsilon \leq \varepsilon_1$.*

PROOF. Because of the assumptions, $A_m \in U(p^{m-1}\varepsilon)$ for all m. Also $A_m A_{m+1} A_{m+2} \cdots A_{m+\ell} \in U(p^{m-1}\varepsilon)$ if $A_{m+1} A_{m+2} \cdots A_{m+\ell} \in U(p^m\varepsilon)$. Hence utilizing induction on ℓ, we can verify that $A_m A_{m+1} \cdots A_{m+\ell}$ is contained in $U(p^{m-1}\varepsilon)$ for all pairs (m, ℓ) of positive integers. In particular

$$A_1 A_2 \cdots A_m \in U(\varepsilon) \tag{2.1.5}$$

holds for every m. Moreover, because of Lemma 2.1.3, we can derive the inequalities

$$|\log(A_1 A_2 \cdots A_{m+\ell}) - \log(A_1 A_2 \cdots A_m)| \leq k_2 p^m \varepsilon. \tag{2.1.6}$$

Therefore, $\log(A_1 A_2 \cdots A_m)$ converge in g as $m \to \infty$. If we denote by E the limit of $\log(A_1 A_2 \cdots A_m)$, then $A_1 A_2 \cdots A_m$ converge to $\exp E$ in G. □

LEMMA 2.1.6. *If three matrices A, B, and C satisfy the two conditions below:*

(i) $A \in U(\delta_1)$, $B \in U(\delta_2)$, $C \in U(\varepsilon_1)$,
(ii) $\log C = \log A + \log B$,

then $A^{-1}CB^{-1}$ is contained in $U(k_3\delta_1\delta_2)$, where $0 < \delta_1 \leq \varepsilon_1$, $0 < \delta_2 \leq \varepsilon_2$, and k_3 is a positive number independent of δ_1 and δ_2.

PROOF. From assumption (i), it follows that $A^{-1} \in U(\delta_1)$ and $B^{-1} \in U(\delta_2)$. Therefore $A^{-1}CB^{-1}$ is contained in $U(\varepsilon_0)$. Also, from assumption (ii), we can derive the inequality

$$|\log(A^{-1}CB^{-1})| \leq K|\log A|\,|\log B|, \tag{2.1.7}$$

where K is a positive number independent of A and B. Utilizing inequality (2.1.7) and definition (2.1.1) of $U(\varepsilon)$, we can complete the proof of the lemma. □

§2.2. A decomposition theorem of Cartan

This section and the next section are very similar to Section E (pp. 192–201), Chapter VI of Gunning–Rossi [2].

We denote by C^ℓ the complex ℓ-dimensional space $C \times C \times \cdots \times C$ (the direct product of ℓ copies of C). Fixing four real numbers $a_1 < a_2 < a_3 < a_4$ and two real numbers $b_1 < b_2$, we define three domains in C, i.e.,

$$K_1 = \{z = x + iy;\ \ a_2 < x < a_3,\ \ b_1 < y < b_2\},$$
$$K_1' = \{z = x + iy;\ \ a_1 < x < a_3,\ \ b_1 < y < b_2\},$$
$$K_1'' = \{z = x + iy;\ \ a_2 < x < a_4,\ \ b_1 < y < b_2\}.$$

The three domains satisfy the condition $K_1' \cap K_1'' = K_1$. Moreover, utilizing bounded and simply connected domains $K_2, K_3, \cdots, K_\ell$, we define three polycylinders in C^ℓ, i.e.,

$$K = K_1 \times K_2 \times \cdots \times K_\ell,$$
$$K' = K_1' \times K_2 \times \cdots \times K_\ell,$$
$$K'' = K_1'' \times K_2 \times \cdots \times K_\ell.$$

Clearly the condition $K' \cap K'' = K$ is satisfied. For a subset $\mathscr{D}$ of the space C, we denote by $\mathscr{D}(\delta)$ the δ-neighborhood of $\mathscr{D}$. Then

$$K(\delta) = K_1(\delta) \times K_2(\delta) \times \cdots \times K_\ell(\delta)$$

is a neighborhood of the domain K in C^ℓ. Hereafter a positive number δ and a linear group G are fixed. We denote also by z a point in C^ℓ.

THEOREM 2.2.1 (Cartan). *If a matrix $F(z)$ is a holomorphic map: $K(\delta) \to G$, there exist two matrices $F'(z)$ and $F''(z)$ satisfying the following conditions*:

(i) *$F'(z)$ is a holomorphic map : $K' \to G$,*
(ii) *$F''(z)$ is a holomorphic map : $K'' \to G$,*
(iii) *$F(z) = F'(z)F''(z)$ at every point z of K.*

This theorem is a particular case of the Grauert lemma which will be explained in §2.4. However this particular case is an important key to the proofs in more general cases. We shall explain the main part of the proof of Theorem 2.2.1 in the next section. Before doing this, we summarize various remarks as lemmas. Hereafter we assume that a matrix $F(z)$ satisfies the conditions given in Theorem 2.2.1.

Since the domain $K(\delta)$ can be shrunk to a point continuously, we can prove the lemma below.

LEMMA 2.2.2. *Given a sufficiently small positive number ε, there exists a finite number of matrices $F_1(z), F_2(z), \cdots, F_r(z)$ such that*

(i) *$F_1(z), \cdots, F_r(z)$ are holomorphic maps : $K(\delta) \to G$,*
(ii) *$F_1(z), \cdots, F_r(z)$ are contained in $U(\varepsilon)$ at every point z of $K(\frac{1}{2}\delta)$,*
(iii) *$F(z) = F_1(z)F_2(z)\cdots F_r(z)$ at every point of $K(\delta)$.*

Every entry of $\log F_j(z)$ is a holomorphic function of z in the domain $K(\frac{1}{2}\delta)$. Further, $\log F_j(z)$ belongs to the Lie algebra g. Expressing $\log F_j(z)$ in terms of a basis $\{E_1, \cdots, E_N\}$ for g and utilizing an approximation of a holomorphic function by a polynomial, we can prove the lemma below.

LEMMA 2.2.3. *Given a positive number ρ, there exist r matrices $P_1(z), \cdots, P_r(z)$ satisfying the following conditions:*

(i) *the entries of $P_1(z), \cdots, P_r(z)$ are polynomials in z,*
(ii) *$P_1(z), \cdots, P_r(z)$ are contained in g at every point z of C^ℓ,*
(iii) *the inequalities $|\log F_j(z) - P_j(z)| < \rho \quad (j = 1, \cdots, r)$ hold at every point z of $K(\frac{1}{4}\delta)$.*

The matrix $\exp P_j(z)$ is contained in G at every point z of C^ℓ, and

$$\exp P_j(z) = \left[\exp\left[\frac{1}{k}P_j(z)\right]\right]^k$$

for every positive integer k. Hence the matrix $\exp P_j(z)$ is a holomorphic map : $C^\ell \to G$. Consequently

$$Q(z) = \exp P_1(z) \exp P_2(z) \cdots \exp P_r(z)$$

is a holomorphic map : $C^\ell \to G$. Furthermore, at every point $z \in K(\frac{1}{4}\delta)$, $Q(z)$ tends to $F(z) = F_1(z)\cdots F_r(z)$ in G as ρ of Lemma 2.2.3 tends to 0. Thus we proved the lemma below.

LEMMA 2.2.4. *Given a sufficiently small positive number ρ, there exists a matrix $Q(z)$ such that*

(i) *$Q(z)$ is a holomorphic map : $C^\ell \to G$,*
(ii) *$F(z)Q(z)^{-1}$ is contained in $U(\rho)$ at every point z of $K(\frac{1}{4}\delta)$.*

If the conclusion of Theorem 2.2.1 holds for the matrix $F(z)Q(z)^{-1}$, then $F(z)Q(z)^{-1}$ can be written in the form $F'(z)F''(z)$. Consequently $F(z) = F'(z)(F''(z)Q(z)^{-1})$. Therefore Theorem 2.2.1 is also verified for

$F(z)$. Because of this reason, in order to prove Theorem 2.2.1, it suffices to consider the case when, given a positive number δ, we can find a positive number ρ so that the matrix $F(z)$ is contained in $U(\rho)$ at every point z of $K(\delta)$.

§2.3. Proof of the Cartan theorem

We assume that a positive number p satisfies the condition given in Lemma 2.1.4. For every positive integer k, we define three polycylinders as follows:

$$V_k = K_1(\frac{1}{2}p^k\delta) \times K_2(\frac{1}{2}\delta) \times \cdots \times K_\ell(\frac{1}{2}\delta),$$
$$V_k' = K_1'(\frac{1}{2}p^k\delta) \times K_2(\frac{1}{2}\delta) \times \cdots \times K_\ell(\frac{1}{2}\delta),$$
$$V_k'' = K_1''(\frac{1}{2}p^k\delta) \times K_2(\frac{1}{2}\delta) \times \cdots \times K_\ell(\frac{1}{2}\delta).$$

We also denote by $\bar{V}_k, \bar{V}_k'$ and $\bar{V}_k''$ the closures of these three polycylinders in C^ℓ respectively.

LEMMA 2.3.1. *If every entry of a matrix $H(z)$ is holomorphic in z in a neighborhood of $\bar{V}_k$ and if $H(z)$ is contained in $U(\varepsilon)$ at every point z of $\bar{V}_k$, there exist two matrices $H'(z)$ and $H''(z)$ such that*

(i) *every entry of $H'(z)$ is holomorphic in z on V_k', and $H'(z)$ is contained in $U(\frac{L}{p^k}\varepsilon)$ at every point z of $\bar{V}_{k+1}'$,*
(ii) *every entry of $H''(z)$ is holomorphic in z on V_k'', and $H''(z)$ is contained in $U(\frac{L}{p^k}\varepsilon)$ at every point z of $\bar{V}_{k+1}''$,*
(iii) $\log H(z) = \log H'(z) + \log H''(z)$ *holds at every point z of V_k.*

Here $0 < \varepsilon \leq \varepsilon_0$ and L is a positive number depending only on the domain $K_1(\delta)$.

PROOF. If we denote by $(z_1, z_2, \cdots, z_\ell)$ the ℓ coordinates of a point z, and by γ the boundary of $K_1(\frac{1}{2}p^k\delta)$ in C, the Cauchy integral formula

$$\log H(z) = \frac{1}{2\pi i}\oint_\gamma \frac{1}{\zeta - z_1} \log H(\zeta, z_2, \cdots, z_\ell)d\zeta \tag{2.3.1}$$

holds at every point z of V_k. Fixing a real number a which satisfies the inequality $a_2 < a < a_3$, we divide γ into two parts γ' and γ'', γ' being on the right side and γ'' on the left side of the line $\mathrm{Re}(z) = a$ which is parallel to the imaginary axis, where $\mathrm{Re}(z)$ denotes the real part of a complex number z. Utilizing the division of γ into γ' and γ'', we divide the integral on the right-hand side of (2.3.1) into two parts

$$E'(z) = \frac{1}{2\pi i}\int_{\gamma'} \frac{1}{\zeta - z_1} \log H(\zeta, z_2, \cdots, z_\ell)d\zeta$$

and

$$E''(z) = \frac{1}{2\pi i} \int_{\gamma''} \frac{1}{\zeta - z_1} \log\, H(\zeta, z_2, \cdots, z_\ell) d\zeta.$$

The entries of the matrix $E'(z)$ are holomorphic in z on V_k', and the entries of the matrix $E''(z)$ are holomorphic in z on V_k''. Also, since $\log\, H(z)$ is contained in the Lie algebra g at every point z of $\bar{V}_k$, the matrix $E'(z)$ is contained in g at every point of V_k', and similarly the matrix $E''(z)$ is contained in g at every point of V_k''. If M is the length of the boundary of the domain $K_1(\delta)$ in C, the inequality

$$|E'(z)| \leq \frac{M}{\pi p^k (1-p)\delta} \operatorname{Max}_{\tilde{z} \in \bar{V}_k} |\log\, H(\tilde{z})|$$

holds at every point z of $\bar{V}_{k+1}'$. Similarly, we can prove the inequality

$$|E''(z)| \leq \frac{M}{\pi p^k (1-p)\delta} \operatorname{Max}_{\tilde{z} \in \bar{V}_k} |\log\, H(\tilde{z})|$$

at every point z of $\bar{V}_{k+1}''$. If we define two matrices $H'(z)$ and $H''(z)$ by $\exp\, E'(z)$ and $\exp E''(z)$ respectively, then conditions (i), (ii) and (iii) of Lemma 2.3.1 are satisfied. □

Let L be the positive number given in Lemma 2.3.1 and let k_3 be the positive number in Lemma 2.1.6. Choose a positive number ρ and a positive integer N so that the inequalities

$$0 < \rho \leq \varepsilon_1, \quad k_3 \rho \leq 1, \quad L p^{N-1} \leq 1 \tag{2.3.2}$$

hold. We shall prove Theorem 2.2.1 only in the case when

$$F(z) \in U(p^{2N} \rho) \tag{2.3.3}$$

holds on $K(\delta)$. To prove Theorem 2.2.1 for the general case, it suffices to consider this particular case only (cf. the end of §2.2).

Utilizing induction on nonnegative integers k, we construct matrices $F_k(z)$, $F_k'(z)$, and $F_k''(z)$ successively so that the conditions below are satisfied:

(i) $F_0(z) = F(z)$,
(ii) $F_k(z) \in U(p^{2(N+k)}\rho) \qquad (z \in \bar{V}_k)$,
(iii) $F_k'(z) \in U(p^{N+k+1}\rho) \qquad (z \in \bar{V}_{k+1}')$,
(iv) $F_k''(z) \in U(p^{N+k+1}\rho) \qquad (z \in \bar{V}_{k+1}'')$,
(v) $F_{k+1}(z) = F_k'(z)^{-1} F_k(z) F_k''(z)^{-1} \qquad (z \in V_k)$.

Assumption (2.3.3) implies that condition (ii) is satisfied if $k = 0$. If we construct $F_0'(z)$ and $F_0''(z)$ by applying Lemma 2.3.1 to the matrix $F_0(z)$, we derive

$$F_0'(z) \in U(Lp^{2N}\rho) \qquad (z \in \bar{V}_1')$$

and

$$F_0''(z) \in U(Lp^{2N}\rho) \qquad (z \in \bar{V}_1'').$$

Since $Lp^{2N} = Lp^{N-1}p^{N+1} \leq p^{N+1}$, condition (iii) and (iv) hold for $k = 0$. If we apply Lemma 2.1.6 to the matrix $F_1(z) = F_0'(z)^{-1}F_0(z)F_0''(z)^{-1}$, we can verify

$$F_1(z) \in U(k_3 p^{2(N+1)}\rho^2) \qquad (z \in \bar{V}_1),$$

and the inequality $k_3 p^{2(N+1)}\rho^2 \leq p^{2(N+1)}\rho$ implies that condition (ii) holds for $k = 1$. Similarly, we can construct F_k, F_k' and F_k'' successively.

If we examine the proof of Lemma 2.1.5 carefully, it becomes clear that the products $F_0'(z)F_1'(z)\cdots F_k'(z)$ converge uniformly in K' as $k \to \infty$ and that its limit is contained in $U(\varepsilon_1)$. Let us denote this limit by $F'(z)$. Similarly we define $F''(z)$ by taking limit of $F_0''(z)F_1''(z)\cdots F_k''(z)$ as $k \to \infty$. Condition (v) implies that

$$F_{k+1}(z) = (F_0'(z)F_1'(z)\cdots F_k'(z))^{-1}F(z)(F_0''(z)F_1''(z)\cdots F_k''(z))^{-1}.$$

Hence by taking limits of the both sides, we have $F'(z)F''(z) = F(z)$ $(z \in K)$, and we complete the proof of Theorem 2.2.1.

§2.4. A lemma of Grauert

In order to improve the Cartan theorem, we define some notations. We utilize the notations K_1, K_1', K_1'', K_2, $K_3, \cdots, K_\ell, K$, K', K'', and $K(\delta)$ as defined in §2.2. In addition to those, we introduce two polycylinders

$$K'(\delta) = K_1'(\delta) \times K_2(\delta) \times \cdots \times K_\ell(\delta),$$

and

$$K''(\delta) = K_1''(\delta) \times K_2(\delta) \times \cdots \times K_\ell(\delta).$$

Evidently $K(\delta) = K'(\delta) \cap K''(\delta)$. Also, for a positive number δ, we fix a subset $\mathscr{D}_0$ of $K'(\delta) \cup K''(\delta)$ as follows:

(i) $\mathscr{D}_0$ is closed as a subset of $K'(\delta) \cup K''(\delta)$,

(ii) $\mathscr{D}_0$ is holomorphically embedded into $K'(\delta) \cup K''(\delta)$ as a complex analytic manifold (i.e., the topology induced on $\mathscr{D}_0$ by $K'(\delta) \cup K''(\delta)$ and the inner topology of $\mathscr{D}_0$ as a manifold are identical).

We do not assume that $\mathscr{D}_0$ is connected.

Next, we set

$$\mathscr{D} = \mathscr{D}_0 \cap K, \quad \mathscr{D}' = \mathscr{D}_0 \cap K' \quad \mathscr{D}'' = \mathscr{D}_0 \cap K'', \quad \mathscr{D}(\rho) = \mathscr{D}_0 \cap K(\rho),$$

where $0 < \rho < \delta$. Evidently $\mathscr{D} = \mathscr{D}' \cap \mathscr{D}''$ and $\mathscr{D} \subset \mathscr{D}(\rho)$. The closure of $K(\rho)$ in C^ℓ is contained in $K(\delta)$.

Denote by ζ a point on the manifold $\mathscr{D}_0$. Hereafter we assume that a matrix $H(\zeta)$ is a holomorphic map : $\mathscr{D}(\rho) \to G$.

DEFINITION 2.4.1. Let J be the interval $0 \leq t \leq 1$ of a real variable t. Assume that a matrix $H(\zeta, t)$ satisfies the conditions below:

(i) $H(\zeta, t)$ is a continuous map : $\mathscr{D}(\rho)\times J \to G$,
(ii) $H(\zeta, t)$ is a holomorphic map : $\mathscr{D}(\rho) \to G$ for each fixed $t \in J$,
(iii) $H(\zeta, 1) = H(\zeta)$, $H(\zeta, 0) = I$.

Then we say that the matrix $H(\zeta)$ can be deformed holomorphically to the identity matrix I on $\mathscr{D}(\rho)$.

LEMMA 2.4.2 (Grauert). *If a matrix $H(\zeta)$ is a holomorphic map: $\mathscr{D}(\rho) \to G$ which can be deformed holomorphically to the identity matrix I on $\mathscr{D}(\rho)$, then there exist two matrices $H'(\zeta)$ and $H''(\zeta)$ such that*

(i) *$H'(\zeta)$ is a holomorphic map : $\mathscr{D}' \to G$ which can be deformed holomorphically to I on $\mathscr{D}'$,*
(ii) *$H''(\zeta)$ is a holomorphic map : $\mathscr{D}'' \to G$ which can be deformed holomorphically to I on $\mathscr{D}''$,*
(iii) *$H(\zeta) = H'(\zeta)H''(\zeta)$ at every point of $\mathscr{D}$.*

PROOF. Given a sufficiently small positive number ε, there exists a finite number of matrices $H_1(\zeta), H_2(\zeta), \cdots, H_r(\zeta)$, satisfying the following conditions:

(i) $H_1(\zeta), \cdots, H_r(\zeta)$ are holomorphic maps : $\mathscr{D}(\rho) \to G$,
(ii) $H_1(\zeta), H_2(\zeta), \cdots, H_r(\zeta)$ are contained in $U(\varepsilon)$ at every point of $\mathscr{D}(\frac{1}{2}\rho)$,
(iii) $H(\zeta) = H_1(\zeta)H_2(\zeta)\cdots H_r(\zeta)$ at every point of $\mathscr{D}(\rho)$.

The matrices $\log H_j(\zeta)$ are contained in the Lie algebra g at every point of $\mathscr{D}(\frac{1}{2}\rho)$. Utilizing a basis $\{E_1, \cdots, E_N\}$ for g, we write the $\log H_j(\zeta)$ in the form

$$\log H_j(\rho) = \sum_{h=1}^{N} f_{j,h}(\zeta)E_h.$$

Then the $f_{j,k}(\zeta)$ are complex-valued holomorphic functions of ζ on $\mathscr{D}(\frac{1}{2}\rho)$. Since $\mathscr{D}(\frac{1}{2}\rho)$ is a nice subset of $K(\frac{1}{2}\rho)$, the $f_{j,h}(\zeta)$ can be extended to holomorphic functions $\hat{f}_{j,h}(z)$ on $K(\frac{1}{2}\rho)$. Hence, if we set

$$L_j(z) = \sum_{h=1}^{N} \hat{f}_{j,h}(z)E_h,$$

the entries of $L_j(z)$ are holomorphic functions of z on $K(\frac{1}{2}\rho)$ and $L_j(\zeta) = \log H_j(\zeta)$ on $\mathscr{D}(\frac{1}{2}\rho)$. Moreover the matrices $L_j(z)$ are contained in g at every point of $K(\frac{1}{2}\rho)$. Consequently, if we set

$$F_j(z) = \exp L_j(z),$$

the $F_j(z)$ are holomorphic maps : $K(\frac{1}{2}\rho) \to G$ and $F_j(\zeta) = H_j(\zeta)$ on $\mathscr{D}(\frac{1}{2}\rho)$. Furthermore if we set

$$F(z) = F_1(z)F_2(z)\cdots F_r(z),$$

$F(z)$ is a holomorphic map : $K(\frac{1}{2}\rho) \to G$ and $F(\zeta) = H(\zeta)$ on $\mathscr{D}(\frac{1}{2}\rho)$. Applying the decomposition theorem of Cartan to $F(z)$, we write F in the form $F(z) = F'(z)F''(z)$. If we restrict this result to $\mathscr{D}_0$, we can complete the proof of Lemma 2.4.2. □

In the proof given above, we used an extension of a holomorphic function on $\mathscr{D}(\frac{1}{2}\rho)$ to a holomorphic function on $K(\frac{1}{2}\rho)$. This is the reason why we assumed two properties (i) and (ii) of the domain $\mathscr{D}_0$ at the beginning of this section. (We can use the same argument even in the case when $\mathscr{D}_0$ is an analytic set instead of a manifold.) To understand these extension theorems precisely, we must study the theory of functions of several complex variables. For example, it is desirable for the reader to understand Theorem 18 on p. 245 of Gunning–Rossi [2]. This means that we must be familiar with the concept of Stein spaces. However, it would suffice to understand Theorem 6 on p. 212 of the same book of Gunning–Rossi if we restrict ourselves to the results concerning polycylinders. The basic argument is to solve the Cousin problem in terms of the theory of sheaf cohomology. In doing this, we make essential use of the concept of coherent sheaves. We would understand these situations more concretely if we read Corollary 4 on p. 35 and the succeeding remarks in the same book. Anyway, the author hopes that the reader would find some supplements concerning those materials in other suitable books.

In the proof given above, we extended $H(\zeta)$ to $F(z)$ and applied the Cartan theorem to $F(z)$. If we look into the proof of the Cartan theorem again, we notice that, if $F(z)$ is too "large", we change $F(z)$ by means of Lemma 2.2.4 so that $F(z)$ is contained in a neighborhood $U(\rho)$ of the identity matrix I. The most important part of the proof of Lemma 2.2.4 is an approximation of a holomorphic function by a polynomial (cf. Lemma 2.2.3).

Now, it is also plausible to interchange the order of these two processes. Namely, we first change $H(\zeta)$ on $\mathscr{D}(\frac{1}{2}\rho)$ so that $H(\zeta)$ is contained in a neighborhood of I and then extend $H(\zeta)$ to $F(z)$ on $K(\frac{1}{2}\rho)$. Next, apply the Cartan theorem in a neighborhood of I. Actually this idea was used by Grauert [18]. To change $H(\zeta)$ on $\mathscr{D}(\frac{1}{2}\rho)$, we need a more sophisticated method than approximation of holomorphic functions by polynomials. Moreover in this process we utilize the fact that, as we extend $f_{j,h}(\zeta)$ to $\hat{f}_{j,h}(z)$, we can make $|\hat{f}_{j,h}(z)|$ sufficiently small if $|f_{j,h}(\zeta)|$ is sufficiently small. This fact is verified by showing that, if we denote by $\mathscr{H}_1$ (resp. $\mathscr{H}_2$) the Frechet space consisting of all functions holomorphic on $K(\frac{1}{2}\rho)$ (resp. $\mathscr{D}(\frac{1}{2}\rho)$), then the map : $\mathscr{H}_1 \to \mathscr{H}_2$ which restricts every function in $\mathscr{H}_1$ to $\mathscr{D}(\frac{1}{2}\rho)$ is continuous and surjective. Hence this map is open (cf. Lemmas 2.6.4 and 2.6.5). On the other hand, it suffices to prove the Cartan theorem only in a neighborhood of I as in §2.3. In Grauert [18], an object more complex than a group G was treated. In such a case perhaps it would be better to follow the argument of Grauert.

To apply Lemma 2.4.2, we do not need the connectedness of the set $\mathscr{D}$. Instead, it is required that the matrix $H(\zeta)$ can be deformed holomorphically to the identity matrix I on $\mathscr{D}(\rho)$. Let us look at an example of such a situation. We consider in C three domains

$$S = \{\ z = x + iy;\ \ |x| < 2,\ \ |y| < 2\ \},$$
$$\Omega = \{\ z;\ \ |z| > 1\ \},$$
$$\Omega_0 = \{\ z:\ \ |z| < 1\ \}.$$

The map $f_1(z) = z : S \to S$ is conformal and bijective. The map $f_2(z) = \frac{1}{z} : \Omega \to \Omega_0$ is also conformal and "bijective". (Assume that ∞ is in Ω if you please.) Next, if we consider the polycylinder $S \times \Omega_0$ in C^2, we can define a map : $S \cap \Omega \to S \times \Omega_0$ by assigning $(f_1(z), f_2(z))$ to z. In this way we can identify

$$D_0 = S \cap \Omega$$

and

$$\mathscr{D}_0 = \left\{ \left(z, \frac{1}{z} \right);\qquad z \in D_0 \right\}.$$

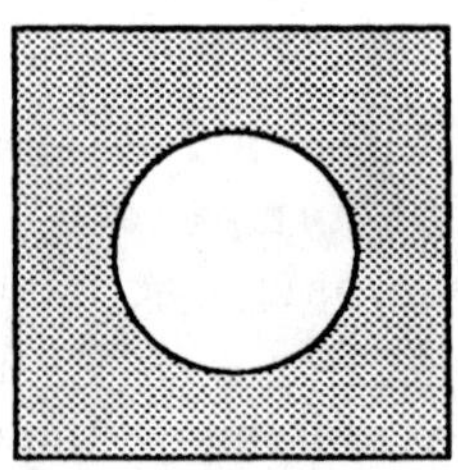

$S \cap \Omega$

FIGURE 1

However, we must have a clear idea of what this identification means. The subset D_0 of C has a natural structure as a complex analytic manifold. We regard $\mathscr{D}_0$ as a complex analytic manifold by bringing this natural structure of D_0 into $\mathscr{D}_0$. Then we can show that $\mathscr{D}_0$ is a closed subset of $S \times \Omega_0$ and is embedded holomorphically into $S \times \Omega_0$ as a submanifold.

Next, fixing two real numbers α and β so that $-2 < \alpha < \beta < 2$, we set

$$\begin{aligned}
K_1 &= \{ z = x + iy;\ \ \alpha < x < \beta,\ |y| < 2 \},\\
K_1' &= \{ z = x + iy;\ \ -2 < x < \beta,\ |y| < 2 \},\\
K_1'' &= \{ z = x + iy;\ \ \alpha < x < 2,\ |y| < 2 \},\\
K &= K_1 \times \Omega_0, \quad K' = K_1' \times \Omega_0, \quad K'' = K_1'' \times \Omega_0,\\
\mathscr{D} &= \mathscr{D}_0 \cap K, \quad \mathscr{D}' = \mathscr{D}_0 \cap K', \quad \mathscr{D}'' = \mathscr{D}_0 \cap K''.
\end{aligned}$$

We can regard $\mathscr{D}$ as a subset of D_0 as well as a subset of $\mathscr{D}_0$. We denote this subset of D_0 by D. Then there are several possibilities as shown by Figure 2.

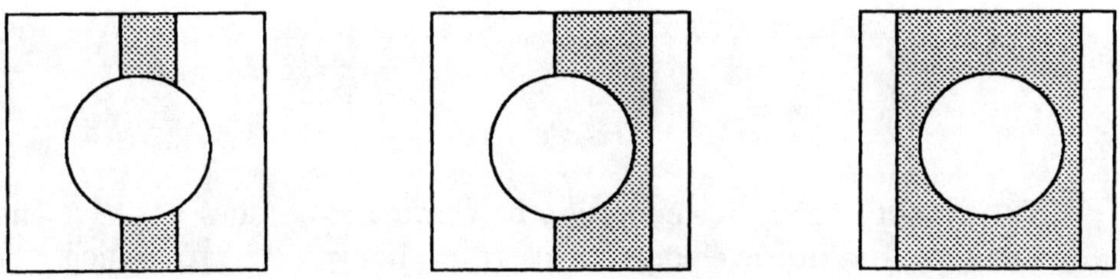

A B C

FIGURE 2

In the case of Figure 2A shown above, D is not connected. To apply Lemma 2.4.2 to this case, we can give a matrix $H(z)$ on each connected component in any way we like. Since each connected component can be shrunk to a point continuously, the requirement that $H(z)$ can be deformed to the identity matrix holomorphically is satisfied. To understand the conclusion of Lemma 2.4.2, we exhibit the subsets D' and D'' of D_0 corresponding to the subsets $\mathscr{D}'$ and $\mathscr{D}''$ of $\mathscr{D}_0$ by Figure 3A on page 44.

In the case of Figure 2B, we can also give $H(z)$ in any way we want. A general idea of the shapes of D' and D'' are given by Figure 3B on page 44. (We can ignore the hole in D'. We may cover it up by pasting a disk over it.)

In the case of Figure 2C, if we wish to utilize Lemma 2.4.2, the matrix $H(z)$ may not be arbitrary. We must assume in this case that $H(z)$ can be deformed holomorphically to the identity matrix.

Case of Figure A:

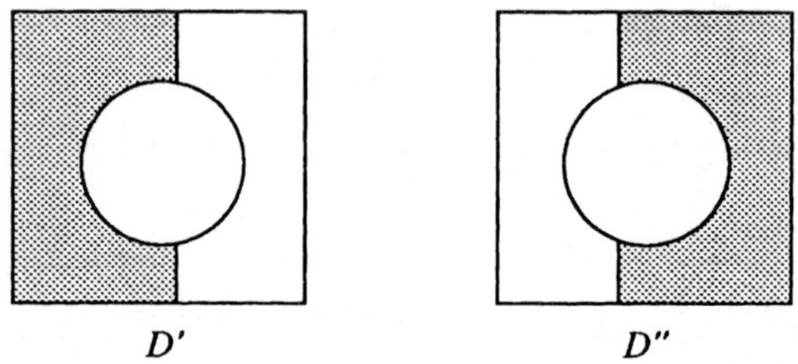

Case of Figure B:

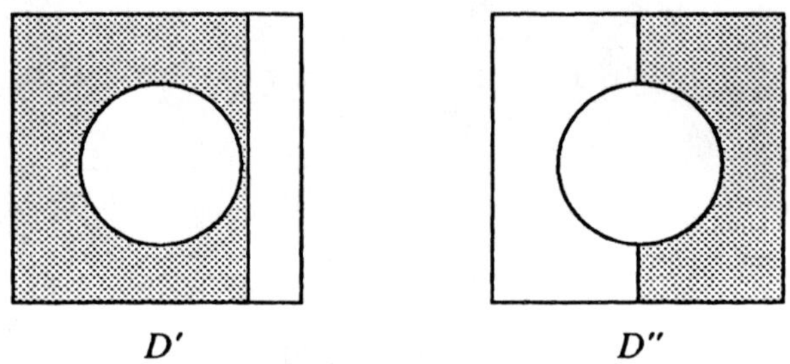

FIGURE 3

If $\mathscr{D}$ is empty, then the conclusion of Lemma 2.4.2 does not have any meaning. We did not exclude such cases explicitly. Hereafter such evident facts will not be emphasized unless it is absolutely necessary. In the conclusion of Lemma 2.4.2, we also included the statement that $H'(\zeta)$ and $H''(\zeta)$ can be deformed holomorphically to the identity matrix. This fact will not be used hereafter.

§2.5. A deformation theorem of Grauert

In order to utilize Lemma 2.4.2, $H(\zeta)$ must be deformed to the identity matrix holomorphically on $\mathscr{D}(\rho)$. In this section, we explain how to treat this requirement. In the case when $H(\zeta)$ does not meet this requirement, we might be able to change $H(\zeta)$ so that the requirement is satisfied. To do this, for example, it suffices to find two matrices $T'(\zeta)$ and $T''(\zeta)$ such that

(A) $T'(\zeta)$ is a holomorphic map : $\mathscr{D}'(\tilde{\rho}) \to G$,
(B) $T''(\zeta)$ is a holomorphic map : $\mathscr{D}''(\tilde{\rho}) \to G$,
(C) $T'(\zeta)H(\zeta)T''(\zeta)$ can be deformed to the identity matrix holomorphically on $\mathscr{D}(\tilde{\rho})$,

where $\mathscr{D}'(\tilde{\rho}) = \mathscr{D}_0 \cap K'(\tilde{\rho})$, $\mathscr{D}''(\tilde{\rho}) = \mathscr{D}_0 \cap K''(\tilde{\rho})$, and $\tilde{\rho}$ is a positive number satisfying the inequality $0 < \tilde{\rho} < \rho$.

To explain a method of finding such T' and T'', we introduce the following definition.

DEFINITION 2.5.1. Let us denote by J the interval $0 \leq t \leq 1$ of a real variable t and assume that a matrix $S(\zeta, t)$ satisfies the conditions below:

(i) $S(\zeta, t)$ is a continuous map : $\mathscr{D}(\rho) \times J \rightarrow G$,
(ii) $S(\zeta, 1) = S(\zeta)$, $\quad S(\zeta, 0) = I$.

Then we say that the matrix $S(\zeta)$ can be deformed to the identity matrix I continuously on $\mathscr{D}(\rho)$. ($S(\zeta)$ does not need to be holomorphic.)

To construct T' and T'', the following two lemmas are important.

LEMMA 2.5.2. *If a matrix $S(\zeta)$ is a continuous map* : $\mathscr{D}(\rho) \rightarrow G$, *then for every fixed positive number $\tilde{\rho}$ such that* $0 < \tilde{\rho} < \rho$, *there exists a matrix $T(\zeta)$ satisfying the two conditions below*:

(i) *$T(\zeta)$ is a holomorphic map* : $\mathscr{D}(\tilde{\rho}) \rightarrow G$,
(ii) *$T(\zeta)S(\zeta)$ can be deformed to the identity matrix I continuously on* $\mathscr{D}(\tilde{\rho})$.

LEMMA 2.5.3. *If a matrix $H(\zeta)$ is a holomorphic map* : $\mathscr{D}(\rho) \rightarrow G$ *and if $H(\zeta)$ can be deformed to the identity matrix I continuously on $\mathscr{D}(\rho)$, then $H(\zeta)$ can be deformed to I holomorphically on $\mathscr{D}(\tilde{\rho})$ for every fixed positive number $\tilde{\rho}$ such that* $0 < \tilde{\rho} < \rho$.

These two lemmas are important results due to Grauert [18]. They can be utilized also on $\mathscr{D}'(\rho)$ and $\mathscr{D}''(\rho)$. We shall prove them later. Instead, we shall now prove the lemma below.

LEMMA 2.5.4. *Suppose that a matrix $H(\zeta)$ is a holomorphic map*: $\mathscr{D}(\rho) \rightarrow G$ *and that $S'(\zeta)$ (resp. $S''(\zeta)$) is a continuous map* : $\mathscr{D}'(\rho) \rightarrow G$ (*resp.* : $\mathscr{D}''(\rho) \rightarrow G$). *Further, assume that*

$$H(\zeta) = S'(\zeta)S''(\zeta) \tag{2.5.1}$$

at every point ζ of $\mathscr{D}(\rho)$. Then for every fixed positive number $\tilde{\rho}$ such that $0 < \tilde{\rho} < \rho$, *there exist two matrices $T'(\zeta)$ and $T''(\zeta)$ satisfying conditions* (A), (B), *and* (C) *given at the beginning of this section.*

PROOF. Applying Lemma 2.5.2 to $S'(\zeta)$ and $S''(\zeta)$, we construct two matrices $T'(\zeta)$ and $T''(\zeta)$ respectively. Precisely speaking, for every fixed positive number $\hat{\rho}$ such that $\tilde{\rho} < \hat{\rho} < \rho$, we can find two matrices $T'(\zeta)$ and $T''(\zeta)$ satisfying the conditions below:

(A′) $T'(\zeta)$ is a holomorphic map : $\mathscr{D}'(\hat{\rho}) \rightarrow G$, and $T'(\zeta)S'(\zeta)$ can be deformed to the identity matrix I continuously on $\mathscr{D}'(\hat{\rho})$,

(B′) $T''(\zeta)$ is a holomorphic map : $\mathscr{D}''(\hat{\rho}) \to G$, and $S''(\zeta)T''(\zeta)$ can be deformed to the identity matrix I continuously on $\mathscr{D}''(\hat{\rho})$.

Then, $T'(\zeta)H(\zeta)T''(\zeta) = T'(\zeta)S'(\zeta)S''(\zeta)T''(\zeta)$ is a holomorphic map : $\mathscr{D}(\hat{\rho}) \to G$ and can be deformed to I continuously on $\mathscr{D}(\hat{\rho})$. Therefore condition (C) is also satisfied because of Lemma 2.5.3. □

Leaving it to the reader to explore condition (2.5.1) further, we shall explain our plan to prove Lemmas 2.5.2 and 2.5.3. By replacing $T(\zeta)$ by $T(\zeta)^{-1}$ we can restate Lemma 2.5.2 as follows.

LEMMA 2.5.2′. *If a matrix $S(\zeta)$ is a continuous map* : $\mathscr{D}(\rho) \to G$, *then for every fixed positive number $\tilde{\rho}$ such that $0 < \tilde{\rho} < \rho$, there exists a matrix $T(\zeta)$ satisfying the conditions below*:

(i) *$T(\zeta)$ is a holomorphic map* : $\mathscr{D}(\tilde{\rho}) \to G$,
(ii) *$S(\zeta)$ can be deformed to $T(\zeta)$ continuously on $\mathscr{D}(\tilde{\rho})$.*

Evidently condition (ii) of this lemma means that we replace condition (ii) of Definition 2.5.1 by $S(\zeta, 1) = S(\zeta)$, $S(\zeta, 0) = T(\zeta)$. In order to prove Lemmas 2.5.2′ and 2.5.3, we shall solve a general problem which contains these two results as particular cases (cf. Grauert [17,18]). To explain those situations, we first give the definition below.

DEFINITION 2.5.5. Let $\mathscr{M}$ be a complex analytic manifold, and let $\mathscr{T}$ be a compact Hausdorff space. Denote by ζ and τ points of $\mathscr{M}$ and $\mathscr{T}$ respectively. Suppose that $\mathscr{G}$ and $\mathscr{H}$ be compact subsets of $\mathscr{T}$ such that $\mathscr{G} \subset \mathscr{H}$. The manifold $\mathscr{M}$ does not need to be connected. The sets $\mathscr{G}$ and $\mathscr{H}$ may be empty. In this situation, a matrix $P(\zeta, \tau)$ is said to be a $(\mathscr{G}, \mathscr{H})$-map : $\mathscr{M} \times \mathscr{T} \to G$ if the following conditions are satisfied:

(i) $P(\zeta, \tau)$ is a continuous map : $\mathscr{M} \times \mathscr{T} \to G$,
(ii) for each fixed τ in $\mathscr{H}$, $P(\zeta, \tau)$ is a holomorphic map : $\mathscr{M} \to G$,
(iii) $P(\zeta, \tau) = I$ on $\mathscr{M} \times \mathscr{G}$.

For example, let $\mathscr{M} = \mathscr{D}(\rho)$, let $\mathscr{T}$ be a set consisting of a point, and let $\mathscr{G}$ and $\mathscr{H}$ be empty. Then the matrix $S(\zeta)$ of Lemma 2.5.2′ is a $(\mathscr{G}, \mathscr{H})$-map. In the case of Lemma 2.5.3, there exists a matrix $P(\zeta, t)$ such that

(i) $P(\zeta, t)$ is a continuous map : $\mathscr{D}(\rho) \times J \to G$,
(ii) $P(\zeta, 1) = H(\zeta)$, $\quad P(\zeta, 0) = I$.

Therefore if we set

$$\mathscr{M} = \mathscr{D}(\rho), \quad \mathscr{T} = J, \quad \mathscr{H} = \{0, 1\}, \quad \mathscr{G} = \{0\},$$

then $P(\zeta, t)$ is a $(\mathscr{G}, \mathscr{H})$-map.

Noting that a $(\mathscr{G}, \mathscr{T})$-map is always a $(\mathscr{G}, \mathscr{H})$-map, we give the following definition.

DEFINITION 2.5.6. Let J be the interval $0 \leq t \leq 1$ and assume that a matrix $P(\zeta, \tau, t)$ satisfies the following conditions:

(i) $P(\zeta, \tau, t)$ is a continuous map : $\mathscr{M} \times \mathscr{T} \times J \to G$,
(ii) for each fixed t in J, $P(\zeta, \tau, t)$ is a $(\mathscr{G}, \mathscr{H})$-map : $\mathscr{M} \times \mathscr{T} \to G$,
(iii) $P(\zeta, \tau, 0)$ is a $(\mathscr{G}, \mathscr{T})$-map.

Then we say that the matrix $P(\zeta, \tau, 1)$ can be deformed to a $(\mathscr{G}, \mathscr{T})$-map $P(\zeta, \tau, 0)$ on $\mathscr{M} \times \mathscr{T}$ through a $(\mathscr{G}, \mathscr{H})$-deformation.

If we regard the matrix $S(\zeta)$ of Lemma 2.5.2′ as a $(\mathscr{G}, \mathscr{H})$-map as mentioned above, then the matrix $T(\zeta)$ of the same lemma is a $(\mathscr{G}, \mathscr{T})$-map. Hence Lemma 2.5.2′ will be proved if we show that $S(\zeta)$ can be deformed to a $(\mathscr{G}, \mathscr{T})$-map $T(\zeta)$ on $\mathscr{D}(\rho) \times J$ through a $(\mathscr{G}, \mathscr{H})$-deformation.

In the case of Lemma 2.5.3, as explained above, we regard the matrix $P(\zeta, t)$ as a $(\mathscr{G}, \mathscr{H})$-map. Note that, in this case, a matrix $Q(\zeta, t)$ is a $(\mathscr{G}, \mathscr{T})$-map if and only if

(a) $Q(\zeta, t)$ is a holomorphic map : $\mathscr{D}(\rho) \to G$ for each fixed t in J,
(b) $Q(\zeta, 0) = I$.

Therefore, $Q(\zeta, 1)$ can be deformed to I on $\mathscr{D}(\rho)$ holomorphically. If $P(\zeta, t)$ can be deformed to $Q(\zeta, t)$ through a $(\mathscr{G}, \mathscr{H})$-deformation, there exists a matrix $P(\zeta, t, s)$ such that

(i) $P(\zeta, t, s)$ is a continuous map : $\mathscr{D}(\rho) \times J \times J \to G$,
(ii) $P(\zeta, t, s)$ is a $(\mathscr{G}, \mathscr{H})$-map : $\mathscr{D}(\rho) \times J \to G$ for each fixed s,
(iii) $P(\zeta, t, 1) = P(\zeta, t), \quad P(\zeta, t, 0) = Q(\zeta, t)$.

Since $\mathscr{H} = \{0, 1\}$, $P(\zeta, 1, s)$ is a holomorphic map : $\mathscr{D}(\rho) \to G$ for each fixed s. Consequently, $H(\zeta) = P(\zeta, 1) = P(\zeta, 1, 1)$ can be deformed to $Q(\zeta, 1) = P(\zeta, 1, 0)$ holomorphically. In this way we can prove Lemma 2.5.3.

In either case, a key to the proof is to deform a given $(\mathscr{G}, \mathscr{H})$-map to a suitable $(\mathscr{G}, \mathscr{T})$-map through a $(\mathscr{G}, \mathscr{H})$-deformation. To solve this general problem, it would be not a good idea to treat each individual space $\mathscr{T}$ separately. We must solve the problem for the most general $\mathscr{T}$ all at once. We shall explain those situations in §2.7. As preparation for doing that, in §2.6, we shall generalize the decomposition theorem of Cartan and the Grauert Lemma 2.4.2 for $(\mathscr{G}, \mathscr{H})$-maps.

§2.6. A lemma of Grauert (continued)

We utilize the same notations K, K', K'', $K(\delta)$ as in §2.2 and $\mathscr{T}$, $\mathscr{G}$, $\mathscr{H}$ as in §2.5. We first generalize the decomposition theorem of Cartan (Theorem 2.2.1) for $(\mathscr{G}, \mathscr{H})$-maps.

LEMMA 2.6.1 (Grauert). *If a matrix $F(z,\tau)$ is a $(\mathscr{G},\mathscr{H})$-map : $K(\delta)\times\mathscr{T}\to G$ and if $F(z,\tau)$ can be deformed to the identity matrix on $K(\delta)\times\mathscr{T}$ through a $(\mathscr{G},\mathscr{H})$-deformation, then there exist two matrices $F'(z,\tau)$ and $F''(z,\tau)$ such that*

(i) *$F'(z,\tau)$ is a $(\mathscr{G},\mathscr{H})$-map : $K'\times\mathscr{T}\to G$ which can be deformed to the identity matrix on $K'\times\mathscr{T}$ through a $(\mathscr{G},\mathscr{H})$-deformation,*
(ii) *$F''(z,\tau)$ is a $(\mathscr{G},\mathscr{H})$-map : $K''\times\mathscr{T}\to G$ which can be deformed to the identity matrix on $K''\times\mathscr{T}$ through a $(\mathscr{G},\mathscr{H})$-deformation,*
(iii) *$F(z,\tau)=F'(z,\tau)F''(z,\tau)$ on $K\times\mathscr{T}$.*

To prove This lemma in a way similar to the proof of Theorem 2.2.1, we shall examine each of the lemmas given in §§2.2 and 2.3 again. First of all, a result similar to Lemma 2.2.2 can be derived from the properties of $F(z,\tau)$. To obtain a result similar to Lemma 2.2.3, it suffices to derive a result similar to the approximation by polynomials.

LEMMA 2.6.2. *Assume that $f(z,\tau)$ is a complex-valued function defined on $K(\delta)\times\mathscr{T}$ which satisfies the conditions below:*

(i) *$f(z,\tau)$ is continuous on $K(\delta)\times\mathscr{T}$,*
(ii) *$f(z,\tau)$ is a holomorphic function of z for each fixed τ in $\mathscr{H}$,*
(iii) *$f(z,\tau)=0$ on $K(\delta)\times\mathscr{G}$.*

Then, given a positive number ρ, we can find a complex-valued function $\hat{f}(z,\tau)$ on $C^\ell\times\mathscr{T}$ so that

(i) *$\hat{f}(z,\tau)$ is continuous on $C^\ell\times\mathscr{T}$,*
(ii) *$\hat{f}(z,\tau)$ is a polynomial in z for each fixed τ in $\mathscr{H}$,*
(iii) *$\hat{f}(z,\tau)=0$ on $C^\ell\times\mathscr{G}$,*
(iv) *$|f(z,\tau)-\hat{f}(z,\tau)|<\rho$ on $K(\frac{1}{2}\delta)\times\mathscr{T}$.*

PROOF. Because of uniform continuity of $f(z,\tau)$ we can find open subsets $V_1, V_2,\cdots, V_q$ of $\mathscr{T}$ such that

(i) $\mathscr{T}=\cup_{q\geq j\geq 1} V_j$,
(ii) if τ and τ' are in V_j, then

$$|f(z,\tau)-f(z,\tau')|<\frac{1}{2}\rho,$$

at every point z of $K(\frac{1}{2}\delta)$,
(iii) we can construct a partition of unity $\sigma_1(\tau), \sigma_2(\tau),\cdots, \sigma_q(\tau)$ associated with $V_1, V_2,\cdots, V_q$; precisely speaking the $\sigma_j(\tau)$ are continuous on $\mathscr{T}$, $\sigma(\tau)\geq 0$ for every τ in $\mathscr{T}$, the support of $\sigma_j(\tau)$ is contained in V_j, and $\sum_{q\geq j\geq 1}\sigma_j(\tau)=1$ at every point τ of $\mathscr{T}$.

For each j we construct $\hat{f}_j(z)$ as follows:

(a) $\hat{f}_j(z) = 0$ on C^ℓ if $V_j \cap \mathscr{G} \neq \varnothing$,

(b) $\hat{f}_j(z)$ is a polynomial in z such that $|f(z, \tau_j) - \hat{f}_j(z)| < \frac{1}{2}\rho$ on $K(\frac{1}{2}\delta)$ for a fixed point τ_j in $V_j \cap \mathscr{H}$ if $V_j \cap \mathscr{G} = \varnothing$ and $V_j \cap \mathscr{H} \neq \varnothing$,

(c) if $V_j \cap \mathscr{H} = \varnothing$, $\hat{f}_j(z)$ is a continuous function on C^ℓ such that $f(z, \tau_j) = \hat{f}_j(z)$ on $K(\frac{1}{2}\delta)$ for a fixed τ_j in V_j.

Utilizing the functions $\hat{f}_j(z)$, set

$$\hat{f}(z, \tau) = \sum_{q \geq j \geq 1} \hat{f}_j(z)\, \sigma_j(\tau).$$

Then the conditions of Lemma 2.6.2 are satisfied. □

If we use Lemma 2.6.2, we can construct a matrix corresponding to the matrix $Q(z)$ of Lemma 2.2.4. Furthermore, if we examine how $Q(z)$ was constructed, it becomes clear that such a matrix can be deformed to the identity matrix on $C^\ell \times \mathscr{T}$ through a $(\mathscr{G}, \mathscr{H})$-deformation.

Next, let us look at Lemma 2.3.1 again. The main part of the proof of this lemma is to write log $H(z)$ as a sum of $E'(z)$ and $E''(z)$. By means of the same idea, if $H(z, \tau)$ is a $(\mathscr{G}, \mathscr{H})$-map, we can construct $E'(z, \tau)$ and $E''(z, \tau)$ on $V_k' \times \mathscr{H}$ and $V_k'' \times \mathscr{H}$ respectively. Then, shrinking V_k'' to $\tilde{V}_k''$ slightly ($\tilde{V}_{k+1}'' \subset \tilde{V}_k''$), we extend $E''(z, \tau)$ to $\tilde{V}_k'' \times \mathscr{T}$ in a suitable way. Furthermore, if we determine $E'(z, \tau)$ on $(V_k \cap \tilde{V}_k'') \times \mathscr{T}$ by

$$E'(z, \tau) = \log\, H(z, \tau) - E''(z, \tau),$$

this definition matches nicely with $E'(z, \tau)$ defined on $V_k' \times \mathscr{H}$. Then shrinking V_k' to $\tilde{V}_k'$ slightly, we can define $E'(z, \tau)$ on $\tilde{V}_k' \times \mathscr{T}$ nicely.

Having seen how Lemma 2.6.1 can be proved in this manner, we leave the details of the proof to the reader, and we shall advance toward a generalization of Lemma 2.4.2. We shall use the same notations as in §§2.4 and 2.5.

LEMMA 2.6.3 (Grauert). *If a matrix $H(\zeta, \tau)$ is a $(\mathscr{G}, \mathscr{H})$-map : $\mathscr{D}(\rho) \times \mathscr{T} \to G$ which can be deformed to the identity matrix on $\mathscr{D}(\rho) \times \mathscr{T}$ through a $(\mathscr{G}, \mathscr{H})$-deformation, there exist two matrices $H'(\zeta, \tau)$ and $H''(\zeta, \tau)$ satisfying the conditions below:*

(i) *$H'(\zeta, \tau)$ is a $(\mathscr{G}, \mathscr{H})$-map : $\mathscr{D}' \times \mathscr{T} \to G$ which can be deformed to the identity matrix on $\mathscr{D}' \times \mathscr{T}$ through a $(\mathscr{G}, \mathscr{H})$-deformation,*

(ii) *$H''(\zeta, \tau)$ is a $(\mathscr{G}, \mathscr{H})$-map : $\mathscr{D}'' \times \mathscr{T} \to G$ which can be deformed to the identity matrix on $\mathscr{D}'' \times \mathscr{T}$ through a $(\mathscr{G}, \mathscr{H})$-deformation,*

(iii) *$H(\zeta, \tau) = H'(\zeta, \tau) H''(\zeta, \tau)$ on $\mathscr{D} \times \mathscr{T}$.*

To prove this lemma in the same way as we proved Lemma 2.4.2, it suffices to examine how we extended the function $f_{j,k}$ to $\hat{f}_{j,k}$. In fact we utilize the following result.

LEMMA 2.6.4. *Suppose that $f(\zeta, \tau)$ is a complex-valued function defined on $\mathscr{D}(\rho) \times \mathscr{T}$ and satisfying the conditions below:*

(i) *$f(\zeta, \tau)$ is continuous on $\mathscr{D}(\rho) \times \mathscr{T}$,*

(ii) *$f(\zeta, \tau)$ is a holomorphic function of ζ on $\mathscr{D}(\rho)$ for each fixed τ in $\mathscr{H}$,*

(iii) *$f(\zeta, \tau) = 0$ on $\mathscr{D}(\rho) \times \mathscr{G}$.*

Then we can construct a complex-valued function $\hat{f}(z, \tau)$ on $K(\rho) \times \mathscr{T}$ such that

(i) *$\hat{f}(\zeta, \tau)$ is continuous on $K(\rho) \times \mathscr{T}$,*

(ii) *$\hat{f}(\zeta, \tau)$ is a holomorphic function of z on $K(\rho)$ for each fixed τ in $\mathscr{H}$,*

(iii) *$\hat{f}(\zeta, \tau) = 0$ on $K(\rho) \times \mathscr{G}$,*

(iv) *$\hat{f}(\zeta, \tau) = f(\zeta, \tau)$ on $\mathscr{D}(\rho) \times \mathscr{T}$.*

To prove this lemma, as in the proof of Lemma 2.6.2, we utilize a partition of unity associated with a suitable open covering of $\mathscr{T}$. However, this is not enough. It is the key to the proof that we can make $|\hat{f}(z, \tau)|$ small if $|f(\zeta, \tau)|$ is small. Even though we explained briefly in §2.4 how to prove this fact, we shall look into this situation more in detail here. We denote by $\mathscr{F}$, $\mathscr{C}$, $\hat{\mathscr{F}}$, and $\hat{\mathscr{C}}$ the set of all holomorphic functions of ζ on $\mathscr{D}(\rho)$, the set of all continuous functions of ζ on $\mathscr{D}(\rho)$, the set of all holomorphic functions of z on $K(\rho)$, and the set of all continuous functions of z on $K(\rho)$, respectively. If we introduce a metric into each of these spaces in terms of the uniform convergence on each compact subset, these spaces become Frechet spaces. Further, $\mathscr{F}$ is a closed subspace of $\mathscr{C}$, while $\hat{\mathscr{F}}$ is a closed subspace of $\hat{\mathscr{C}}$. If we set

$$\phi(\hat{f}) = \hat{f}|_{\mathscr{D}(\rho)} \quad (= \text{the restriction of} \hat{f} \text{ on } \mathscr{D}(\rho))$$

for $\hat{f}$ in $\hat{\mathscr{C}}$, then ϕ is a continuous and surjective linear operator: $\hat{\mathscr{C}} \to \mathscr{C}$. Moreover, ϕ maps $\hat{\mathscr{F}}$ onto $\mathscr{F}$. Consequently ϕ is an open map not only $: \hat{\mathscr{C}} \to \mathscr{C}$ but also $: \hat{\mathscr{F}} \to \mathscr{F}$. Utilizing the given function $f(\zeta, \tau)$, we set

$$\psi(\tau) = f(\cdot, \tau).$$

Then $\psi(\tau)$ is a continuous map $: \mathscr{T} \to \mathscr{C}$ and satisfies the two conditions below:

(i) $\psi(\mathscr{H}) \subset \mathscr{F}$,

(ii) $\psi(\mathscr{G}) = 0$.

To prove Lemma 2.6.4, it suffices to construct a continuous map : $\mathscr{T} \to \hat{\mathscr{C}}$ such that

(i) $\hat{\psi}(\mathscr{H}) \subset \hat{\mathscr{F}}$,
(ii) $\hat{\psi}(\mathscr{G}) = 0$,
(iii) $\phi \circ \hat{\psi} = \psi$.

Since $\hat{\psi}$ is given on $\mathscr{G}$, it suffices to extend $\hat{\psi}$ to $\mathscr{H}$ and then to $\mathscr{T}$. To do this, we can apply the following lemma of Cartan.

LEMMA 2.6.5. (Cartan). *Assume that*

(i) $\mathscr{F}$ *and* $\hat{\mathscr{F}}$ *are Frechet spaces,*
(ii) $\mathscr{A}$ *is a compact Hausdorff space,*
(iii) $\mathscr{B}$ *is a compact subset of* $\mathscr{A}$; $\mathscr{B}$ *may be empty,*
(iv) ϕ *is a continuous and surjective linear operator* : $\hat{\mathscr{F}} \to \mathscr{F}$,
(v) ψ_1 *is a continuous map* : $\mathscr{A} \to \mathscr{F}$,
(vi) ψ_2 *is a continuous map:* $\mathscr{B} \to \hat{\mathscr{F}}$,
(vii) $\phi \circ \psi_2 = \psi_1$ *on* $\mathscr{B}$.

Then, we can construct a continuous map : $\mathscr{A} \to \hat{\mathscr{F}}$ *such that*

(a) $\psi_3 = \psi_2$ *on* $\mathscr{B}$,
(b) $\phi \circ \psi_3 = \psi_1$ *on* $\mathscr{A}$.

We can prove this lemma by utilizing the fact that ϕ is an open map and a suitable partition of unity on $\mathscr{A}$ as we remarked above. Asking the reader to read a proof in detail which is appended to the paper [17] of Grauert, we declare that Lemmas 2.6.1 and 2.6.3 were proved.

§2.7. A deformation theorem of Grauert (continued)

To prove Lemmas 2.5.2 and 2.5.3, we must deform a given $(\mathscr{G}, \mathscr{H})$-map to a suitable $(\mathscr{G}, \mathscr{T})$-map through a $(\mathscr{G}, \mathscr{H})$-deformation. In this section we shall explain this problem in detail. To simplify the exposition, we shall introduce some notations. We write the coordinates z_j of a point $z = (z_1, \cdots, z_\ell)$ of C^ℓ in the form:

$$z_j = x_{2j-1} + ix_{2j} \qquad (i = \sqrt{-1}),$$

where $(x_1, x_2, \cdots, x_{2\ell-1}, x_{2\ell})$ is a point of the real 2ℓ-dimensional space $R^{2\ell}$. In the case when two points $a = (a_1, \cdots, a_{2\ell})$ and $b = (b_1, \cdots, b_{2\ell})$ satisfy the inequalities

$$a_j < b_j \qquad (j = 1, \cdots, 2\ell)$$

we write

$$a < b.$$

We also set

$$\Delta(a,b) = \{z;\ \ a_j < x_j < b_j \qquad (j = 1, \cdots, 2\ell)\}.$$

If $a < \tilde{a} < \tilde{b} < b$, then the closure of $\Delta(\tilde{a}, \tilde{b})$ in C^ℓ is contained in $\Delta(a,b)$. Being a direct product of ℓ rectangles, $\Delta(a,b)$ is a polycylinder. The main result of this section is the theorem below.

THEOREM 2.7.1. (Grauert). *Let us assume that*

(a) *$\mathscr{D}_0$ is closed as a subset of a polycylinder $\Delta(a,b)$,*
(b) *$\mathscr{D}_0$ is embedded holomorphically into $\Delta(a,b)$ as a complex analytic submanifold; $\mathscr{D}_0$ may not be connected,*
(c) *a matrix $P(\zeta,\tau)$ is a $(\mathscr{G},\mathscr{H})$-map : $\mathscr{D}_0 \times \mathscr{T} \to G$,*
(d) *$\tilde{a}$ and $\tilde{b}$ satisfy the inequality $a < \tilde{a} < \tilde{b} < b$.*

Then, if we set $\mathscr{D} = \mathscr{D}_0 \cap \Delta(\tilde{a}, \tilde{b})$, we can find a $(\mathscr{G},\mathscr{T})$-map $Q(\zeta,\tau)$: $\mathscr{D} \times \mathscr{T} \to G$ such that $P(\zeta,\tau)$ can be deformed to $Q(\zeta,\tau)$ on $\mathscr{D} \times \mathscr{T}$ through a $(\mathscr{G},\mathscr{H})$-deformation.

The idea of the proof is induction on the real dimension. First of all, fixing a point ζ_0 in $\mathscr{D}_0$, we show that the conclusion of the theorem is true in a sufficiently small neighborhood of ζ_0. Denote by J the interval $0 \le t \le 1$ of a real variable t. For a suitably chosen neighborhood V of ζ_0 in $\mathscr{D}_0$, we can construct a map $\phi(\zeta, t) : V \times J \to V$ in such a way that

(i) $\phi(\zeta,t)$ is continuous on $V \times J$,
(ii) $\phi(\zeta,t)$ is a holomorphic function of ζ for each fixed t,
(iii) $\phi(\zeta,1) = \zeta, \quad \phi(\zeta,0) = \zeta_0$.

Utilizing this function $\phi(\zeta,t)$ we set

$$P(\zeta,\tau,t) = P(\phi(\zeta,t),\tau).$$

Then,

(i) $P(\zeta,\tau,t)$ is a continuous map : $V \times \mathscr{T} \times J \to G$,
(ii) $P(\zeta,\tau,t)$ is a $(\mathscr{G},\mathscr{H})$-map : $V \times \mathscr{T} \to G$ for each fixed t in J,
(iii) $P(\zeta,\tau,1) = P(\zeta,\tau), \quad P(\zeta,\tau,0) = P(\zeta_0,\tau)$.

Since the matrix $P(\zeta,\tau,0)$ is independent of ζ for each fixed τ, this matrix is a holomorphic function of ζ. Therefore, $P(\zeta,\tau,0)$ is a $(\mathscr{G},\mathscr{T})$-map : $V \times \mathscr{T} \to G$. This means that $P(\zeta,\tau)$ can be deformed to a $(\mathscr{G},\mathscr{T})$-map on $V \times \mathscr{T}$ through a $(\mathscr{G},\mathscr{H})$-deformation. This result is true for all $(\mathscr{G},\mathscr{H})$-maps no matter how we choose $\mathscr{G}$, $\mathscr{H}$, and $\mathscr{T}$.

To state the assumption of induction, we introduce several notations. Fixing $\xi = (\xi_1, \xi_2, \cdots, \xi_{2\ell})$ so that $a < \xi < b$, we denote by $\Delta_k(\tilde{a}, \tilde{b}, \varepsilon, \xi)$ the set of all z in C^ℓ satisfying the conditions

$$\tilde{a}_j - \varepsilon < x_j < \tilde{b}_j + \varepsilon \qquad (j = 1, \cdots, k-1)$$

and

$$|x_j - \xi_j| < \varepsilon \qquad (j = k, \cdots, 2\ell)$$

for a sufficiently small positive number ε, where $k = 1, \cdots, 2\ell + 1$. The Δ_k are polycylinders. Furthermore, Δ_1 is a neighborhood of the point $(\xi_1 + i\xi_2, \cdots, \xi_{2\ell-1} + i\xi_{2\ell})$ of $\Delta(a, b)$, and $\Delta_{2\ell+1}$ is a neighborhood of $\Delta(\tilde{a}, \tilde{b})$. Also set

$$\mathscr{D}_k(\tilde{a}, \tilde{b}, \varepsilon, \xi) = \mathscr{D}_0 \cap \Delta_k(\tilde{a}, \tilde{b}, \varepsilon, \xi) \qquad (k = 1, \cdots, 2\ell + 1).$$

ASSUMPTION OF INDUCTION. *Fixing ε_0 and ξ, we assume that a matrix $P(\zeta, \tau)$ is a $(\mathscr{G}, \mathscr{H})$-map : $\mathscr{D}_k(\tilde{a}, \tilde{b}, \varepsilon_0, \xi) \times \mathscr{T} \to G$. Then, if $\mathscr{D}_k(\tilde{a}, \tilde{b}, \varepsilon, \xi)$ is not empty no matter how small an ε we choose, there exists a positive number ε not greater than ε_0 such that the matrix $P(\zeta, \tau)$ can be deformed to a $(\mathscr{G}, \mathscr{T})$-map on $\mathscr{D}_k(\tilde{a}, \tilde{b}, \varepsilon, \xi) \times \mathscr{T}$ through a $(\mathscr{G}, \mathscr{H})$-deformation.*

This is a recursive assumption on k, where we may choose ε_0, ξ, $\mathscr{G}$, $\mathscr{H}$, and $\mathscr{T}$ in any way we want. In particular, this assumption is satisfied for $k = 1$. In order to verify this assumption for $k + 1$ in the case when the assumption is satisfied for k, we utilize a lemma which we shall explain later. Before doing that, we introduce several notations again. Let an arbitrary ξ be given. Also, let us fix four numbers α_1, α_2, β_1 and β_2 such that $\alpha_1 < \alpha_2 < \beta_1 < \beta_2$. Denote by $\Delta'(\varepsilon)$ the set of all points z in C^ℓ satisfying the conditions

$$\begin{cases} \tilde{a}_j - \varepsilon < x_j < \tilde{b}_j + \varepsilon & (j = 1, \cdots, k - 1), \\ \alpha_1 < x_k < \beta_1, & \\ |x_j - \xi_j| < \varepsilon & (j = k + 1 \cdots, 2\ell), \end{cases}$$

and by $\Delta''(\varepsilon)$ the set of all z satisfying the conditions

$$\begin{cases} \tilde{a}_j - \varepsilon < x_j < \tilde{b}_j + \varepsilon & (j = 1, \cdots, k - 1), \\ \alpha_2 < x_k < \beta_2, & \\ |x_j - \xi_j| < \varepsilon & (j = k + 1 \cdots, 2\ell). \end{cases}$$

The set $\Delta'(\varepsilon) \cup \Delta''(\varepsilon)$ consists of all z satisfying the conditions

$$\begin{cases} \tilde{a}_j - \varepsilon < x_j < \tilde{b}_j + \varepsilon & (j = 1, \cdots, k - 1), \\ \alpha_1 < x_k < \beta_2, & \\ |x_j - \xi_j| < \varepsilon & (j = k + 1 \cdots, 2\ell). \end{cases}$$

Further, we set

$$\mathscr{D}'(\varepsilon) = \mathscr{D}_0 \cap \Delta'(\varepsilon), \qquad \mathscr{D}''(\varepsilon) = \mathscr{D}_0 \cap \Delta''(\varepsilon)$$

and denote $\mathscr{D}'(\varepsilon) \cup \mathscr{D}''(\varepsilon)$ by $\tilde{\mathscr{D}}(\varepsilon)$.

LEMMA 2.7.2. *Assume that*

(a) *the assumption of induction is satisfied for* k,

(b) *a matrix* $P(\zeta,\tau)$ *is a* $(\mathscr{G},\mathscr{H})$*-map* : $\tilde{\mathscr{D}}(\varepsilon_0)\times\mathscr{T} \to G$,

(c) $P(\zeta,\tau)$ *can be deformed to a* $(\mathscr{G},\mathscr{T})$*-map through* $(\mathscr{G},\mathscr{H})$*-deformation on* $\mathscr{D}'(\varepsilon_0)\times\mathscr{T}$ *and* $\mathscr{D}''(\varepsilon_0)\times\mathscr{T}$ *respectively.*

Then, if we choose a suitable positive number ε *not greater than* ε_0, *the matrix* $P(\zeta,\tau)$ *can be deformed to a* $(\mathscr{G},\mathscr{T})$*-map on* $\tilde{\mathscr{D}}(\varepsilon)\times\mathscr{T}$ *through a* $(\mathscr{G},\mathscr{H})$*-deformation.*

PROOF. Let us state assumption (c) more precisely. If we denote by J the interval $0\le t\le 1$ of a real variable t, there exist two matrices $P'(\zeta,\tau,t)$ and $P''(\zeta,\tau,t)$ satisfying the conditions below:

(i) $P'(\zeta,\tau,t)$ is a continuous map : $\mathscr{D}'(\varepsilon_0)\times\mathscr{T}\times J \to G$ and a $(\mathscr{G},\mathscr{H})$-map : $\mathscr{D}'(\varepsilon_0)\times\mathscr{T} \to G$ for each fixed t; $P'(\zeta,\tau,0)$ is a $(\mathscr{G},\mathscr{T})$-map and $P'(\zeta,\tau,1)=P(\zeta,\tau)$;

(ii) $P''(\zeta,\tau,t)$ is a continuous map : $\mathscr{D}''(\varepsilon_0)\times\mathscr{T}\times J \to G$ and a $(\mathscr{G},\mathscr{H})$-map : $\mathscr{D}''(\varepsilon_0)\times\mathscr{T} \to G$ for each fixed t; $P''(\zeta,\tau,0)$ is a $(\mathscr{G},\mathscr{T})$-map and $P''(\zeta,\tau,1)=P(\zeta,\tau)$;

We used only $\xi_{k+1},\cdots,\xi_{2\ell}$ to define $\Delta'(\varepsilon)$ and $\Delta''(\varepsilon)$. Therefore if we define $\mathscr{D}_k(\tilde{a},\tilde{b},\varepsilon,\xi)$ by utilizing a real number ξ_k such that $\alpha_2<\xi_k<\beta_1$ in addition to those $\xi_{k+1},\cdots,\xi_{2\ell}$, we have

$$\mathscr{D}_k(\tilde{a},\tilde{b},\varepsilon,\xi)\subset\mathscr{D}'(\varepsilon_0)\cap\mathscr{D}''(\varepsilon_0)$$

provided that ε is sufficiently small. If $\mathscr{D}(\tilde{a},\tilde{b},\varepsilon,\xi)$ is empty, $\tilde{\mathscr{D}}(\varepsilon)$ splits into two disjoint parts which are subsets of $\mathscr{D}'(\varepsilon)$ and $\mathscr{D}''(\varepsilon)$ respectively. Therefore, in this case, Lemma 2.7.2 holds.

Assume that $\mathscr{D}_k(\tilde{a},\tilde{b},\varepsilon,\xi)$ is not empty for all ε. Then if we set

$$R(\zeta,\tau,t)=P'(\zeta,\tau,t)P''(\zeta,\tau,t)^{-1}, \tag{2.7.1}$$

(1) $R(\zeta,\tau,t)$ is a continuous map : $\mathscr{D}_k(\tilde{a},\tilde{b},\varepsilon,\xi)\times\mathscr{T}\times J \to G$,

(2) for each fixed t, $R(\zeta,\tau,t)$ is a $(\mathscr{G},\mathscr{H})$-map : $\mathscr{D}_k(\tilde{a},\tilde{b},\varepsilon,\xi)\times\mathscr{T} \to G$ and $R(\zeta,\tau,0)$ is a $(\mathscr{G},\mathscr{T})$-map,

(3) $R(\zeta,\tau,1)=I$ on $\mathscr{D}_k(\tilde{a},\tilde{b},\varepsilon,\xi)\times\mathscr{T}$.

Therefore, if we set

$$\tilde{\mathscr{T}}=\mathscr{T}\times J,$$
$$\tilde{\mathscr{H}}=\mathscr{H}\times J\cup\mathscr{T}\times\{0,1\},$$
$$\tilde{\mathscr{G}}=\mathscr{G}\times J\cup\mathscr{T}\times\{1\},$$

$R(\zeta,\tau,t)$ is a $(\tilde{\mathscr{G}},\tilde{\mathscr{H}})$-map : $\mathscr{D}_k(\tilde{a},\tilde{b},\varepsilon,\xi)\times\tilde{\mathscr{T}} \to G$. Therefore, because of the assumption of induction, $R(\zeta,\tau,t)$ can be deformed to a $(\tilde{\mathscr{G}},\tilde{\mathscr{T}})$-map

$Q(\zeta,\tau,t)$ on $\mathscr{D}_k(\tilde{a},\tilde{b},\varepsilon,\xi)\times\tilde{\mathscr{T}}$ through a $(\tilde{\mathscr{G}},\mathscr{H})$-deformation, provided that we choose ε sufficiently small.

Note that $Q(\zeta,\tau,t)$ is a $(\tilde{\mathscr{G}},\tilde{\mathscr{T}})$-map on $\mathscr{D}_k(\tilde{a},\tilde{b},\varepsilon,\xi)\times\tilde{\mathscr{T}}$ if the conditions below are satisfied:

(i) Q is continous on $\mathscr{D}_k(\tilde{a},\tilde{b},\varepsilon,\xi)\times\tilde{\mathscr{T}}$,
(ii) for each fixed (τ,t), Q is a holomorphic function of ζ on $\mathscr{D}_k(\tilde{a},\tilde{b},\varepsilon,\xi)$,
(iii) $Q(\zeta,\tau,t)=I$ if τ is in $\mathscr{G}$,
(iv) $Q(\zeta,\tau,1)=I$.

Therefore, if we set

$$Q(\zeta,\tau,t,s)=Q(\zeta,\tau,1-s(1-t)),$$

then $Q(\zeta,\tau,t,s)$ is also a $(\tilde{\mathscr{G}},\tilde{\mathscr{T}})$-map for each fixed s, and

$$Q(\zeta,\tau,t,1)=Q(\zeta,\tau,t)$$
$$Q(\zeta,\tau,t,0)=I.$$

This means that $Q(\zeta,\tau,t)$ can be deformed to the identity matrix through a $(\tilde{\mathscr{G}},\tilde{\mathscr{T}})$-deformation. Consequently, $R(\zeta,\tau,t)$ can be deformed to the identity matrix on $\mathscr{D}_k(\tilde{a},\tilde{b},\varepsilon,\xi)\times\tilde{\mathscr{T}}$ through a $(\tilde{\mathscr{G}},\mathscr{H})$-deformation.

We can shrink $\mathscr{D}'(\varepsilon)$ and $\mathscr{D}''(\varepsilon)$ without changing $\mathscr{D}'(\varepsilon)\cup\mathscr{D}''(\varepsilon)$ so that

$$\mathscr{D}_k(\tilde{a},\tilde{b},\varepsilon,\xi)=\mathscr{D}'(\varepsilon)\cap\mathscr{D}''(\varepsilon).$$

To do this, it suffices to replace α_2 and β_1 by $\xi_k-\varepsilon$ and $\xi_k+\varepsilon$ respectively. Now, we apply Lemma 2.6.3 to $R(\zeta,\tau,t)$. If we choose ε much smaller, there exist two matrices $R'(\zeta,\tau,t)$ and $R''(\zeta,\tau,t)$ satisfying the following conditions:

(i) $R'(\zeta,\tau,t)$ is a $(\tilde{\mathscr{G}},\mathscr{H})$-map : $\mathscr{D}'(\varepsilon)\times\tilde{\mathscr{T}}\to G$,
(ii) $R''(\zeta,\tau,t)$ is a $(\tilde{\mathscr{G}},\mathscr{H})$-map : $\mathscr{D}''(\varepsilon)\times\tilde{\mathscr{T}}\to G$,
(iii)

$$R(\zeta,\tau,t)=R'(\zeta,\tau,t)R''(\zeta,\tau,t)\quad\text{on}\quad\mathscr{D}(\tilde{a},\tilde{b},\varepsilon,\xi)\times\tilde{\mathscr{T}}. \tag{2.7.2}$$

From (2.7.1) and (2.7.2), it follows that

$$R'(\zeta,\tau,t)^{-1}P'(\zeta,\tau,t)=R''(\zeta,\tau,t)P''(\zeta,\tau,t)\quad\text{on}\,\mathscr{D}_k(\tilde{a},\tilde{b},\varepsilon,\xi)\times\tilde{\mathscr{T}}. \tag{2.7.3}$$

The left-hand side of (2.7.3) is defined on $\mathscr{D}'(\varepsilon)\times\tilde{\mathscr{T}}$, while the right-hand side of (2.7.3) is defined on $\mathscr{D}''(\varepsilon)\times\tilde{\mathscr{T}}$. Hence (2.7.3) defines a matrix $P(\zeta,\tau,t)$ on $\tilde{\mathscr{D}}(\varepsilon)\times\tilde{\mathscr{T}}$ which satisfies the conditions below:

(i) $P(\zeta,\tau,t)$ is a continuous map : $\tilde{\mathscr{D}}(\varepsilon)\times\tilde{\mathscr{T}}\to G$,
(ii) for each fixed t, $P(\zeta,\tau,t)$ is a $(\mathscr{G},\mathscr{H})$-map : $\tilde{\mathscr{D}}(\varepsilon)\times\mathscr{T}\to G$,
(iii) $P(\zeta,\tau,1)=P(\zeta,\tau)$,
(iv) $P(\zeta,\tau,0)$ is a $(\mathscr{G},\mathscr{T})$-map. □

If we use Lemma 2.7.2 in a suitable way, we can complete induction on k and the proof of Theorem 2.7.1. The details are left to the reader. The reader will find much deeper results in Grauert [18] (cf. in particular the neighborhood of Theorem 5 on p. 455 of [18]). Our proof of Theorem 2.7.1 followed the same plan as given in [18].

§2.8. An existence theorem

Utilizing what we have written so far as a preparation, we shall explain a rough outline of various aspects of an existence theorem of Grauert (cf. Theorem 2.9.4). We shall introduce some notations. First of all, let us fix 4ℓ real numbers a_ν, a'_ν, b_ν, b'_ν ($\nu = 1, 2, \cdots, \ell$) so that the inequalities $a_\nu < a'_\nu$ and $b_\nu < b'_\nu$ are satisfied. Then, for a positive number ε, we define ℓ rectangular domains in C by

$$V_\nu(\varepsilon) = \{\ z = x + iy\ ;\ a_\nu - \varepsilon < x < a'_\nu + \varepsilon,\ b_\nu - \varepsilon < y < b'_\nu + \varepsilon\ \}.$$

Further, utilizing these rectangular domains, we construct a polycylinder

$$V(\varepsilon) = V_1(\varepsilon) \times V_2(\varepsilon) \times \cdots \times V_\ell(\varepsilon)$$

in C^ℓ. Next, for a positive integer N, we set

$$\begin{cases} \alpha_{\nu,k} = a_\nu + \dfrac{k}{N}(a'_\nu - a_\nu), \\ \beta_{\nu,k} = b_\nu + \dfrac{k}{N}(b'_\nu - b_\nu), \end{cases} \qquad (k = 0, 1, \cdots, N),$$

and

$$U_{\nu,jk}(\varepsilon) = \{\ z = x+iy\ ; \alpha_{\nu,j}-\varepsilon < x < \alpha_{\nu,j+1}+\varepsilon,\ \beta_{\nu,k}-\varepsilon < y < \beta_{\nu,k+1}+\varepsilon\ \},$$

where $j, k = 0, 1, \cdots, N-1$. Then, for a 2ℓ-tuple $\mu = \{j_1 \cdots, j_\ell; k_1, \cdots, k_\ell\}$ of integers such that $0 \le j_\nu \le N-1$ and $0 \le k_\nu \le N-1$, we define a polycylinder

$$K_\mu(\varepsilon) = U_{1,j_1k_1}(\varepsilon) \times \cdots \times U_{\ell,j_\ell k_\ell}(\varepsilon)$$

in C^ℓ. Thus we have

$$V(\varepsilon) = \bigcup_\mu K_\mu(\varepsilon).$$

In general, we fix N and choose ε small.

Fixing a sufficiently small positive number ε_0, we assume that a subset $\mathscr{D}_0$ of $V(\varepsilon_0)$ satisfies the two conditions below:

(i) $\mathscr{D}_0$ is closed as a subset of $V(\varepsilon_0)$,

(ii) $\mathscr{D}_0$ is embedded holomorphically in $V(\varepsilon_0)$ as a complex analytic submanifold; $\mathscr{D}_0$ may not be connected.

For a positive number ε not greater than ε_0, we set

$$\mathscr{D}_\mu(\varepsilon) = \mathscr{D}_0 \cap K_\mu(\varepsilon),$$
$$\mathscr{D}_{\lambda\mu}(\varepsilon) = \mathscr{D}_\lambda(\varepsilon) \cap \mathscr{D}_\mu(\varepsilon) = \mathscr{D}_0 \cap K_\lambda(\varepsilon) \cap K_\mu(\varepsilon),$$

where we distinguish (λ, μ) from (μ, λ).

LEMMA 2.8.1. (Grauert). *Assume that*

(1) *a continuous map* $S_\mu(\zeta) : \mathscr{D}_\mu(\varepsilon_0) \to G$ *is given for every* μ *satisfying* $\mathscr{D}_\mu(\varepsilon_0) \neq \varnothing$,

(2) *a holomorphic map* $F_{\lambda\mu}(\zeta) : \mathscr{D}_{\lambda\mu}(\varepsilon_0) \to G$ *is given for every* (λ, μ) *satisfying* $\mathscr{D}_{\lambda\mu}(\varepsilon_0) \neq \varnothing$,

(3) *we have*

$$F_{\lambda\mu}(\zeta) = S_\lambda(\zeta)^{-1} S_\mu(\zeta) \tag{2.8.1}$$

on $\mathscr{D}_{\lambda\mu}(\varepsilon_0)$ *for every* (λ, μ) *satisfying* $\mathscr{D}_{\lambda\mu}(\varepsilon_0) \neq \varnothing$.

In this case, if we fix a positive number ε *less than* ε_0, *then for each* μ *satisfying* $\mathscr{D}_\mu(\varepsilon) \neq \varnothing$, *we can find a holomorphic map* $F_\mu(\zeta) : \mathscr{D}_\mu(\varepsilon) \to G$ *such that*

$$F_{\lambda\mu}(\zeta) = F_\lambda(\zeta)^{-1} F_\mu(\zeta) \tag{2.8.2}$$

on $\mathscr{D}_{\lambda\mu}(\varepsilon)$ *for every* (λ, μ) *satisfying* $\mathscr{D}_{\lambda\mu}(\varepsilon) \neq \varnothing$. *Also, if we set*

$$S(\zeta) = S_\mu(\zeta) F_\mu(\zeta)^{-1} \quad \text{on } \mathscr{D}_\mu(\varepsilon) \tag{2.8.3}$$

for every μ *satisfying* $\mathscr{D}_\mu(\varepsilon) \neq \varnothing$, *then* $S(\zeta)$ *is a continuous map* : $\mathscr{D}_0 \cap V(\varepsilon) \to G$. *Moreover, if we choose* $F_\mu(\varepsilon)$ *in a suitable way,* $S(\zeta)$ *can be continuously deformed to the identity matrix on* $\mathscr{D}_0 \cap V(\varepsilon)$.

PROOF. We use induction on the total number of the μ satisfying the condition:

$$\mathscr{D}_\mu(\varepsilon_0) \neq \varnothing. \tag{2.8.4}$$

Since (2.8.1) implies that $F_{\mu\mu} = I$, we can choose F_μ by using Lemma 2.5.2 if only one μ satisfies (2.8.4).

Now, assuming that the conclusion of Lemma 2.8.1 is true in the case when the number of the μ satisfying (2.8.4) is q, we look into the case when the number of such μ is $q + 1$. To do this we divide $V(\varepsilon_0)$ into two parts in such a way that we can use Lemma 2.8.1 on each part. For example, if we set

$$V_1'(\varepsilon) = \{\ z = x + iy\ ; a_1 - \varepsilon < x < \alpha_{1,k_0} + \varepsilon,\ b_1 - \varepsilon < y < b_1' + \varepsilon\ \},$$
$$V_1''(\varepsilon) = \{\ z = x + iy\ ; \alpha_{1,k_0} - \varepsilon < x < a_1' + \varepsilon,\ b_1 - \varepsilon < y < b_1' + \varepsilon\ \},$$
$$V'(\varepsilon) = V_1'(\varepsilon) \times V_2(\varepsilon) \times \cdots \times V_\ell(\varepsilon),$$
$$V''(\varepsilon) = V_1''(\varepsilon) \times V_2(\varepsilon) \times \cdots \times V_\ell(\varepsilon)$$

for a fixed positive integer k_0 less than N, we have either

$$K_\mu(\varepsilon_0) \subset V'(\varepsilon_0) \tag{2.8.5$'$}$$

or

$$K_\mu(\varepsilon_0) \subset V''(\varepsilon_0) \tag{2.8.5$''$}$$

for every μ. We denote by M' (resp. M'') the set of all μ satisfying conditions (2.8.4) and (2.8.5$'$) (resp. (2.8.5$''$)). We also assume that each M' and M'' contains at most q elements so that we can apply Lemma 2.8.1 on $V'(\varepsilon_0)$ and $V''(\varepsilon_0)$.

Under the situation given above, if we fix a positive number $\tilde{\varepsilon}$ so that $0 < \varepsilon < \tilde{\varepsilon} < \varepsilon_0$, we can derive the results below from the assumption of induction:

(a) for each μ satisfying $\mathscr{D}_\mu(\tilde{\varepsilon}) \neq \varnothing$, we can find a holomorphic map $\tilde{F}_\mu(\zeta) : \mathscr{D}_\mu(\tilde{\varepsilon}) \to G$ such that

$$\begin{cases} F_{\lambda\mu}(\zeta) = \tilde{F}_\lambda(\zeta)^{-1}\tilde{F}_\mu(\zeta) & (\lambda, \mu \in M'), \\ F_{\lambda\mu}(\zeta) = \tilde{F}_\lambda(\zeta)^{-1}\tilde{F}_\mu(\zeta) & (\lambda, \mu \in M'') \end{cases}$$

on $\mathscr{D}_{\lambda\mu}(\tilde{\varepsilon})$ $(\neq \varnothing)$,

(b) if we set

$$\begin{cases} S'(\zeta) = S_\mu(\zeta)\tilde{F}_\mu(\zeta)^{-1} & (\mu \in M'), \\ S''(\zeta) = S_\mu(\zeta)\tilde{F}_\mu(\zeta)^{-1} & (\mu \in M'') \end{cases}$$

on $\mathscr{D}_\mu(\tilde{\varepsilon})$ $(\neq \varnothing)$, then $S'(\zeta)$ (resp. $S''(\zeta)$) is a continuous map: $\mathscr{D}_0 \cap V'(\tilde{\varepsilon}) \to G$ (resp. : $\mathscr{D}_0 \cap V''(\tilde{\varepsilon}) \to G$) which can be continuously deformed to the identity matrix on $\mathscr{D}_0 \cap V'(\tilde{\varepsilon})$ (resp. $\mathscr{D}_0 \cap V''(\tilde{\varepsilon})$).

To simplify the exposition hereafter, we set

$$\mathscr{D}'(\tilde{\varepsilon}) = \mathscr{D}_0 \cap V'(\tilde{\varepsilon}), \qquad \mathscr{D}''(\tilde{\varepsilon}) = \mathscr{D}_0 \cap V''(\tilde{\varepsilon}).$$

If $\mathscr{D}'(\tilde{\varepsilon}) \cap \mathscr{D}''(\tilde{\varepsilon})$ is empty, the proof of Lemma 2.8.1 has been completed. Let us assume that

$$\mathscr{D}'(\tilde{\varepsilon}) \cap \mathscr{D}''(\tilde{\varepsilon}) \neq \varnothing. \tag{2.8.6}$$

This means that we are assuming that

$$\mathscr{D}_{\lambda\mu}(\tilde{\varepsilon}) \neq \varnothing \tag{2.8.7}$$

if we choose $\lambda \in M'$ and $\mu \in M''$ in a suitable way. Then, if we set

$$F(\zeta) = S'(\zeta)^{-1}S''(\zeta), \tag{2.8.8}$$

we have

$$\begin{aligned} F(\zeta) &= \tilde{F}_\lambda(\zeta) S_\lambda(\zeta)^{-1} S_\mu(\zeta) \tilde{F}_\mu(\zeta)^{-1} \\ &= \tilde{F}_\lambda(\zeta) F_{\lambda\mu}(\zeta) \tilde{F}_\mu(\zeta)^{-1} \end{aligned} \tag{2.8.9}$$

for $\lambda \in M'$ and $\mu \in M''$ satisfying (2.8.7). Hence $F(\zeta)$ is a holomorphic map : $\mathscr{D}'(\tilde{\varepsilon}) \cap \mathscr{D}''(\tilde{\varepsilon}) \to G$. Furthermore, since the right-hand side of (2.8.8) can be deformed to the identity matrix continuously on $\mathscr{D}'(\tilde{\varepsilon}) \cap \mathscr{D}''(\tilde{\varepsilon})$, $F(\zeta)$ can be deformed to the identity matrix holomorphically on $\mathscr{D}'(\hat{\varepsilon}) \cap \mathscr{D}''(\hat{\varepsilon})$ provided that $0 < \varepsilon < \hat{\varepsilon} < \tilde{\varepsilon}$ (cf. Lemma 2.5.3).

The situation explained above allows us to apply Lemma 2.4.2 to $F(\zeta)$. This means that if we fix a positive number ε' so that $0 < \varepsilon < \varepsilon' < \varepsilon_0$, we can find a holomorphic map $F'(\zeta) : \mathscr{D}'(\varepsilon') \to G$ and a holomorphic map $F''(\zeta) : \mathscr{D}''(\varepsilon') \to G$ such that

$$F(\zeta) = F'(\zeta)^{-1} F''(\zeta) \tag{2.8.10}$$

on $\mathscr{D}'(\varepsilon') \cap \mathscr{D}''(\varepsilon')$. Then, if we set

$$\hat{F}_\mu(\zeta) = \begin{cases} F'(\zeta)\tilde{F}_\mu(\zeta) & (\mu \in M'), \\ F''(\zeta)\tilde{F}_\mu(\zeta) & (\mu \in M'') \end{cases}$$

for every μ satisfying $\mathscr{D}_\mu(\varepsilon') \neq \varnothing$, we have $F_{\lambda\mu}(\zeta) = \hat{F}_\lambda(\zeta)^{-1}\hat{F}_\mu(\zeta)$ for every (λ, μ) such that $\mathscr{D}_{\lambda\mu}(\varepsilon') \neq \varnothing$ (cf. in particular (2.8.9) and (2.8.10)). Moreover, if we define $\hat{S}(\zeta)$ by

$$\hat{S}(\zeta) = S_\mu(\zeta)\hat{F}_\mu(\zeta)^{-1}, \tag{2.8.11}$$

we have

$$\hat{S}(\zeta) = \begin{cases} S'(\zeta)F'(\zeta)^{-1} & (\zeta \in \mathscr{D}'(\varepsilon')), \\ S''(\zeta)F''(\zeta)^{-1} & (\zeta \in \mathscr{D}''(\varepsilon')), \end{cases}$$

and hence $\hat{S}(\zeta)$ is a continuous map : $\mathscr{D}'(\varepsilon') \cup \mathscr{D}''(\varepsilon') \to G$ because of (2.8.8) and (2.8.10). Note that

$$\mathscr{D}'(\varepsilon') \cup \mathscr{D}''(\varepsilon') = \mathscr{D}_0 \cap V(\varepsilon').$$

Applying Lemma 2.5.2 to $\hat{S}(\zeta)$, we can find a holomorphic map $T(\zeta)$: $\mathscr{D}_0 \cap V(\varepsilon) \to G$ so that

$$S(\zeta) = \hat{S}(\zeta)T(\zeta)^{-1}$$

can be continuously deformed to the identity matrix on $\mathscr{D}_0 \cap V(\varepsilon)$. Hence if we set

$$F_\mu(\zeta) = T(\zeta)\hat{F}_\mu(\zeta),$$

we can complete the proof of Lemma 2.8.1. □

Since this section has already become long, we shall continue the discussion in the next section.

§2.9. An existence theorem (continued)

Let M be a complex analytic manifold. We denote by $\mathscr{O}(M)$ the set of all complex-valued functions which are holomorphic on M. Also, let us set

$$\hat{K} = \bigcap_{f \in \mathscr{O}(M)} \{\, \zeta \in M \,;\, |f(\zeta)| \leq \max_{\xi \in K} |f(\xi)| \,\}$$

for a compact subset K of M. We say that M is **holomorphically convex** if $\hat{K}$ is compact for every compact subset K of M.

DEFINITION 2.9.1. A connected complex analytic manifold M is called a **Stein manifold** if the conditions below are satisfied:

(i) points of M can be separated by functions in $\mathscr{O}(M)$,
(ii) M is holomorphically convex,
(iii) M admits an enumerable base of open sets,
(iv) at each point of M a system of local coordinates is given by suitable functions in $\mathscr{O}(M)$.

As in §2.8, choosing ℓ rectangular domains $V_1, V_2, \cdots, V_\ell$ in C, we define a polycylinder

$$V = V_1 \times V_2 \times \cdots \times V_\ell$$

in C^ℓ. Next, let M be a Stein manifold. If we choose ℓ functions $f_1, \cdots, f_\ell$ from $\mathscr{O}(M)$ and set

$$\mathscr{F}(\zeta) = (f_1(\zeta), \cdots, f_\ell(\zeta)) \tag{2.9.1}$$

at every point ζ of M, then $\mathscr{F}$ is a holomorphic map : $M \to C^\ell$. If D_0 is an open subset of M, D_0 also becomes a complex analytic manifold with the structure of a manifold induced by the structure of M. Hereafter we assume that the closure of D_0 in M is compact in M. We denote by $\mathscr{D}_0$ the image of D_0 by map (2.9.1). If $\mathscr{F}$ is a bijection : $D_0 \to \mathscr{D}_0$, we can induce a complex analytic structure on $\mathscr{D}_0$ from D_0. If such a complex analytic manifold $\mathscr{D}_0$ satisfies the following conditions, the open set D_0 is called an **Oka–Weil domain** on M defined by functions in $\mathscr{O}(M)$:

(a) $\mathscr{D}_0$ is closed as a subset of V,
(b) $\mathscr{D}_0$ is embedded holomorphically into V as a submanifold.

For a given D_0, the choice of V and $\mathscr{F}$ is not unique. Furthermore, ℓ is not necessarily the same for two Oka–Weil domains. For an Oka–Weil domain, the following approximation theorem holds (cf. Gunning–Rossi [2], Chap.VII, Sec. A, Theorem 6, p. 213).

LEMMA 2.9.2.. *Let M be a Stein manifold, D_0 an Oka–Weil domain and K a compact subset of M such that $K \subset D_0$. Then, if ϕ is a complex-valued holomorphic function defined on D_0 and if ε is a given positive number, there exists a function f in $\mathscr{O}(M)$ such that*

$$|f(\zeta) - \phi(\zeta)| \leq \varepsilon \tag{2.9.2}$$

at every point ζ of K.

PROOF. Assume that D_0 is mapped onto $\mathscr{D}_0$ by the bijection $\mathscr{F}$ defined by (2.9.1) and that $\mathscr{D}_0$ satisfies conditions (a) and (b). Since $\phi \circ \mathscr{F}^{-1}$ is a holomorphic function on $\mathscr{D}_0$, there exists a holomorophic function $F(z)$ on V such that $\phi \circ \mathscr{F}^{-1}$ is the restriction of F on $\mathscr{D}_0$. If, utilizing a polynomial approximation $P(z)$ of F, we set $f(\zeta) = P(\mathscr{F}(\zeta))$, we have (2.9.2). □

One of the most important properties of a Stein Manifold is that it is approximated by Oka–Weil domains, i.e.,

LEMMA 2.9.3. *For a given Stein manifold M, we can choose a sequence $D_1, D_2, \cdots$ of Oka–Weil domains on M such that*

(1) $\bar{D}_m \subset D_{m+1}$ *($\bar{D}_m$ is the closure of D_m in M),*
(2) $M = \bigcup_{m \geq 1} D_m$.

For the proof of this lemma, look at Gunning–Rossi [2], Chap. VII, Sec. A, Proposition 3, Corollary 4, pp. 211-212.

Utilizing what we have explained above as a preparation, we state an existence theorem of Grauert which is the main subject of this chapter.

THEOREM 2.9.4. (Grauert). *Assume that*

(I) *M is a Stein manifold,*
(II) *$\{\Omega_j\}$ is a covering of M consisting of open subsets of M,*
(III) *a continuous map $S_j(\zeta) : \Omega_j \to G$ is given for every j,*
(IV) *a holomorphic map $F_{jk} : \Omega_j \cap \Omega_k \to G$ is given whenever $\Omega_j \cap \Omega_k \neq \varnothing$ so that*

$$F_{jk}(\zeta) = S_j(\zeta)^{-1} S_k(\zeta) \tag{2.9.3}$$

at every point ζ of $\Omega_j \cap \Omega_k$.

Then for every j we can find a holomorphic map $F_j(\zeta) : \Omega_j \to G$ such that

$$F_{jk}(\zeta) = F_j(\zeta)^{-1} F_k(\zeta) \tag{2.9.4}$$

on $\Omega_j \cap \Omega_k$ $(\neq \varnothing)$.

PROOF. First, we shall prove this theorem for a fixed Oka–Weil domain D_0 on M. To do this, we utilize the results of §2.8. Identifying D_0 with

$\mathscr{D}_0$ of §2.8, we denote by D_μ the subset of D_0 corresponding to $\mathscr{D}_\mu$ of §2.8. If we choose N of §2.8 sufficiently large, we can find $j(\mu)$ for D_μ ($\neq \varnothing$) in such a way that

$$D_\mu \subset \Omega_{j(\mu)} \cap D_0.$$

Now, we set

$$S_\mu(\zeta) = S_{j(\mu)}(\zeta) \quad \text{on } D_\mu,$$

and

$$F_{\lambda\mu}(\zeta) = F_{j(\lambda)j(\mu)}(\zeta) \quad \text{on } D_\lambda \cap D_\mu$$

for every (λ, μ) satisfying $D_\lambda \cap D_\mu \neq \varnothing$. As we can use Lemma 2.8.1 in this way, we can find a holomorphic map $F_\mu(\zeta) : D_\mu$ ($\neq \varnothing$) $\to G$ so that

$$F_{\lambda\mu}(\zeta) = F_\lambda(\zeta)^{-1}F_\mu(\zeta)$$

on $D_\lambda \cap D_\mu$ ($\neq \varnothing$). Then, if we set

$$\tilde{F}_j(\zeta) = F_\mu(\zeta)F_{j(\mu)j}(\zeta) \qquad (\zeta \in \Omega_j \cap D_\mu),$$

$\tilde{F}_j(\zeta)$ is a holomorphic map : $\Omega_j \cap D_0 \to G$ since

$$F_\mu F_{j(\mu)j} = F_\lambda F_{\lambda\mu} F_{j(\mu)j} = F_\lambda F_{j(\lambda)j(\mu)} F_{j(\mu)j} = F_\lambda F_{j(\lambda)j}.$$

Also we have

$$\begin{aligned}\tilde{F}_j(\zeta)^{-1}\tilde{F}_k(\zeta) &= F_{jj(\lambda)}(\zeta)F_\lambda(\zeta)^{-1}F_\mu(\zeta)F_{j(\mu)k}(\zeta)\\ &= F_{jj(\lambda)}(\zeta)F_{j(\lambda)j(\mu)}(\zeta)F_{j(\mu)k}(\zeta)\\ &= F_{jk}(\zeta).\end{aligned}$$

Furthermore, if we set

$$\tilde{S}(\zeta) = S_j(\zeta)\tilde{F}_j(\zeta)^{-1}$$

on $\Omega_j \cap D_0$, $\tilde{S}(\zeta)$ is a continuous map : $D_0 \to G$. Hence we can modify $\tilde{S}(\zeta)$ by utilizing Lemma 2.5.2 so that $\tilde{S}(\zeta)$ can be deformed to the identity matrix continuously on D_0. (In order to use Lemma 2.8.1 (resp. Lemma 2.5.2), we must choose ε_0 and ε (resp. ρ and $\tilde{\rho}$). Such details are left to the reader.)

Utilizing Lemma 2.9.3 we choose a sequence $\{D_m\}$ of Oka–Weil domains and apply the results given above to each of the D_m. To do this, if we set

$$\Omega_j^{(m)} = \Omega_j \cap D_m,$$

there exists a holomorphic map $F_j^{(m)}(\zeta) : \Omega_j^{(m)}$ ($\neq \varnothing$) $\to G$ such that

$$F_{jk}(\zeta) = F_j^{(m)}(\zeta)^{-1}F_k^{(m)}(\zeta)$$

on $\Omega_j^{(m)} \cap \Omega_k^{(m)}$ $(\neq \emptyset)$. Furthermore if we set

$$S^{(m)}(\zeta) = S_j(\zeta)F_j^{(m)}(\zeta)^{-1}$$

on $\Omega_j^{(m)}$ $(\neq \emptyset)$, then $S^{(m)}(\zeta)$ is a continuous map : $D_m \to G$ which can be deformed to the identity matrix continuously on D_m.

Next, we modify $F_j^{(m)}(\zeta)$ in such a way that the sequence $\{F_j^{(m)};\ m = 1, 2, \cdots\}$ converges on Ω_j as $m \to \infty$. To do this, we set

$$F^{(m)}(\zeta) = S^{(m+1)}(\zeta)^{-1}S^{(m)}(\zeta) \tag{2.9.5}$$

on D_m. Since

$$F^{(m)}(\zeta) = F_j^{(m+1)}(\zeta)F_j^{(m)}(\zeta)^{-1} \tag{2.9.6}$$

on $\Omega_j^{(m)}$ $(\neq \emptyset)$, $F^{(m)}(\zeta)$ is a holomorphic map : $D_m \to G$. Since the right-hand side of (2.9.5) can be deformed to the identity matrix continuously on D_m, we can assume that $F^{(m)}(\zeta)$ can be deformed to the identity matrix holomorphically on D_m because of Lemma 2.5.3. Now choosing a positive number ε sufficiently small, we write $F^{(m)}(\zeta)$ on D_m in the form

$$F^{(m)}(\zeta) = H_1(\zeta)H_2(\zeta)\cdots H_N(\zeta), \qquad H_k(\zeta) \in U(\varepsilon)$$

(cf. notations of §2.1). Then applying Lemma 2.9.2 to $\log H_k(\zeta)$ $(\in g)$, we can find a holomorphic map $H(\zeta) : M \to G$ such that

$$H(\zeta)F^{(m)}(\zeta) \in U(\varepsilon) \quad \text{on } \bar{D}_{m-1}.$$

Hence if we replace $F_j^{(m+1)}$ by $HF_j^{(m+1)}$, $F^{(m)}(\zeta)$ is contained in $U(\varepsilon)$ on $\bar{D}_{m-1}$. Since ε is arbitrary, we can modify $F^{(m)}$ successively so that we can use Lemma 2.1.5. Hence the sequence $\{F_j^{(m)}\ ;\ m = 1, 2, \cdots\}$ converges on Ω_j. □

To prove results such as the theorem given above, it will be more convenient to define a Stein manifold by utilizing the conclusion of Lemma 2.9.3.

§2.10. The case of one independent variable (Theorem 2.0.1)

In order to derive Theorem 2.0.1 (cf. the beginning of this chapter) from Theorem 2.9.4, it suffices to verify the following two facts:

(A) an open Riemann surface is a Stein manifold,

(B) condition (V) of Theorem 2.0.1 implies the existence of the S_j satisfying condition (IV) (i.e., (2.9.3)) of Theorem 2.9.4.

As to (A), it is sufficient to understand Theorem 10 of Chapter IX on page 270 of Gunning–Rossi [2]. Actually we must verify only conditions (i) and (ii) of Stein manifolds (cf. Definition 2.9.1). Hence it is easy to show that every connected open set of C is a Stein manifold. In particular, if a domain $\mathscr{D}$ is constructed by removing some isolated points from C, then $\mathscr{D}$ is a Stein manifold. To study such a Stein manifold, it is helpful to know that a domain which is constructed by removing a finite number of disks from a large disk is an Oka–Weil domain (cf. examples of §2.4).

As to (B), we use an idea based on a fibre bundle. To do this, we construct a fibre bundle by utilizing the assumptions of Theorem 2.0.1. (For detail, look at Part I, §3, pp. 14–16 of N. Steenrod [10].) First of all, let us denote by N the set of all indices j of the covering $\{\Omega_j\}$ of M. We introduce the discrete topology on N. Then, constructing a topological space $M \times G \times N$, we set

$$T = \bigcup_{j \in N} \Omega_j \times G \times \{j\}.$$

A relative topology is induced on T from $M \times G \times N$. We identify two points (z, g, j) and (z', g', k) of T whenever

$$z = z' \in \Omega_j \cap \Omega_k, \qquad F_{kj}(z)g = g'.$$

Then by virtue of condition (V) of Theorem 2.0.1, we can construct a topological space B. We denote by $[z, g, j]$ the point of B corresponding to the point (z, g, j) of T. We introduce a topology on B so that a map $q(z, g, j) : T \to B$ defined by

$$q(z, g, j) = [z, g, j]$$

becomes continuous. If we set

$$p([z, g, j]) = z$$

then p is also continuous as a map : $B \to M$, and if we set

$$\phi_j(z, g) = [z, g, j] \qquad (z \in \Omega_j)$$

then ϕ_j is a homeomorphism : $\Omega_j \times G \to p^{-1}(\Omega_j)$. Furthermore, if we set

$$\phi_{j,z}(g) = \phi_j(z, g) \qquad (z \in \Omega_j),$$

we have

$$\phi_{j,z}^{-1} \circ \phi_{k,z}(g) = F_{jk}(z)g$$

for $z \in \Omega_j \cap \Omega_k$. This follows from $[z, g, k] = [z, F_{jk}(z)g, j]$. In this way, we can construct a fibre bundle with the base space M, the fibre G and the coordinate transformation $F_{jk}(z)$ on M.

A continuous map $S(z) : M \to B$ is called a (global) section if $pS(z) = z$ for all points z of M. In the case when $S(z)$ is a section,

$$\tilde{S}_j(z) = \phi_{j,z}^{-1}(S(z)) \qquad (z \in \Omega_j)$$

is a continuous map : $\Omega_j \rightarrow G$ and we have

$$\tilde{S}_j(z) = \phi_{j,z}^{-1} \circ \phi_{k,z}(\tilde{S}_k(z)) = F_{jk}(z)\tilde{S}_k(z)$$

for $z \in \Omega_j \cap \Omega_k$. Therefore, if we set $S_j(z) = \tilde{S}_j(z)^{-1}$, condition (2.9.3) is satisfied. Therefore it suffices to construct a continuous map $S(z) : M \rightarrow B$ satisfying the condition $pS(z) = z$ for all $z \in M$.

A detailed explanation of constructing a section is found in Part III of N. Steenrod [10]. Roughly speaking, we utilize a theory of cohomologies concerning "obstruction cocycle" (cf. §§29.4, 29.7, 29.8 and in particular §§34.3 and 34.4 on p. 176 of [10]). Since the fibre G is a connected topological group (a Lie group), the fundamental group of G is commutative. This fact is also proved in §§16.6–16.9 on pp. 88–89 of [10]. Then, by virtue of the theory of cohomologies concerning an open Riemann surface, we can construct a global section $S(z)$. Therefore, we can use Theorem 2.9.4 without assuming the existence of the S_j. (In the general case of Stein manifolds, we must assume the existence of the S_j.)

We explain how we can construct a section $S(z)$ in a simpler but less rigorous manner. First of all we consider a "triangle" σ on M which is totally contained in M so that σ is simply connected in M. We denote by B_σ the restriction of the fibre bundle B on σ (i.e., $B_\sigma = p^{-1}(\sigma)$). Since σ is simply connected, we can regard B_σ as $\sigma \times G$. Therefore, it is not difficult to construct a section on σ. If $S(z)$ is a continuous map : $\sigma \rightarrow B_\sigma$ then we can construct a section on σ by deforming $S(z)$ in a suitable way (cf. §29.5, p. 170 of [10]). If we fix values of $S(z)$ on the three sides of σ, then a closed curve is given in B_σ. If this closed curve cannot be shrunk to a point continuously in B_σ, we cannot define $S(z)$ with this boundary condition. However, if the boundary values are given on two sides or one side of σ, we can define $S(z)$ because of the connectedness of $\sigma \times G$. Now, subdividing M into triangles, let us construct a section on a compact subset K which is a union of a finite number of triangles. Let us assume that it is evident that K admits a nonempty boundary, since M is an open Riemann surface. We use induction on the total number q of triangles of which K consists. We took care of the case $q = 1$. Supposing that the case of q triangles has been taken care of, we look at the case of $q + 1$ triangles. To do this, removing from K a triangle σ on the boundary of K, we construct a section $S(z)$ on $K - \sigma$. Then the values of $S(z)$ are fixed on two sides or one side of σ, but not on three sides, since σ is on the boundary of K. Hence we can extend the section $S(z)$ on $K - \sigma$ to σ. In this way, we can eventually construct a global section of B by repeating this process successively.

Now we have finished the explanation of the existence theorem of Grauert. In the next section, we shall explain a simple application of this theorem, combining Theorems 1.11.3 and 2.0.1.

§2.11. Improvements and applications of Theorem 1.11.3

In order to combine Theorems 1.11.3 and 2.0.1 in a useful way, we prove the lemma below.

LEMMA 2.11.1. *Assume that $\mathscr{D}$ is an open set on the z-plane and that a matrix $W(z)$ is a holomorphic map : $\mathscr{D} \to G$, where G is a linear group. Define a matrix $A(z)$ by*

$$A(z) = \frac{dW(z)}{dz} W(z)^{-1} \qquad (z \in \mathscr{D}).$$

Then $A(z)$ belongs to $gl(n, \mathscr{O}(\mathscr{D}))$ and is contained in the Lie algebra g of G at every point z of $\mathscr{D}$.

PROOF. If we fix a point z in $\mathscr{D}$ and if we set

$$V(t, z) = W(t)W(z)^{-1},$$

then $V(t, z)$ tends to the identity matrix in G as t approaches z. Consequently $V(t, z)$ can be expressed in terms of the system of standard coordinates as t approaches z (cf. §2.1). The standard coordinates are holomorphic functions of t and z. Hence if we differentiate V with respect to t and if we set $t = z$, we can prove the lemma. □

Combining Lemmas 2.11.1 and 1.11.4, we derive the following lemma.

LEMMA 2.11.2. *Let $\mathscr{D}$ be a connected open set on the z-plane and let $\{\Delta(z_j, r_j)\}$ be a covering of $\mathscr{D}$. Also, for every j, let $W_j(z)$ be a holomorphic map : $\Delta(z_j, r_j) \to G$, where G is a linear group. Then, if*

$$C_{jk} = W_j(z)^{-1} W_k(z) \tag{2.11.1}$$

is a constant matrix independent of z whenever $\Delta(z_j, r_j) \cap \Delta(z_k, r_k) \neq \varnothing$, the matrix

$$A(z) = \frac{dW_j(z)}{dz} W_j(z)^{-1} \qquad (z \in \Delta(z_j, r_j)) \tag{2.11.2}$$

belongs to $gl(n, \mathscr{O}(\mathscr{D}))$ and is contained in the Lie algebra g of G at every point z of $\mathscr{D}$. Moreover $\{\Delta(z_j, r_j), W_j(z)\}$ is a system of fundamental solutions of the differential equation

$$\frac{dW}{dz} = A(z)W. \tag{2.11.3}$$

The converse of Lemma 2.11.2 is also true. We summarize such a result as follows.

LEMMA 2.11.3. *Let $\mathscr{D}$ be a connected open set on the z-plane. Assume that a matrix $A(z)$ belongs to $gl(n,\mathscr{O}(\mathscr{D}))$ and is contained in the Lie algebra g of a linear group G at every point z of $\mathscr{D}$. Then, we can construct a system $\{\Delta(z_j, r_j), W_j(z)\}$ of fundamental solutions of differential equation* (2.11.3) *in such a way that $W_j(z)$ is a holomorphic map : $\Delta(z_j, r_j) \to G$ for every j.*

Let us fix a neighborhood U of the identity matrix in G. To prove Lemma 2.11.3, it suffices to construct an approximate solution by means of the Cauchy polygon so that the approximate solution is contained in U in a neighborhood of a point of $\mathscr{D}$. The details are left to the reader. Utilizing what we have explained so far as a preparation, we explain the theorem below.

THEOREM 2.11.4. *Let $\mathscr{D}$ be a connected open set on the z-plane. Assume that $\mathscr{C}$ is a structure of analytic continuation on $\mathscr{D}$ and that G is a linear group. Then, in order that we can find a matrix $A(z)$ in $gl(n,\mathscr{O}(\mathscr{D}))$ such that*

(i) *$A(z)$ is contained in the Lie algebra g of G at every point z of $\mathscr{D}$,*
(ii) *differential equation* (2.11.3) *defines the structure $\mathscr{C}$ of analytic continuation,*

it is necessary and sufficient that we find a system $\{\Delta(z_j, r_j), C_{jk}\}$ of connection matrices belonging to $\mathscr{C}$ such that every matrix C_{jk} is contained in G.

The necessity of the condition of this theorem follows from Lemma 2.11.3. To verify the sufficiency, we can apply Lemma 2.11.2 since we can construct $W_j(z)$ satisfying (2.11.1) if we set $M = \mathscr{D}$, $\Omega_j = \Delta(z_j, r_j)$ and $F_{jk}(z) = C_{jk}$ in Theorem 2.0.1.

The theorem below is a consequence of Theorems 2.11.4 and 1.11.1.

THEOREM 2.11.5. *Let $\mathscr{D}$ be a connected open set on the z-plane. Assume that a matrix $A(z)$ belongs to $gl(n,\mathscr{O}(\mathscr{D}))$ and that G is a linear group. Then, in order that there exists a matrix $P(z)$ in $GL(n,\mathscr{O}(\mathscr{D}))$ such that the transformation*

$$W = P(z)V \tag{2.11.4}$$

changes differential equation (2.11.3) *to*

$$\frac{dV}{dz} = B(z)V \tag{2.11.5}$$

with a matrix $B(z)$ contained in the Lie algebra g of G at every point z of $\mathscr{D}$, it is necessary and sufficient that the structure $\mathscr{C}$ of analytic continuation defined by (2.11.3) *on $\mathscr{D}$ satisfies the condition of Theorem* 2.11.4.

The proof of this theorem is left to the reader as an exercise.

In the two theorems above, all conditions were given in terms of a structure $\mathscr{C}$ of analytic continuation. Those conditions can be also given in terms of the structure $\mathscr{M}(\mathscr{C})$ of monodromy determined by $\mathscr{C}$ on $\mathscr{D}$. The following example shows such a situation.

EXAMPLE 2.11.6. Let us denote by G the set of all upper triangular matrices in $GL(n, C)$. Then G is a linear group. If we denote by g the set of all upper triangular matrices in $gl(n, C)$, the Lie algebra of G is g. Applying Theorem 2.11.5 to this group G, we can derive the following result:

Let $\mathscr{D}$ be a connected open set on the z-plane. Assume that a matrix $A(z)$ belongs to $gl(n, \mathscr{O}(\mathscr{D}))$. Then, in order that there exists a matrix $P(z)$ in $GL(n, \mathscr{O}(\mathscr{D}))$ such that transformation (2.11.4) changes (2.11.3) to (2.11.5) with a matrix $B(z)$ which is upper triangular at every point z of $\mathscr{D}$, it is necessary and sufficient that we find a system $\{\Delta(z_j, r_j), C_{jk}\}$ of connection matrices in the structure $\mathscr{C}$ of analytic continuation defined by (2.11.3) on $\mathscr{D}$ such that every matrix C_{jk} is upper triangular.

The structure $\mathscr{C}$ of analytic continuation satisfies the condition above if we can find a representation $\rho(\sigma)$ of $\pi_1(\mathscr{D}, z_0)$ in C^n in the structure $\mathscr{M}(\mathscr{C})$ of monodromy such that $\rho(\sigma)$ is upper triangular for every $\sigma \in \pi_1(\mathscr{D}, z_0)$. (To prove this, it suffices to utilize what we explained in §1.7.) In such a situation, there exists a matrix Γ in $GL(n, C)$ for each representation $\rho(\sigma)$ in $\mathscr{M}(\mathscr{C})$ such that $\Gamma^{-1}\rho(\sigma)\Gamma$ is upper triangular for every $\sigma \in \pi_1(\mathscr{D}, z_0)$.

CHAPTER III

A Theorem of G. D. Birkhoff

§3.0. Preliminaries

Theorem 1.11.1 states that the classification of linear ordinary differential equations by means of transformations with coefficients in $\mathcal{O}(\mathcal{D})$ and the classification by means of structures of analytic continuation defined on $\mathcal{D}$ by such differential equations are identical. However, if we restrict coefficients of differential equations and of transformations to a set smaller than $\mathcal{O}(\mathcal{D})$, the situation changes. Let $\mathcal{D}$ be the domain derived from a disk $\Delta(a, r)$ by deleting its center a (i.e., $\mathcal{D} = \{ z \ ; \ 0 < |z-a| < r\}$). In this chapter, utilizing Theorem 2.11.5 on $\mathcal{D}$, we shall first find standard differential equations by means of transformations with coefficients in $\mathcal{O}(\mathcal{D})$ (cf. §3.1). Next, we shall explain a theorem of G. D. Birkhoff (cf. §3.3) which is one of the most significant results concerning standard differential equations in the case when coefficients are meromorphic at the center a of the disk $\Delta(a, r)$. The Birkhoff theorem is based on several lemmas (cf. §§3.4–3.7) which are by themselves important results concerning characterization of complex analytic vector bundles over the Riemann sphere. The existence theorem (i.e., Theorem 2.0.1) of Grauert is a result on an open Riemann surface. In order to study an ordinary differential equation at singular points, we need also some results on a closed Riemann surface. The lemmas mentioned above are such examples. We shall come back to some situations on a closed Riemann surface again in Chapter 4.

§3.1. Standard forms of linear ordinary differential equations at singular points

Setting

$$\mathcal{D} = \{ z \ ; \ 0 < |z - a| < R_0\}$$

we apply Theorem 2.11.5 given at the end of Chapter 2, where a is a fixed point on the z-plane and R_0 is a fixed positive number. Given a matrix $A(z) \in gl(n, \mathcal{O}(\mathcal{D}))$, we shall find a standard form (in the sense of Theorem 2.11.5) for the differential equation

$$\frac{dW}{dz} = A(z)W. \tag{3.1.1}$$

To do this, let $\rho_0(\sigma)$ be a representation of $\pi_1(\mathscr{D}, z_0)$ in C^n which belongs to the structure $\mathscr{M}(\mathscr{C})$ of monodromy corresponding to the structure $\mathscr{C}$ of analytic continuation defined on $\mathscr{D}$ by differential equation (3.1.1), where $z_0 \in \mathscr{D}$ and $\sigma \in \pi_1(\mathscr{D}, z_0)$.

Let ℓ_0 be the circle with the center at a, passing through z_0 and oriented counterclockwise. Then $\pi_1(\mathscr{D}, z_0)$ is a cyclic group generated by $\sigma_0 = [\ell_0]$. Consequently, if we set

$$\rho_0(\sigma_0) = C_0$$

and if G is a linear group containing C_0, we have

$$\rho_0(\sigma) \in G$$

for every σ in $\pi_1(\mathscr{D}, z_0)$. Therefore, if we utilize what we explained in §1.7, we can choose a system $\{\Delta(z_j, r_j), C_{jk}\}$ of connection matrices from the structure $\mathscr{C}$ of analytic continuation in such a way that every C_{jk} belongs to G. If we regard differential equation (2.11.5) of Theorem 2.11.5 as a standard form, the matrix $B(z)$ of the standard form belongs to the Lie algebra g of G at every point z of $\mathscr{D}$.

In order to find a linear group containing C_0, it suffices to construct a one-parameter subgroup of $GL(n, C)$ containing C_0. To do this, fixing a matrix A_0 in $gl(n, C)$ by the condition

$$C_0 = \exp\ A_0,$$

we set

$$G = \{\exp\ zA_0\ ; \quad z \in C\ \}.$$

The Lie algebra g of G is the subalgebra of $gl(n, C)$ generated by A_0, i.e.,

$$g = \{\ \alpha A_0\ ; \quad \alpha \in C\ \}.$$

Hence, the matrix $B(z)$ of standard form (2.11.5) is given by

$$B(z) = \alpha(z)A_0 \tag{3.1.2}$$

where $\alpha(z) \in \mathscr{O}(\mathscr{D})$.

Fixing a function $\alpha(z)$ in $\mathscr{O}(\mathscr{D})$, let us study the structure of analytic continuation defined on $\mathscr{D}$ by the differential equation

$$\frac{dV}{dz} = \alpha(z)A_0V. \tag{3.1.3}$$

To do this, if we set

$$V_j(z) = \exp\left[\int_{z_j}^{z} \alpha(\xi)d\xi A_0\right],$$

then $\{\Delta(z_j, r_j), V_j(z)\}$ is a system of fundamental solutions of (3.1.3). Therefore, if we set

$$C_{jk} = \exp\left[\int_{z_k}^{z_j} \alpha(\xi)d\xi A_0\right], \tag{3.1.4}$$

then $\{\Delta(z_j, r_j), C_{jk}\}$ is a system of connection matrices defined by $\{\Delta(z_j, r_j), V_j(z)\}$, where the path of integration on the right-hand side of (3.1.4) is a line segment joining z_k to z_j. If we construct a representation ρ_h of $\pi_1(\mathscr{D}, z_0)$ by applying Theorem 1.7.2 to the system $\{\Delta((z_j, r_j), C_{jk}\}$ of connection matrices, we obtain

$$\rho_h(\sigma_0) = \exp\left[-\int_{\ell_0} \alpha(\xi)d\xi A_0\right].$$

Hence if $\alpha(z)$ satisfies the condition

$$\int_{\ell_0} \alpha(\xi)d\xi = -1, \tag{3.1.5}$$

ρ_h and ρ_0 agree with each other exactly. This means that the structure of analytic continuation defined by (3.1.3) and the structure defined by (3.1.1) are identical.

THEOREM 3.1.1. *Let $\rho(\sigma)$ be a representation of $\pi_1(\mathscr{D}, z_0)$ belonging to the structure of monodromy defined on $\mathscr{D}$ by differential equation* (3.1.1) *and let ℓ_0 be the circle with the center at a, passing through z_0 and oriented counterclockwise. Fix a matrix A_0 in $gl(n, C)$ by the condition*

$$\rho([\ell_0]) = \exp\ A_0$$

and choose a function $\alpha(z)$ in $\mathscr{O}(\mathscr{D})$ satisfying condition (3.1.5). *Then there exists a matrix $P(z)$ in $GL(n, \mathscr{O}(\mathscr{D}))$ such that the transformation*

$$W = P(z)V \tag{3.1.6}$$

changes (3.1.1) *to* (3.1.3).

There are many functions $\alpha(z)$ satisfying condition (3.1.5). The simplest $\alpha(z)$ is given by

$$\alpha(z) = \frac{-1}{2\pi i(z-a)}. \tag{3.1.7}$$

Customarily, differential equation (3.1.3) with $\alpha(z)$ given by (3.1.7) is regarded as the standard form of (3.1.1) on $\mathscr{D}$. However we must note that there are also many choices of A_0 (cf. Lemma 3.3.3).

§3.2. The case when coefficients are restricted to a set smaller than $\mathscr{O}(\mathscr{D})$

The classification of linear ordinary differential equations by means of transformations with coefficients in $\mathscr{O}(\mathscr{D})$ and the classification by means of structures of analytic continuation defined on $\mathscr{D}$ by such differential equations are identical (cf. Theorem 1.11.1). However, in some cases, $\mathscr{O}(\mathscr{D})$ is too large as a set of functions. In this section, we consider a smaller set. As in §3.1, we set

$$\mathscr{D} = \{\, z \;;\; 0 < |z-a| < R_0 \}, \tag{3.2.1}$$

and

$$\mathscr{m}(\mathscr{D}) = \{\, (z-a)^k f(z) \;;\; f \in \mathscr{O}(\Delta(a, R_0)),\;\; k \in Z \,\}, \tag{3.2.2}$$

where Z is the ring of all integers. This means that $\mathscr{m}(\mathscr{D})$ is the set of all functions which are holomorphic on $\mathscr{D}$ and admit at worst a pole at $z = a$.

We denote by $gl(n, \mathscr{m}(\mathscr{D}))$ the set of all $n \times n$ matrices whose entries are in $\mathscr{m}(\mathscr{D})$. The inverse of a matrix in $gl(n, \mathscr{m}(\mathscr{D}))$ belongs also to $gl(n, \mathscr{m}(\mathscr{D}))$. We denote by $GL(n, \mathscr{m}(\mathscr{D}))$ the set of all invertible matrices in $gl(n, \mathscr{m}(\mathscr{D}))$. Since $\mathscr{m}(\mathscr{D})$ is contained in $\mathscr{O}(\mathscr{D})$, we have

$$gl(n, \mathscr{m}(\mathscr{D})) \subset gl(n, \mathscr{O}(\mathscr{D}))$$
$$GL(n, \mathscr{m}(\mathscr{D})) = gl(n, \mathscr{m}(\mathscr{D})) \cap GL(n, \mathscr{O}(\mathscr{D})).$$

Fixing a matrix $A(z)$ in $gl(n, \mathscr{m}(\mathscr{D}))$, we apply Theorem 3.1.1 to the differential equation

$$\frac{dW}{dz} = A(z)W. \tag{3.2.3}$$

This means that there exist a matrix A_0 in $gl(n, C)$ and a matrix $P(z)$ in $GL(n, \mathscr{O}(\mathscr{D}))$ such that the transformation

$$W = P(z)V \tag{3.2.4}$$

changes (3.2.3) to the differential equation

$$\frac{dV}{dz} = \frac{-1}{2\pi i(z-a)} A_0 V. \tag{3.2.5}$$

The matrix A_0 is determined by the structure of monodromy defined on $\mathscr{D}$ by (3.2.3), while $P(z)$ is determined by systems of fundamental solutions of (3.2.3) and (3.2.5). If $P(z)$ belongs to $GL(n, \mathcal{m}(\mathscr{D}))$, we say that differential equation (3.2.3) admits at worst **a regular singular point** at $z = a$. We shall explain regular singular points in detail in Chapters 4 and 5. In order to construct an example of a singular point at $z = a$ which is not a regular singular point, it suffices to study the relation between differential equations (3.1.3) and (3.2.5) by utilizing various functions $\alpha(z)$.

In this chapter, considering the general case when the matrix $A(z)$ of differential equation (3.2.3) belongs to $gl(n, \mathcal{m}(\mathscr{D}))$, we shall explain standard forms by means of transformations (3.2.4) with $P(z)$ in $GL(n, \mathcal{m}(\mathscr{D}))$. The main result is a theorem of G. D. Birkhoff (Theorem 3.3.1) which will be given in the next section. We shall also verify that (3.2.3) admits at worst a regular singular point at $z = a$ if every entry of $A(z)$ admits at worst a simple pole at $z = a$ (cf. Theorem 3.9.4 (Hukuhara)). In §3.10 (the last section), we shall explain a theorem of H. L. Turrittin which is an improvement of the Birkhoff theorem.

§3.3. A theorem of G. D. Birkhoff

To explain a theorem of Birkhoff [12] which is the main concern of this chapter, it is convenient to place the singular point at $z = \infty$. Therefore we set

$$\mathscr{D} = \{ z \, ; \, |z| > R_0 \} \tag{3.3.1}$$

and

$$\mathcal{m}(\mathscr{D}) = \left\{ z^k f\left(\frac{1}{z}\right) \, ; \, f \in \mathscr{O}\left(\Delta\left(0, \frac{1}{R_0}\right)\right), \; k \in Z \right\}. \tag{3.3.2}$$

The definitions of $gl(n, \mathcal{m}(\mathscr{D}))$ and $GL(n, \mathcal{m}(\mathscr{D}))$ are the same as in §3.2. If A is a matrix in $gl(n, \mathscr{O}(\Delta(0, \frac{1}{R_0})))$ and if k is an integer, then $z^k A(\frac{1}{z})$ is in $gl(n, \mathcal{m}(\mathscr{D}))$. Conversely, every matrix in $gl(n, \mathcal{m}(\mathscr{D}))$ can be written in such a form. Using those facts as a preparation, we state a result of Birkhoff [12].

THEOREM 3.3.1. (G. D. Birkhoff). *Given an integer k and a matrix $A(\zeta) \in gl(n, \mathscr{O}(\Delta(0, \frac{1}{R_0})))$, we consider the differential equation*

$$\frac{dW}{dz} = z^k A\left(\frac{1}{z}\right) W. \tag{3.3.3}$$

Then, there exists a matrix $P(\zeta) \in GL(n, \mathscr{O}(\Delta(0, \frac{1}{R_0})))$ such that the transformation

$$W = P\left(\frac{1}{z}\right) V \tag{3.3.4}$$

changes (3.3.3) *to the differential equation*

$$\frac{dV}{dz} = z^k B\left(\frac{1}{z}\right) V \tag{3.3.5}$$

with a matrix $B(\frac{1}{z})$ *whose entries are polynomials in* $\frac{1}{z}$ *with complex constant coefficients. Moreover we can choose such a* $P(\zeta)$ *so that* $z = 0$ *is at worst a regular singular point of* (3.3.5).

If we set $\zeta = \frac{1}{z}$, differential equation (3.3.3) becomes

$$\frac{dW}{d\zeta} = -\zeta^{-k-2} A(\zeta) W. \tag{3.3.6}$$

If $k \leq -2$, (3.3.6) is a differential equation on $\Delta(0, \frac{1}{R_0})$. Since $\Delta(0, \frac{1}{R_0})$ is simply connected, we can derive Theorem 3.3.1 by utilizing Theorem 1.11.2. Therefore the main problem is the case when $k \geq -1$.

Since the fundamental group of $\mathscr{D}$ is cyclic, we can use Theorem 3.1.1. We summarize such a result as a lemma.

LEMMA 3.3.2. *There exist a matrix* $A_0 \in gl(n, C)$ *and a matrix* $T(z) \in GL(n, \mathscr{O}(\mathscr{D}))$ *such that the transformation*

$$W = T(z)V \tag{3.3.7}$$

changes differential equation (3.3.3) *to*

$$\frac{dV}{dz} = \frac{1}{z} A_0 V. \tag{3.3.8}$$

Fixing a point z_0 on $\mathscr{D}$, we denote by ℓ_0 the circle with the center at $z = 0$, passing through z_0 and oriented clockwise. Let $\rho(\sigma)$ be a representation of $\pi_1(\mathscr{D}, z_0)$ in C^n which belongs to the structure of monodromy defined on $\mathscr{D}$ by differential equation (3.3.3). Then we can determine A_0 by the condition

$$\rho([\ell_0]) = \exp[2\pi i A_0]. \tag{3.3.9}$$

The choice of A_0 is not unique. However for every A_0 satisfying the condition of Lemma 3.3.2, there exists a representation ρ satisfying condition (3.3.9). Since these representations belong to the structure of monodromy defined on $\mathscr{D}$ by (3.3.3), they are mutually equivalent (cf. §1.6). Hence we obtain the lemma below.

LEMMA 3.3.3. *If two matrices* A_0 *and* $\tilde{A}_0$ *in* $gl(n, C)$ *satisfy the condition of Lemma* 3.3.2, *there exists a matrix* Q *in* $GL(n, C)$ *such that*

$$\exp[2\pi i \tilde{A}_0] = \exp[2\pi i Q^{-1} A_0 Q]. \tag{3.3.10}$$

Conversely, if (3.3.10) *is satisfied by some* Q *and if* A_0 *satisfies the condition of Lemma* 3.3.2, *then* $\tilde{A}_0$ *also satisfies the condition of Lemma* 3.3.2.

We remark that A_0 (resp. $\tilde{A}_0$) satisfies the condition of Lemma 3.3.2 if there exists $T(z)$ (resp. $\tilde{T}(z)$) such that the conclusion of Lemma 3.3.2 holds. The two matrices $T(z)$ and $\tilde{T}(z)$ are not necessarily the same. The details of the proof of Lemma 3.3.3 are left to the reader. The main part of the proof of Theorem 3.3.1 is to study the structure of the matrix $T(z)$. In the next section we shall explain this problem in detail.

§3.4. A lemma of G. D. Birkhoff

To prove Theorem 3.3.1, the result given below is the most important lemma.

LEMMA 3.4.1. *Assume that a matrix* $T(z)$ *belongs to* $GL(n,\mathcal{O}(\mathcal{D}))$, *where*

$$\mathcal{D} = \{\, z\,;\ |z| > R_0 \,\}. \tag{3.4.1}$$

Then $T(z)$ *can be written in the following form:*

$$T(z) = P\left(\frac{1}{z}\right)E(z)\Lambda(z), \tag{3.4.2}$$

where

(i) $P(\zeta) \in GL(n,\mathcal{O}(\Delta(0,\frac{1}{R_0})))$,
(ii) $E(z) \in GL(n,\mathcal{O}(\mathbb{C}))$,
(iii) $\Lambda(z)$ *is a diagonal matrix*

$$\Lambda(z) = \begin{bmatrix} z^{k_1} & 0 & 0 & \cdots & 0 \\ 0 & z^{k_2} & 0 & \cdots & 0 \\ \cdots & \cdots & \cdots & \cdots & \cdots \\ 0 & 0 & 0 & \cdots & z^{k_n} \end{bmatrix}, \tag{3.4.3}$$

and $k_1, \cdots, k_n$ *are integers.*

Lemma 3.4.1 will be used to prove Theorem 3.3.1. However, we modify the lemma slightly for various purposes.

LEMMA 3.4.2. *Assume that a matrix* $T(z)$ *belongs to* $GL(n,\mathcal{O}(\mathcal{D}))$, *where*

$$\mathcal{D} = \{\, z\,;\ |z| > R_0 \,\}.$$

Then $T(z)$ *can be written in the following form:*

$$T(z) = P\left(\frac{1}{z}\right)\Lambda(z)E(z), \tag{3.4.4}$$

where P, E, and Λ satisfy conditions (i), (ii), *and* (iii) *of Lemma* 3.4.1. *However P, E, and Λ of this lemma are not identical with those of Lemma* 3.4.1. *In particular, the integers $k_1, \cdots, k_n$ of this lemma can be arranged in the ascending order $k_1 \leq k_2 \leq \cdots \leq k_n$ (actually in any order).*

If we denote by $\mathscr{D}_0$ the set obtained by adding ∞ to $\mathscr{D}$, then $\mathscr{D}_0 \cup C$ is the Riemann sphere and $\mathscr{D}_0 \cap C = \mathscr{D}$. If the Riemann sphere were an open Riemann surface, the formula $T(z) = P(\frac{1}{z})E(z)$ would hold because of the existence theorem of Grauert (i.e., Theorem 2.0.1). Note that, in this formula, the matrix $\Lambda(z)$ of (3.4.4) is replaced by the identity matrix. However, since the Riemann sphere is a closed Riemann surface, we cannot utilize the existence theorem of Grauert.

Given a matrix $T(z)$, we can construct an analytic vector bundle over the Riemann sphere whose fibre is C^n and whose structure group is $GL(n, C)$ (cf. §2.10). Similarly, given a matrix $\Lambda(z)$, we can construct another vector bundle over the Riemann sphere. Formula (3.4.4) asserts that these two vector bundles are analytically isomorphic (cf. N. Steenrod [10], Lemma 2.10 on p. 12). Since $\Lambda(z)$ is diagonal, we can prove by utilizing Theorem 2.0.1 and Lemma 3.4.2 that every complex analytic vector bundle over the Riemann sphere is analytically isomorphic to a direct sum of line bundles.

Generally speaking, every complex analytic vector bundle over a closed Riemann surface admits a meromorphic section. For a proof of this fact in the case of a line bundle, see for example R. Gunning [3] (Theorem 12 on p.170) or Hurwitz–Courant [5] (Appendix, Chapter 2, Section 5). The author does not know where we can find any proof for the general case, so we leave that to the reader. Anyway, such a result over the Riemann sphere can be proved by means of what we explained above. Actually Lemma 3.4.2 is a much deeper result.

To prove Lemmas 3.4.1 and 3.4.2, we shall utilize the lemma below.

LEMMA 3.4.3. *Assume that a matrix $T(z)$ belongs to $GL(n, \mathscr{O}(\mathscr{D}))$, where*

$$\mathscr{D} = \{ z \; ; \; |z| > R_0 \}.$$

Then $T(z)$ can be written in the form:

$$T(z) = Q(z)E(z), \tag{3.4.5}$$

where

(i) $Q(z) \in GL(n, \mathscr{m}(\mathscr{D}))$,
(ii) $E(z) \in GL(n, \mathscr{O}(C))$.

In particular, if $T(z)$ is upper (resp. lower) triangular, then Q and E can be chosen to be upper (resp. lower) triangular.

This lemma also has a deeper meaning than the existence of a meromorphic section.

In the situation above, the structure group of the vector bundle is $GL(n, C)$. In Lemmas 3.4.1 and 3.4.2 we cannot replace $GL(n, C)$ by a linear group G. However, Lemma 3.4.3 is different. For example, we can derive the same conclusion within the group of triangular matrices as we emphasized above.

§3.5. Proof of Lemma 3.4.3

Birkhoff gave two proofs of Lemma 3.4.3. In one of the two proofs, he identifies $T(z)$ with its Laurent series, regards (3.4.5) as an equation in the infinite dimensional space consisting of all Laurent series, and solves such an equation (cf. G. D. Birkhoff [12], §1, Lemma on Analytic Functions, on pp. 438–443; see also Y. Sibuya [28], (II), §5, Lemmas I and I′ on pp. 11–14). The other proof is somewhat similar to the proof of the decomposition theorem of Cartan (Theorem 2.2.1). Therefore we shall explain that proof here. The idea of the method is to transform (3.4.5) into an integral equation of the Riemann–Hilbert type (cf. G. D. Birkhoff [13], pp. 123–127).

Considering a domain

$$\mathscr{D} = \{ z \,;\, |z| > R_0 \} \tag{3.5.1}$$

on the z-plane, we denote by ℓ_0 a circle with the center at $z = 0$, contained in $\mathscr{D}$ and oriented counterclockwise. We also denote by Δ the disk bounded by ℓ_0. If we use the Laurent series of a function $f(z)$ in $\mathscr{O}(\mathscr{D})$, we can write $f(z)$ in the form

$$f(z) = \frac{1}{z}\,\alpha\left(\frac{1}{z}\right) + \beta(z), \tag{3.5.2}$$

where

(i) $\alpha(\zeta) \in \mathscr{O}(\Delta(0, \frac{1}{R_0}))$,
(ii) $\beta(z) \in \mathscr{O}(C)$.

By utilizing this expression, we can derive

$$\frac{1}{2\pi i}\int_{\ell_0} \frac{f(\tau)}{\tau - z}\,d\tau = -\frac{1}{z}\,\alpha\left(\frac{1}{z}\right) \tag{3.5.3}$$

for z not contained in the closure of Δ. This means that, if we operate on f using the integral operator on the left-hand side of (3.5.3), the essential singularity of f at $z = \infty$ disappears. This situation is similar to the situation in which we divided a path γ of integration into γ' and γ'' in §2.3. If we use (3.5.3), we can write (3.5.2) as

$$f(z) = -\frac{1}{2\pi i}\int_{\ell_0} \frac{f(\tau)}{\tau - z} d\tau + \beta(z)$$

and hence

$$\beta(z) = f(z) + \frac{1}{2\pi i}\int_{\ell_0} \frac{f(\tau)}{\tau - z} d\tau. \tag{3.5.4}$$

By virtue of (3.5.4) we can extract $\beta(z)$ from $f(z)$.

Let us write a matrix $F(z)$ in $gl(n, \mathscr{O}(\mathscr{D}))$ in the form

$$F(z) = \frac{1}{z} A\left(\frac{1}{z}\right) + B(z), \tag{3.5.5}$$

where

(1) $A(\zeta) \in gl(n, \mathscr{O}(\Delta(0, \frac{1}{R_0})))$,
(2) $B(z) \in gl(n, \mathscr{O}(C))$.

Then, because of (3.5.4), we have

$$B(z) = F(z) + \frac{1}{2\pi i}\int_{\ell_0} \frac{F(\tau)}{\tau - z} d\tau. \tag{3.5.6}$$

If there exists a matrix $B(z)$ in $gl(n, \mathscr{O}(C))$ such that

$$T(z)B(z) \in GL(n, \mathscr{m}(\mathscr{D}))$$

for the matrix $T(z)$ of Lemma 3.4.3, we come close to the conclusion of Lemma 3.4.3. To accomplish this, it suffices to find a matrix $F(z)$ in $gl(n, \mathscr{O}(\mathscr{D}))$ such that

$$T(z)\left\{F(z) + \frac{1}{2\pi i}\int_{\ell_0} \frac{F(\tau)}{\tau - z} d\tau\right\} \in GL(n, \mathscr{m}(\mathscr{D})).$$

Thus skipping some arguments in detail, we arrive at the following integral equation:

$$T(z)\left\{F(z) + \frac{1}{2\pi i}\int_{\ell_0} \frac{F(\tau)}{\tau - z} d\tau\right\} = \frac{1}{2\pi i}\int_{\ell_0} \frac{T(\tau)F(\tau)}{\tau - z} d\tau + G(z),$$

where the entries of G are polynomials in z. The integral on the right-hand side does not have an essential singular point at $z = \infty$. If we set

$$K(z, \tau) = T(z)^{-1}T(\tau) - I, \tag{3.5.7}$$

this integral equation can be written in the form

$$\frac{1}{2\pi i}\int_{\ell_0} \frac{K(z, \tau)F(\tau)}{\tau - z} d\tau = F(z) - T(z)^{-1}G(z), \tag{3.5.8}$$

where I is the identity matrix.

The kernel $K(z, \tau)/(\tau - z)$ is holomorphic on $\mathscr{D} \times \mathscr{D}$. If we first find a continuous solution of this equation on ℓ_0, we obtain a holomorphic solution on $\mathscr{D}$ as a natural consequence of the form of (3.5.8). In order that (3.5.8) admits a solution, the matrix $T(z)^{-1}G(z)$ must satisfy a finite number of linear conditions. Since $G(z)$ contains sufficiently many arbitrary constants, it suffices to choose those constants suitably so that the linear conditions are satisfied. Furthermore, if we make the degree of each entry of $G(z)$ with respect to z sufficiently large, the coefficients of terms with the highest degrees can be given arbitrarily. In this manner, we can derive the lemma below.

LEMMA 3.5.1. *Assume that* $\mathscr{D}$ *and* $T(z)$ *are the same as in Lemma* 3.4.3. *Then, setting*

$$\mathscr{D}' = \{ z ; |z| > R \} \tag{3.5.9}$$

for a suitable positive number R *not less than* R_0, *we can find a matrix* $B(z)$ *in* $gl(n, \mathscr{O}(C))$ *such that*

$$T(z)B(z) \in GL(n, \mathscr{m}(\mathscr{D}')). \tag{3.5.10}$$

Furthermore, if T *is upper (resp. lower) triangular, we can choose* B *to be upper (resp. lower) triangular.*

To prove Lemma 3.4.3, we modify the matrix $B(z)$ without changing (3.5.10) so that $B(z)$ will be eventually contained in $GL(n, \mathscr{O}(C))$. Then, $T(z)B(z)$ is automatically contained in $GL(n, \mathscr{m}(\mathscr{D}))$. By virtue of (3.5.10), it can be shown that the determinant of $B(z)$ has at most a finite number of zeros. We eliminate those zeros one by one. To do this, it suffices to modify the column vectors of $B(z)$ in a suitable way by multiplying $B(z)$ from the right by suitable matrices successively. In this way, we can complete the proof of Lemma 3.4.3. The details are left to the reader.

§3.6. Proof of Lemma 3.4.2

To prove Lemma 3.4.2, we rewrite (3.4.5) of Lemma 3.4.3 in the form

$$T(z)B(z) = Q(z), \tag{3.6.1}$$

where

(i) $Q(z) \in GL(n, \mathscr{m}(\mathscr{D}))$,
(ii) $B(z) \in GL(n, \mathscr{O}(C))$.

If we choose a matrix $A(\zeta) \in gl(n, \mathscr{O}(\Delta(0, \frac{1}{R_0})))$ and integers $k_1, k_2, \cdots, k_n$ in a suitable way, we can write the matrix $Q(z)$ in the form

$$Q(z) = A\left(\frac{1}{z}\right)\begin{bmatrix} z^{k_1} & 0 & 0 & \cdots & 0 \\ 0 & z^{k_2} & 0 & \cdots & 0 \\ \cdots & \cdots & \cdots & \cdots & \cdots \\ 0 & 0 & 0 & \cdots & z^{k_n} \end{bmatrix}. \tag{3.6.2}$$

If the determinant of $A(0)$ is not zero, the proof of Lemma 3.4.2 is finished.

In the case when the determinant of $A(0)$ is zero, we first rearrange the column vectors of $Q(z)$ by multiplying $Q(z)$ from the right by a constant matrix so that the inequalities

$$k_1 \le k_2 \le \cdots \le k_n \tag{3.6.3}$$

are satisfied. There exists an upper triangular matrix C_0 in $GL(n, C)$ such that some column vectors of $A(0)C_0$ are zero. Then, setting

$$M(z) = \begin{bmatrix} z^{k_1} & 0 & 0 & \cdots & 0 \\ 0 & z^{k_2} & 0 & \cdots & 0 \\ \cdots & \cdots & \cdots & \cdots & \cdots \\ 0 & 0 & 0 & \cdots & z^{k_n} \end{bmatrix}$$

we write (3.6.2) in the form

$$Q(z) = A\left(\frac{1}{z}\right) C_0 M(z) M(z)^{-1} C_0^{-1} M(z).$$

Since $M(z)^{-1} C_0^{-1} M(z)$ is in $GL(n, \mathscr{O}(C))$, we have

$$T(z)\tilde{B}(z) = \tilde{Q}(z),$$

where

$$\tilde{Q}(z) = A\left(\frac{1}{z}\right) C_0 M(z), \qquad \tilde{B}(z) = B(z) M(z)^{-1} C_0 M(z),$$

and the matrices $\tilde{Q}$ and $\tilde{B}$ satisfy conditions (i) and (ii) respectively. If we set

$$\tilde{A}(\zeta) = A(\zeta) C_0,$$

some column vectors of $\tilde{A}(0)$ are zero. Hence there exist nonnegative integers $h_1, \cdots, h_n$ not all zero such that

$$\tilde{A}(\zeta)\begin{bmatrix} \zeta^{-h_1} & 0 & 0 & \cdots & 0 \\ 0 & \zeta^{-h_2} & 0 & \cdots & 0 \\ \cdots & \cdots & \cdots & \cdots & \cdots \\ 0 & 0 & 0 & \cdots & \zeta^{-h_n} \end{bmatrix} \in gl\left(n, \mathscr{O}\left(\Delta\left(0, \frac{1}{R_0}\right)\right)\right)$$

In this way, we can eliminate the zero of the determinant of $A(\zeta)$ at $\zeta = 0$ in a finite number of steps.

§3.7. Proof of Lemma 3.4.1 and a lemma of T. Kimura

We need some preparation. First, we rewrite (3.4.4) of Lemma 3.4.2 in the form

$$T(z)B(z) = P\left(\frac{1}{z}\right)\Lambda(z), \tag{3.7.1}$$

where

(i) $P(\zeta) \in GL(n, \mathscr{O}(\Delta(0, \frac{1}{R_0})))$,
(ii) $B(z) \in GL(n, \mathscr{O}(C))$,
(iii) $\Lambda(z)$ is a diagonal matrix

$$\Lambda(z) = \begin{bmatrix} z^{k_1} & 0 & 0 & \cdots & 0 \\ 0 & z^{k_2} & 0 & \cdots & 0 \\ \cdots & \cdots & \cdots & \cdots & \cdots \\ 0 & 0 & 0 & \cdots & z^{k_n} \end{bmatrix}, \tag{3.7.2}$$

and $k_1, k_2, \cdots, k_n$ are integers satisfying inequalities

$$k_1 \leq k_2 \leq \cdots \leq k_n. \tag{3.7.3}$$

Let $C(z)$ be an upper triangular matrix whose entry $c_{j\ell}(z)$ on the jth row and ℓth column is a polynomial in z of degree at most $k_\ell - k_j$. Furthermore, assume that the entries $c_{jj}(z)$ on the main diagonal are 1. Then $\Lambda(z)C(z)\Lambda(z)^{-1}$ has the same property as $P(\frac{1}{z})$. Precisely speaking, if we set

$$\tilde{P}(\zeta) = \Lambda\left(\frac{1}{\zeta}\right) C\left(\frac{1}{\zeta}\right) \Lambda\left(\frac{1}{\zeta}\right)^{-1},$$

then $\tilde{P}(\zeta)$ is contained in $GL(n, \mathscr{O}(\Delta(0, \frac{1}{R_0})))$. On the other hand, the matrix $C(z)$ belongs to $GL(n, \mathscr{O}(C))$. Moreover, (3.7.1) implies that

$$T(z)B(z)C(z) = P\left(\frac{1}{z}\right)\tilde{P}\left(\frac{1}{z}\right)\Lambda(z). \tag{3.7.4}$$

Letting $(\alpha_1, \alpha_2, \cdots, \alpha_n)$ be a permutation of $(1, 2, \cdots, n)$, we denote by $B_\ell(z)$ the submatrix constructed by choosing α_1st,α_2nd,$\cdots$,α_ℓth rows and 1st,2nd,$\cdots$,ℓth columns from B. Since $B(0)$ is in $GL(n, C)$, each $B_\ell(0)$ belongs to $GL(\ell, C)$ for a suitable choice of $(\alpha_1, \alpha_2, \cdots, \alpha_n)$. Fixing $(\alpha_1, \alpha_2, \cdots, \alpha_n)$ in this way, we set

$$h_{\alpha_j} = k_j \qquad (j = 1, 2, \cdots, n). \tag{3.7.5}$$

Note that $(h_1, h_2, \cdots, h_n)$ is a permutation of $(k_1, k_2, \cdots, k_n)$.

Let us denote the entries of B and BC on the jth row and the kth column by $b_{j\ell}(z)$ and $\hat{b}_{j\ell}(z)$ respectively. Then

$$\begin{bmatrix} \hat{b}_{\alpha_1\, \ell+1} \\ \hat{b}_{\alpha_2\, \ell+1} \\ \cdot \\ \cdot \\ \cdot \\ \hat{b}_{\alpha_\ell\, \ell+1} \end{bmatrix} = B_\ell \begin{bmatrix} c_{1\, \ell+1} \\ c_{2\, \ell+1} \\ \cdot \\ \cdot \\ \cdot \\ c_{\ell\, \ell+1} \end{bmatrix} + \begin{bmatrix} b_{\alpha_1\, \ell+1} \\ b_{\alpha_2\, \ell+1} \\ \cdot \\ \cdot \\ \cdot \\ b_{\alpha_\ell\, \ell+1} \end{bmatrix} \qquad (\ell = 1, \cdots, n-1).$$

Therefore we can choose $c_{1\, \ell+1}, c_{2\, \ell+1}, \cdots, c_{\ell\, \ell+1}$ so that

$$\hat{b}_{\alpha_j\, \ell+1} = O(z^{k_{\ell+1}-k_j}) \qquad (j = 1, \cdots, \ell)$$

at $z = 0$. Then, from (3.7.3) and (3.7.5) we derive

$$\hat{b}_{j\ell}(z) = O(z^{k_\ell - h_j}) \qquad (j, \ell = 1, \cdots, n)$$

at $z = 0$ for a suitable choice of $C(z)$. Hence, if we set

$$\tilde{b}_{j\ell}(z) = z^{h_j - k_\ell}\hat{b}_{j\ell}(z),$$

every $\tilde{b}_{j\ell}(z)$ is contained in $\mathscr{O}(C)$. Denote by $\tilde{B}(z)$ the matrix whose entry on the jth row and the ℓth column is $\tilde{b}_{j\ell}$. Then

$$B(z)C(z) = \begin{bmatrix} z^{-h_1} & \cdots & 0 \\ \cdots & \cdots & \cdots \\ 0 & \cdots & z^{-h_n} \end{bmatrix} \tilde{B}(z) \begin{bmatrix} z^{k_1} & \cdots & 0 \\ \cdots & \cdots & \cdots \\ 0 & \cdots & z^{k_n} \end{bmatrix}. \qquad (3.7.6)$$

Since the determinants of both sides of (3.7.6) are identical, $\tilde{B}(z)$ is in $GL(n, \mathscr{O}(C))$. Setting

$$\tilde{\Lambda}(z) = \begin{bmatrix} z^{h_1} & 0 & 0 & \cdots & 0 \\ 0 & z^{h_2} & 0 & \cdots & 0 \\ \cdots & \cdots & \cdots & \cdots & \cdots \\ 0 & 0 & 0 & \cdots & z^{h_n} \end{bmatrix},$$

we derive

$$T(z)\tilde{\Lambda}(z)^{-1}\tilde{B}(z) = P\left(\frac{1}{z}\right)\tilde{P}\left(\frac{1}{z}\right)$$

from (3.7.2), (3.7.4), and (3.7.6). Thus we complete the proof of Lemma 3.4.1.

In exactly the same way, we can prove a lemma due to T. Kimura which will play an important role when we solve the Riemann problem over the Riemann sphere (cf. T. Kimura [20], §7, Lemma, p. 225; also see Chapter 4, §4.5 of this book).

LEMMA 3.7.1. (T. Kimura). *Assume that a matrix $T(z)$ is in $GL(n, \mathcal{O}(\mathcal{D}))$, where*

$$\mathcal{D} = \{ z \; ; \; |z| > R_0 \}. \tag{3.7.7}$$

Then we can write $T(z)$ in the form

$$T(z) = \tilde{\Lambda}(z)\tilde{P}\left(\frac{1}{z}\right)\tilde{E}(z), \tag{3.7.8}$$

where

(i) $\tilde{P}(\zeta) \in GL(n, \mathcal{O}(\Delta(0, \frac{1}{R_0})))$,
(ii) $\tilde{E}(z) \in GL(n, \mathcal{O}(C))$,
(iii) $\tilde{\Lambda}(z)$ *is a diagonal matrix*

$$\tilde{\Lambda}(z) = \begin{bmatrix} z^{\ell_1} & 0 & 0 & \cdots & 0 \\ 0 & z^{\ell_2} & 0 & \cdots & 0 \\ \cdots & \cdots & \cdots & \cdots & \cdots \\ 0 & 0 & 0 & \cdots & z^{\ell_n} \end{bmatrix}, \tag{3.7.9}$$

and $\ell_1, \cdots, \ell_n$ are integers.

PROOF. Again we start with (3.7.1). As in the proof of Lemma 3.4.1, we can find a matrix $C(\zeta)$ and a permutation $(\ell_1, \cdots, \ell_n)$ of $(k_1, \cdots, k_n)$ satisfying the conditions below:

(1) $\tilde{\Lambda}(\zeta)P(\zeta)C(\zeta)\Lambda(\zeta)^{-1} \in GL(n, \mathcal{O}(\Delta(0, \frac{1}{R_0})))$,
(2) $\Lambda(\frac{1}{z})C(\frac{1}{z})\Lambda(\frac{1}{z})^{-1} \in GL(n, \mathcal{O}(C))$.

Then if we set

$$\tilde{P}(\zeta) = \tilde{\Lambda}(\zeta)P(\zeta)C(\zeta)\Lambda(\zeta)^{-1}, \qquad \tilde{B}(z) = B(z)\Lambda\left(\frac{1}{z}\right)C\left(\frac{1}{z}\right)\Lambda\left(\frac{1}{z}\right)^{-1},$$

we have

$$\begin{aligned} T(z)\tilde{B}(z) &= P\left(\frac{1}{z}\right)\Lambda(z)\Lambda\left(\frac{1}{z}\right)C\left(\frac{1}{z}\right)\Lambda\left(\frac{1}{z}\right)^{-1} \\ &= P\left(\frac{1}{z}\right)C\left(\frac{1}{z}\right)\Lambda\left(\frac{1}{z}\right)^{-1} \\ &= \tilde{\Lambda}\left(\frac{1}{z}\right)^{-1}\tilde{P}\left(\frac{1}{z}\right) = \tilde{\Lambda}(z)\tilde{P}\left(\frac{1}{z}\right). \quad \square \end{aligned}$$

If we use the fact that $(\ell_1, \cdots, \ell_n)$ of Lemma 3.7.1 is a permutation of $(k_1, \cdots, k_n)$, we can prove rather easily a result due to H. L. Turrittin which will be explained in §3.10. Turrittin's original proof was explained in detail in Y. Sibuya [28]. Compare these two proofs if you please (cf. also H. L. Turrittin [34].)

§3.8. Proof of the Birkhoff theorem

To prove Theorem 3.3.1, we first find the matrix $T(z)$ of (3.3.7) and the matrix A_0 of (3.3.8) by applying Lemma 3.3.2. The relation between the matrix $A(\frac{1}{z})$ of differential equation (3.3.3) and the two matrices $T(z)$ and A_0 is given by

$$z^k A\left(\frac{1}{z}\right) T(z) = \frac{dT(z)}{dz} + T(z)\frac{1}{z}A_0. \tag{3.8.1}$$

Next, utilizing Lemma 3.4.1, we write $T(z)$ in the form

$$T(z) = P\left(\frac{1}{z}\right) E(z)\Lambda(z), \tag{3.8.2}$$

where

(i) $P(\zeta) \in GL(n, \mathscr{O}(\Delta(0, \frac{1}{R_0})))$,
(ii) $E(z) \in GL(n, \mathscr{O}(C))$,
(iii) $\Lambda(z)$ is diagonal matrix (3.4.3).

Now, if the transformation

$$W = P\left(\frac{1}{z}\right) V \tag{3.8.3}$$

changes differential equation (3.3.3) to the differential equation

$$\frac{dV}{dz} = z^k B\left(\frac{1}{z}\right) V, \tag{3.8.4}$$

then

$$z^k A\left(\frac{1}{z}\right) P\left(\frac{1}{z}\right) = \frac{dP(\frac{1}{z})}{dz} + P\left(\frac{1}{z}\right) z^k B\left(\frac{1}{z}\right). \tag{3.8.5}$$

Hence we derive

$$B\left(\frac{1}{z}\right) = P\left(\frac{1}{z}\right)^{-1}\left\{A\left(\frac{1}{z}\right) P\left(\frac{1}{z}\right) - z^{-k}\frac{dP(\frac{1}{z})}{dz}\right\}. \tag{3.8.6}$$

Therefore, if k satisfies the inequality

$$k \geq -2, \tag{3.8.7}$$

we have

$$B(\zeta) \in gl\left(n, \mathscr{O}\left(\Delta\left(0, \frac{1}{R_0}\right)\right)\right). \tag{3.8.8}$$

If we differentiate both sides of (3.8.2) with respect to z, we obtain

$$\frac{dT(z)}{dz} = \frac{dP(\frac{1}{z})}{dz}E(z)\Lambda(z) + P\left(\frac{1}{z}\right)\frac{dE(z)}{dz}\Lambda(z) + P\left(\frac{1}{z}\right)E(z)\frac{d\Lambda(z)}{dz}.$$

Then, by virtue of (3.8.1), we derive

$$\begin{aligned} z^k A\left(\frac{1}{z}\right)P\left(\frac{1}{z}\right) - \frac{dP(\frac{1}{z})}{dz} = P\left(\frac{1}{z}\right)\frac{dE(z)}{dz}E(z)^{-1} \\ + P\left(\frac{1}{z}\right)E(z)\frac{d\Lambda(z)}{dz}\Lambda(z)^{-1}E(z)^{-1} \\ + P\left(\frac{1}{z}\right)E(z)\Lambda(z)\frac{1}{z}A_0\Lambda(z)^{-1}E(z)^{-1}. \end{aligned}$$

Therefore, $B(\frac{1}{z})$ can be written in the form

$$\begin{aligned} B\left(\frac{1}{z}\right) = z^{-k}\left\{\frac{dE(z)}{dz}E(z)^{-1} + E(z)\frac{d\Lambda(z)}{dz}\Lambda(z)^{-1}E(z)^{-1}\right. \\ \left. + E(z)\Lambda(z)\frac{1}{z}A_0\Lambda(z)^{-1}E(z)^{-1}\right\} \end{aligned} \tag{3.8.9}$$

because of (3.8.5). We can derive $B(\zeta) \in gl(n, \mathscr{O}(C))$ from (3.8.8) and (3.8.9). Furthermore, $B(\zeta)$ admits at worst a pole at $\zeta = 0$. This means that $B(\zeta)$ is a polynomial in ζ.

If we set $S(z) = E(z)\Lambda(z)$, (3.8.9) becomes

$$z^k B\left(\frac{1}{z}\right)S(z) = \frac{dS}{dz} + S(z)\frac{1}{z}A_0.$$

Hence, the transformation $V = S(z)U$ changes differential equation (3.8.4) to the differential equation

$$\frac{dU}{dz} = \frac{1}{z}A_0 U.$$

Since each entry of $S(z)$ admits at worst a pole at $z = 0$, (3.8.4) admits at worst a regular singular point at $z = 0$. This completes the proof of Theorem 3.3.1.

In this proof we utilized inequality (3.8.7) as we remarked in §3.3 that it would be sufficient.

§3.9. The case when coefficients admit a simple pole
(a theorem of M. Hukuhara)

Let us look at the theorem of G. D. Birkhoff (Theorem 3.3.1) more closely. We change differential equation (3.3.3) to (3.3.6) by setting $\zeta = \frac{1}{z}$. Then, as we remarked in §3.3, if k satisfies the inequality

$$k \leq -2, \tag{3.9.1}$$

(3.3.6) becomes differential equation on $\Delta(0, \frac{1}{R_0})$. Since $\Delta(0, \frac{1}{R_0})$ is simply connected, the matrix $B(\zeta)$ of differential equation (3.3.5) of Theorem 3.3.1 can be changed to the zero matrix by virtue of Theorem 1.11.2. Thus we are led to the following remark.

REMARK 3.9.1. In the case when k satisfies inequality (3.9.1), we can choose the matrix $B(\zeta)$ of differential equation (3.3.5) of Theorem 3.3.1 to be the zero matrix.

Next, let us consider the case when

$$k = -1. \tag{3.9.2}$$

This means that differential equation (3.3.3) has the form

$$\frac{dW}{dz} = \frac{1}{z} A\left(\frac{1}{z}\right) W, \tag{3.9.3}$$

while (3.3.5) has the form

$$\frac{dV}{dz} = \frac{1}{z} B\left(\frac{1}{z}\right) V, \tag{3.9.4}$$

where every entry of $B(\zeta)$ is a polynomial in ζ (with complex constant coefficients) and $z = 0$ is at worst a regular singular point of (3.9.4). Furthermore, these two differential equations are related by the transformation

$$W = P\left(\frac{1}{z}\right) V, \tag{3.9.5}$$

where

$$A(\zeta) \in gl\left(n, \mathcal{O}\left(\Delta\left(0, \frac{1}{R_0}\right)\right)\right), \tag{3.9.6}$$

and

$$P(\zeta) \in GL\left(n, \mathcal{O}\left(\Delta\left(0, \frac{1}{R_0}\right)\right)\right). \tag{3.9.7}$$

If we set $\zeta = \frac{1}{z}$, (3.9.3), (3.9.4), and (3.9.5) become

$$\frac{dW}{d\zeta} = -\frac{1}{\zeta} A(\zeta) W, \tag{3.9.8}$$

$$\frac{dV}{d\zeta} = -\frac{1}{\zeta} B(\zeta) V, \tag{3.9.9}$$

and

$$W = P(\zeta) V, \tag{3.9.10}$$

respectively. The two differential equations (3.9.8) and (3.9.9) are related by transformation (3.9.10). Consequently

$$\zeta \frac{dP(\zeta)}{d\zeta} = P(\zeta) B(\zeta) - A(\zeta) P(\zeta). \tag{3.9.11}$$

Conversely, if we determine P and B by (3.9.11), then (3.9.8) is changed to (3.9.9) by transformation (3.9.10).

Now, we shall look into the problem of determining P and B by (3.9.11). If we replace in (3.9.11) the matrices A, B, and P, respectively by their Taylor series at $\zeta = 0$, we can regard (3.9.11) as a differential equation in $gl(n, C\{\zeta\})$ (cf. §1.8). More precisely speaking, first we set

$$T(A,0) = \sum_{m\geq 0} A_m\zeta^m, \qquad T(B,0) = \sum_{m\geq 0} B_m\zeta^m.$$

(Note that the matrix A_0 of $T(A,0)$ is not necessarily identical with A_0 of Lemma 3.3.2. We believe that this will not create any confusion here. In the next section, the notation A_0 will be used to denote the matrix A_0 of Lemma 3.3.2.). Next we consider the differential equation

$$\zeta\frac{dX}{dz} = XT(B,0) - T(A,0)X. \tag{3.9.12}$$

If we regard X as a formal power series

$$X = \sum_{m\geq 0} X_m\zeta^m, \tag{3.9.13}$$

differential equation (3.9.12) becomes

$$\begin{cases} X_0B_0 - A_0X_0 = 0, \\ X_mB_0 - A_0X_m - mX_m = \displaystyle\sum_{\ell=0}^{m-1} A_{m-\ell}X_\ell - \sum_{\ell=0}^{m-1} X_\ell B_{m-\ell} \qquad (m \geq 1). \end{cases} \tag{3.9.14}$$

In a way similar to the proof of Theorem 1.8.3, we can prove the lemma below.

LEMMA 3.9.2. *Assume that $A(\zeta)$ and $B(\zeta)$ are in $gl(n, \mathcal{O}(\Delta(0,r_0)))$. Then, if a formal power series* (3.9.13) *satisfies* (3.9.12) (*i.e.* (3.9.14)), *the radius of convergence of every entry of the matrix X is not smaller than r_0.*

Assuming that series (3.9.13) satisfies (3.9.12), we denote by $P(\zeta)$ the sum of the power series X. This means that

$$P = S(X) \tag{3.9.15}$$

in terms of notations in §§1.2 and 1.9.

LEMMA 3.9.3. *Assume that $A(\zeta)$ and $B(\zeta)$ belong to $gl(n, \mathcal{O}(\Delta(0,r_0)))$. Then, if series* (3.9.13) *satisfies differential equation* (3.9.12) *and if X_0 belongs to $GL(n, C)$, the sum $S(X)$ of the power series X is in $GL(n, \mathcal{O}(\Delta(0,r_0)))$.*

PROOF. If we define $P(\zeta)$ by (3.9.15), $P(\zeta)$ belongs to $gl(n,\mathcal{O}(\Delta(0,r_0)))$ by virtue of Lemma 3.9.1. Since X_0 belongs to $GL(n,C)$, the determinant of $P(\zeta)$ does not vanish identically on $\Delta(0,r_0)$. Furthermore $P(\zeta)$ satisfies differential equation (3.9.11) on $\Delta(0,r_0)$. Therefore, if $\{\Delta(z_j,r_j), W_j(\zeta)\}$ and $\{\Delta(z_j,r_j), V_j(\zeta)\}$ are systems of fundamental solutions of (3.9.8) and (3.9.9) on $\mathcal{D} = \{\ \zeta\ ;\ 0 < |\zeta| < r_0\ \}$ respectively, we have

$$\begin{aligned} \frac{d}{d\zeta}\{W_j^{-1}PV_j\} &= \frac{dW_j^{-1}}{d\zeta}PV_j + W_j^{-1}\frac{dP}{d\zeta}V_j + W_j^{-1}P\frac{dV_j}{d\zeta} \\ &= \frac{1}{\zeta}W_j^{-1}APV_j + \frac{1}{\zeta}W_j^{-1}\{PB - AP\}V_j - \frac{1}{\zeta}W_j^{-1}PBV_j \\ &= 0 \end{aligned}$$

on $\Delta(z_j,r_j)$. Hence

$$W_j^{-1}PV_j \in GL(n,C). \quad \square$$

By virtue of Lemmas 3.9.2 and 3.9.3, the problem of determining P and B by (3.9.11) is reduced to the problem of determining X_m and B_m by (3.9.14). To determine X_m and B_m, it is convenient to change A_0 to its Jordan canonical form. This means that

$$A_0 = \begin{bmatrix} \lambda_1 & \delta_2 & 0 & 0 & \cdots & 0 & 0 \\ 0 & \lambda_2 & \delta_3 & 0 & \cdots & 0 & 0 \\ 0 & 0 & \lambda_3 & \delta_4 & \cdots & 0 & 0 \\ \cdots & \cdots & \cdots & \cdots & \cdots & \cdots & \cdots \\ 0 & 0 & 0 & 0 & \cdots & \lambda_{n-1} & \delta_n \\ 0 & 0 & 0 & 0 & \cdots & 0 & \lambda_n \end{bmatrix}, \tag{3.9.16}$$

where $\lambda_1, \lambda_2, \cdots, \lambda_n$ and $\delta_2, \delta_3, \cdots, \delta_n$ are complex numbers satisfying the condition

$$\lambda_k = \lambda_{k-1} \qquad \text{if} \quad \delta_k \neq 0. \tag{3.9.17}$$

We set

$$X_0 = I \qquad \text{(the identity matrix)}, \tag{3.9.18}$$

although there are many choices of X_0. Then, it follows from (3.9.14) that

$$B_0 = A_0.$$

For $m \geq 1$, if we denote by $\vec{x}_1^{(m)}, \cdots, \vec{x}_n^{(m)}$ (resp. $\vec{b}_1^{(m)}, \cdots, \vec{b}_n^{(m)}$) the column vectors of X_m (resp. B_m), system (3.9.14) can be rewritten in the form

$$\begin{cases} [(\lambda_1 - m)I - A_0]\vec{x}_1^{(m)} + \vec{b}_1^{(m)} = \vec{h}_1^{(m)}, \\ [(\lambda_k - m)I - A_0]\vec{x}_k^{(m)} + \vec{b}_k^{(m)} = -\delta_k\vec{x}_{k-1}^{(m)} + \vec{h}_k^{(m)} \qquad (k = 2, \cdots, n), \end{cases} \tag{3.9.19}$$

where $\vec{h}_k^{(m)}$ is a polynomial in the entries of $X_0, \cdots, X_{m-1}$ and $B_0, \cdots, B_{m-1}$. If we denote by $b_{jk}^{(m)}$ the entry on the jth row and the kth column of B_m, we can determine X_m and B_m so that the conditions

$$b_{jk}^{(m)} = 0 \qquad (\lambda_k - \lambda_j \neq m) \tag{3.9.20}$$

are satisfied. This becomes clear if we keep solving (3.9.19) successively. Thus B_m becomes the zero matrix as m becomes sufficiently large. Hence the entries of $B(\zeta)$ are polynomials in ζ. As a matter of fact, $B(\zeta)$ has a more specific form because of (3.9.20).

Let us fix integers $\ell_1, \cdots, \ell_n$ by the following condition:

$$\lambda_k - \lambda_j = \ell_k - \ell_j \quad \text{if } \lambda_k - \lambda_j \text{ is an integer.} \tag{3.9.21}$$

Assume further that the inequalities

$$\ell_1 \leq \ell_2 \leq \cdots \leq \ell_n \tag{3.9.22}$$

are satisfied. Then by virtue of (3.9.20) we must have $k > j$ if $b_{jk}^{(m)} \neq 0$. Hence we can determine a matrix $C \in gl(n, C)$ so that

(a) C is upper triangular,
(b) the entries on the main diagonal of C are zero,
(c) if we determine a diagonal matrix $\Lambda(\zeta)$ by

$$\Lambda(\zeta) = \begin{bmatrix} \zeta^{\ell_1} & 0 & 0 & \cdots & 0 \\ 0 & \zeta^{\ell_2} & 0 & \cdots & 0 \\ \cdots & \cdots & \cdots & \cdots & \cdots \\ 0 & 0 & 0 & \cdots & \zeta^{\ell_n} \end{bmatrix}, \tag{3.9.23}$$

then

$$B(\zeta) = \sum_{m \geq 0} B_m \zeta^m = \Lambda(\zeta)^{-1}\{A_0 + C\}\Lambda(\zeta). \tag{3.9.24}$$

In this way, we derive the theorem below which is due to M. Hukuhara (cf. M. Hukuhara [4], Theorem 41.1, p. 143).

THEOREM 3.9.4. (M.Hukuhara). *Given a matrix* $A(z) \in gl(n, \mathcal{O}(\Delta(a, r_0)))$, *we consider a differential equation*

$$\frac{dW}{dz} = \frac{1}{z-a} A(z) W. \tag{3.9.25}$$

Assuming further that $A(a)$ $(= A_0)$ *is in its Jordan canonical form* (3.9.16), *we choose integers* $\ell_1, \cdots, \ell_n$ *so that*

$$(-\lambda_k) - (-\lambda_j) = \ell_k - \ell_j \qquad \textit{if } \lambda_k - \lambda_j \textit{ is an integer.}$$

We also assume that the inequalities

$$\ell_1 \le \ell_2 \le \cdots \le \ell_n$$

hold. Then, there exists a matrix $P(z) \in GL(n, \mathscr{O}(\Delta(a, r_0)))$ *such that the transformation*

$$W = P(z)V \tag{3.9.26}$$

changes differential equation (3.9.25) *to the differential equation*

$$\frac{dV}{dz} = \frac{1}{z-a}\Lambda(z)^{-1}\{A_0 + C\}\Lambda(z)V, \tag{3.9.27}$$

where

$$\Lambda(z) = \begin{bmatrix} (z-a)^{\ell_1} & 0 & 0 & \cdots & 0 \\ 0 & (z-a)^{\ell_2} & 0 & \cdots & 0 \\ \cdots & \cdots & \cdots & \cdots & \cdots \\ 0 & 0 & 0 & \cdots & (z-a)^{\ell_n} \end{bmatrix}, \tag{3.9.28}$$

and C *is an upper triangular matrix in* $gl(n, C)$ *whose entries on the main diagonal are zero.*

Theorem 3.9.4 contains more information than Theorem 3.3.1 for $k = -1$. In particular, the transformation

$$V = \Lambda(z)^{-1}U \tag{3.9.29}$$

changes differential equation (3.9.27) to

$$\frac{dU}{dz} = \frac{1}{z-a}\{A_0 + L + C\}U, \tag{3.9.30}$$

where

$$L = \begin{bmatrix} \ell_1 & 0 & 0 & \cdots & 0 \\ 0 & \ell_2 & 0 & \cdots & 0 \\ \cdots & \cdots & \cdots & \cdots & \cdots \\ 0 & 0 & 0 & \cdots & \ell_n \end{bmatrix}. \tag{3.9.31}$$

Hence, the transformation

$$W = P(z)\Lambda(z)^{-1}U \tag{3.9.32}$$

changes (3.9.25) to (3.9.30). Since $P(z)\Lambda(z)^{-1}$ belongs to $GL(n, \mathscr{M}(\mathscr{D}))$ on the domain

$$\mathscr{D} = \{\, z \,;\, 0 < |z-a| < r_0 \,\},$$

differential equation (3.9.25) admits at worst a regular singular point at $z = a$. We summarize this result as a theorem.

THEOREM 3.9.5. *If a matrix* $A(z)$ *is in* $gl(n, \mathcal{O}(\Delta(a, r_0)))$, *then the differential equation*

$$\frac{dW}{dz} = \frac{1}{z-a}A(z)W$$

admits at worst a regular singular point at $z = a$.

We remark that the two matrices A_0 and L are commutative.

§3.10. A theorem of H. L. Turrittin

If the degrees of the entries of the matrix $B(\zeta)$ of (3.3.5) with respect to ζ are at most $k+1$, differential equation (3.3.5) admits at worst a regular singular point at $z = 0$ because of Theorem 3.9.5. The converse of Theorem 3.9.5 is not true. Actually, as we stated it in Theorem 3.3.1, $z = 0$ is always at worst a regular singular point of (3.3.5) of Theorem 3.3.1 no matter how large such degrees may be. In fact, the degrees of some entries of $B(\zeta)$ with respect to ζ are larger than $k+1$ for some cases.

If we examine (3.8.9) more closely, we can derive the lemma below.

LEMMA 3.10.1. *Let us write* $T(z)$ *of Lemma* 3.4.1 *in form* (3.4.2). *Then, if the matrix* A_0 *of Lemma* 3.3.2 *and the matrix* $\Lambda(z)$ *are commutative* (*i.e.*, $A_0\Lambda(z) = \Lambda(z)A_0$), *the degrees of the entries of the matrix* $B(\zeta)$ *of Theorem* 3.3.1 *with respect to* ζ *are at most* $k+1$.

In particular, if either A_0 is diagonal or $\Lambda(z)$ is the identity matrix, Lemma 3.10.1 applies. We derive from Lemmas 3.3.3 and 3.10.1 the theorem below.

THEOREM 3.10.2. (Birkhoff). *Assume that the curve* ℓ_0 *and the representation* $\rho(\sigma)$ *are the same as in the explanation of Lemma* 3.3.3. *Then, if the eigenvalues of* $\rho([\ell_0])$ *are mutually distinct, the degrees of the entries of the matrix* $B(\zeta)$ *of Theorem* 3.3.1 *with respect to* ζ *are at most* $k+1$. (*The eigenvalues of the matrix* $\rho([\ell_0])$ *are determined by differential equation* (3.3.3) *and independent of the choice of* ρ.)

The following theorem of H. L. Turrittin [34] concerns a similar problem in a different situation.

THEOREM 3.10.3. (H. L. Turrittin). *Given a matrix*

$$A(\zeta) \in gl\left(n, \mathcal{O}\left(\Delta(0, \frac{1}{R_0}\right)\right)$$

and an integer k, *we consider a differential equation*

$$\frac{dW}{dz} = z^k A\left(\frac{1}{z}\right) W. \tag{3.10.1}$$

Then, if the eigenvalues of $A(0)$ are mutually distinct, there exists a matrix $Q(z)$ in $GL(n, \mathscr{m}(\mathscr{D}))$ on the domain

$$\mathscr{D} = \{ z ; |z| > R_0 \} \tag{3.10.2}$$

such that the transformation

$$W = Q(z)V \tag{3.10.3}$$

changes (3.10.1) *to a differential equation*

$$\frac{dV}{dz} = z^k B\left(\frac{1}{z}\right) V \tag{3.10.4}$$

with a matrix $B(\frac{1}{z})$ whose entries are polynomials in $\frac{1}{z}$ with complex constant coefficients and degrees of at most $k+1$.

PROOF. By virtue of the results of §3.9, we can restrict ourselves to the case when

$$k \geq 0. \tag{3.10.5}$$

We shall apply Lemma 3.3.2 to differential equation (3.10.1). This means that we change (3.10.1) by the transformation

$$W = T(z)U \tag{3.10.6}$$

to the differential equation

$$\frac{dU}{dz} = \frac{1}{z} A_0 U \tag{3.10.7}$$

by choosing a matrix A_0 in $gl(n, C)$ and a matrix $T(z)$ in $GL(n, \mathscr{O}(\mathscr{D}))$ in a suitable way. Next, applying Lemma 3.4.2 to the matrix $T(z)$, we write $T(z)$ in the form

$$T(z) = P\left(\frac{1}{z}\right) \Lambda(z) E(z), \tag{3.10.8}$$

where

(i) $P(\zeta) \in GL(n, \mathscr{O}(\Delta(0, \frac{1}{R_0})))$,
(ii) $E(z) \in GL(n, \mathscr{O}(C))$,
(iii) $\Lambda(z)$ is a diagonal matrix

$$\Lambda(z) = \begin{bmatrix} z^{k_1} & 0 & 0 & \cdots & 0 \\ 0 & z^{k_2} & 0 & \cdots & 0 \\ \cdots & \cdots & \cdots & \cdots & \cdots \\ 0 & 0 & 0 & \cdots & z^{k_n} \end{bmatrix}, \tag{3.10.9}$$

and $k_1, \cdots, k_n$ are integers such that

$$k_1 \leq k_2 \leq \cdots \leq k_n. \tag{3.10.10}$$

Since the eigenvalues of $A(0)$ are mutually distinct, we can find a matrix $P_1(\zeta) \in GL(n, \mathcal{O}(\Delta(0, \frac{1}{R})))$ for a suitable positive number R ($\geq R_0$) such that the transformation

$$W = P_1\left(\frac{1}{z}\right) W_1 \tag{3.10.11}$$

changes (3.10.1) to a differential equation of the form below:

$$\frac{dW_1}{dz} = z^k \left\{ D\left(\frac{1}{z}\right) + z^{-N} A_1\left(\frac{1}{z}\right)\right\} W_1, \tag{3.10.12}$$

where

(1) $D(\zeta)$ and $A_1(\zeta)$ are in $gl(n, \mathcal{O}(\Delta(0, \frac{1}{R})))$,
(2) $D(\zeta)$ is a diagonal matrix,
(3) N is an integer larger than $k_n - k_1$.

If we set

$$S(z) = P_1\left(\frac{1}{z}\right)^{-1} T(z), \tag{3.10.13}$$

the transformation

$$W_1 = S(z)U \tag{3.10.14}$$

changes (3.10.12) to (3.10.7). We also derive

$$S(z) = P_1\left(\frac{1}{z}\right)^{-1} P\left(\frac{1}{z}\right)\Lambda(z)E(z) \tag{3.10.15}$$

from (3.10.8). Starting with (3.10.15), we apply the Kimura lemma (i.e., Lemma 3.7.1) to $S(z)$. Then $S(z)$ can be written in the form

$$S(z) = \tilde{\Lambda}(z)\tilde{P}\left(\frac{1}{z}\right)\tilde{E}(z), \tag{3.10.16}$$

where

(α) $\tilde{P}(\zeta) \in GL(n, \mathcal{O}(\Delta(0, \frac{1}{R})))$,
(β) $\tilde{E}(z) \in GL(n, \mathcal{O}(C))$,
(γ) $\tilde{\Lambda}(z)$ is a diagonal matrix

$$\tilde{\Lambda}(z) = \begin{bmatrix} z^{\ell_1} & 0 & 0 & \cdots & 0 \\ 0 & z^{\ell_2} & 0 & \cdots & 0 \\ \cdots & \cdots & \cdots & \cdots & \cdots \\ 0 & 0 & 0 & \cdots & z^{\ell_n} \end{bmatrix}, \tag{3.10.17}$$

and $(\ell_1, \cdots, \ell_n)$ is a permutation of $(k_1, \cdots, k_n)$.

The transformation

$$W_1 = \tilde{\Lambda}(z)W_2 \tag{3.10.18}$$

changes (3.10.12) to

$$\frac{dW_2}{dz} = z^k \left\{ D\left(\frac{1}{z}\right) + z^{-N}\tilde{\Lambda}(z)^{-1}A_1\left(\frac{1}{z}\right)\tilde{\Lambda}(z) - z^{-k}\tilde{\Lambda}(z)^{-1}\frac{d\tilde{\Lambda}(z)}{dz} \right\} W_2. \tag{3.10.19}$$

If we denote by $A_2(\frac{1}{z})$ the matrix $\{\cdots\}$, inequality (3.10.5) and condition (3) on the integer N imply that $A_2(\zeta)$ belongs to $gl(n, \mathscr{O}(\Delta(0, \frac{1}{R})))$.

The transformation

$$W_2 = \tilde{\Lambda}(z)^{-1}S(z)U \tag{3.10.20}$$

changes (3.10.19) to (3.10.7). If we compare $\tilde{\Lambda}(z)^{-1}S(z) = \tilde{P}(\frac{1}{z})\tilde{E}(z)$ and (3.4.2) of Lemma 3.4.1, we notice that $\tilde{\Lambda}(z)^{-1}S(z)$ is in the case when $\Lambda(z)$ of (3.4.2) is the identity matrix. Hence Lemma 3.10.1 applies to differential equation (3.10.19). This means that if the transformation

$$W_2 = \tilde{P}\left(\frac{1}{z}\right)V \tag{3.10.21}$$

changes (3.10.19) to (3.10.4), the matrix $B(\frac{1}{z})$ satisfies the conclusion of Theorem 3.10.3.

Now if we set

$$Q(z) = P_1\left(\frac{1}{z}\right)\tilde{\Lambda}(z)\tilde{P}\left(\frac{1}{z}\right),$$

transformation (3.10.3) is the composition of three transformations (3.10.11), (3.10.18), and (3.10.21). Therefore the proof of Theorem 3.10.3 will be finished if we show that $Q(z)$ belongs to $GL(n, m(\mathscr{D}))$. If we set

$$\mathscr{D}' = \{\, z\,;\ |z| > R \,\},$$

$Q(z)$ belongs to $GL(n, m(\mathscr{D}'))$ and two differential equations (3.10.1) and (3.10.4) are related by transformation (3.10.3) on $\mathscr{D}'$. Therefore

$$\frac{dQ(z)}{dz} = z^k\left\{A\left(\frac{1}{z}\right)Q(z) - Q(z)B\left(\frac{1}{z}\right)\right\} \tag{3.10.22}$$

on $\mathscr{D}'$. Consequently we can extend $Q(z)$ on $\mathscr{D}$ and show that $Q(z) \in GL(n, m(\mathscr{D}))$ by utilizing systems of fundamental solutions of (3.10.1) and (3.10.4) on $\mathscr{D}$ in the same way as in the proof of Lemma 3.9.3. □

We did not explain how to construct $P_1(\zeta)$ in the proof above. To construct $P_1(\zeta)$, assuming that the entries of $P_1(\zeta)$ are polynomials in ζ, we determine their coefficients by the required condition. We shall come back to similar problems in Chapters 5 and 6 (cf. also the method explained in §3.9).

CHAPTER IV

The Riemann Problem

§4.0. Preliminaries

As we remarked at the beginning of Chapter 3, the classification of linear ordinary differential equations by means of transformations with coefficients in $\mathscr{O}(\mathscr{D})$ and the classification by means of structures of analytic continuation defined on $\mathscr{D}$ by such differential equations are identical (cf. Theorem 1.11.1). However, if we restrict the coefficients to $\mathscr{m}(\mathscr{D})$, the situation is different. (For the definition of $\mathscr{m}(\mathscr{D})$, see §§3.2 and 4.3.) In this chapter we consider the case when the coefficients of transformations and of differential equations are restricted to $\mathscr{m}(\mathscr{D})$.

Speaking more in detail, we shall explain first that, if we consider exclusively such differential equations that all singular points are regular singular points, the classification of linear ordinary differential equations by means of transformations with coefficients in $\mathscr{m}(\mathscr{D})$ and the classification by means of structures of analytic continuation defined on $\mathscr{D}$ by such differential equations are identical (cf. §4.3). Thereafter, by utilizing the existence theorem of Grauert (i.e., Theorem 2.0.1), those lemmas related with Theorem 3.3.1 of Birkhoff and, in particular, the Kimura lemma (i.e., Lemma 3.7.1), we shall prove the existence of a differential equation of (quasi-) Fuchsian type that defines a preassigned structure of analytic continuation on $\mathscr{D}$. The result over the complex plane (i.e., the case of quasi-Fuchsian type) is a slight modification of a theorem due to H. Röhrl [27] (cf. §4.4), while the result over the Riemann sphere (i.e., the case of Fuchsian type) is a theorem due to T. Kimura [20] (cf. §4.5).

At the end, we shall give a necessary and sufficient condition that a differential equation of Fuchsian type can be changed to a triangular form by a transformation with rational coefficients (cf. §4.6). The ideas of proofs of such results will be shown very clearly by means of what we have explained so far, although there are some other ideas.

In the outline given above, we used the concept of structures of analytic continuation. However, we can also utilize structures of monodromy as well (cf. Theorem 1.7.5). If we look at the problem in this way, the main problem of this chapter is the Riemann problem in which we seek a differential equation of Fuchsian type that defines a preassigned structure of monodromy on $\mathscr{D}$. The author notes here with his appreciation that he was helped very much by the lecture notes [7] of Professor T. Kimura concerning the Riemann problem.

§4.1. Transformation of a linear ordinary differential equation at a regular singular point

Set

$$\mathscr{D} = \{ z ; 0 < |z-a| < r_0 \} \tag{4.1.1}$$

and assume that the matrix $A(z)$ of a differential equation

$$\frac{dW}{dz} = A(z)W \tag{4.1.2}$$

on $\mathscr{D}$ belongs to $gl(n, \mathscr{m}(\mathscr{D}))$. If there exist a matrix $P(z) \in GL(n, \mathscr{m}(\mathscr{D}))$ and a matrix $A_0 \in gl(n, \mathbf{C})$ such that the transformation

$$W = P(z)V \tag{4.1.3}$$

changes (4.1.2) to

$$\frac{dV}{dz} = \frac{1}{z-a}A_0 V, \tag{4.1.4}$$

then $z = a$ is said to be (at worst) a regular singular point of differential equation (4.1.2) (cf. §3.2). In this definition, we ought to find such matrices $P(z)$ and A_0. This situation is explained more precisely in Lemma 4.1.1.

LEMMA 4.1.1. *Assume that $z = a$ is at worst a regular singular point of (4.1.2). Then, if there exist a matrix $Q(z) \in GL(n, \mathscr{O}(\mathscr{D}))$ and a matrix $\tilde{A}_0$ in $gl(n, \mathbf{C})$ such that the transformation*

$$W = Q(z)U \tag{4.1.5}$$

changes (4.1.2) to

$$\frac{dU}{dz} = \frac{1}{z-a}\tilde{A}_0 U, \tag{4.1.6}$$

the matrix $Q(z)$ is contained in $GL(n, \mathscr{m}(\mathscr{D}))$.

This lemma is a particular case of the following lemma.

LEMMA 4.1.2. *Given two matrices $A_1(z)$ and $A_2(z) \in gl(n, \mathscr{M}(\mathscr{D}))$ and a matrix $Q(z) \in GL(n, \mathscr{O}(\mathscr{D}))$ such that the transformation*

$$W_1 = Q(z)W_2 \tag{4.1.7}$$

changes the differential equation

$$\frac{dW_1}{dz} = A_1(z)W_1 \tag{4.1.8}$$

to

$$\frac{dW_2}{dz} = A_2(z)W_2, \tag{4.1.9}$$

then, if differential equation (4.1.8) *and* (4.1.9) *admit at worst regular singular points at $z = a$ respectively, the matrix $Q(z)$ is contained in $GL(n, \mathscr{M}(\mathscr{D}))$.*

The proof of this lemma is reduced to that of the following lemma.

LEMMA 4.1.3. *Given two matrices A_0 and $\tilde{A}_0 \in gl(n, \mathbf{C})$ and a matrix $Q(z) \in GL(n, \mathscr{O}(\mathscr{D}))$ such that transformation* (4.1.7) *changes the differential equation*

$$\frac{dW_1}{dz} = \frac{1}{z-a}A_0W_1 \tag{4.1.10}$$

to

$$\frac{dW_2}{dz} = \frac{1}{z-a}\tilde{A}_0W_2, \tag{4.1.11}$$

then every entry of the matrix $Q(z)$ is a rational function of z.

PROOF. Since (4.1.7) changes (4.1.10) to (4.1.11), we have

$$(z-a)\frac{dQ(z)}{dz} = A_0Q(z) - Q(z)\tilde{A}_0.$$

This means that, if $\sum_{-\infty<m<\infty} Q_m(z-a)^m$ is the Laurent series of $Q(z)$ at $z = a$, we have

$$mQ_m = A_0Q_m - Q_m\tilde{A}_0$$

for all integers m. Therefore Q_m must be the zero matrix if $|m|$ is sufficiently large. □

We leave it to the reader to derive Lemma 4.1.2 from Lemma 4.1.3.

In order that the matrix $Q(z)$ of Lemma 4.1.2 exists, it is necessary and sufficient that the structure of analytic continuation defined on $\mathscr{D}$ by (4.1.8) and the structure defined by (4.1.9) are identical (cf. Theorem 1.11.1). This means that, at a regular singular point, the classification of linear ordinary differential equations by means of transformations with coefficients in $\mathscr{M}(\mathscr{D})$ and the classification by means of structures of analytic continuation defined on $\mathscr{D}$ by such differential equations are identical. In §3.2, we introduced a notion of regular singular points in a hasty manner. However, since Lemma 4.1.3 holds, such a notion of regular singular points goes very well with the set $m(\mathscr{D})$ of functions .

§4.2. Normally regular singular points

The definition of regular singular points was given only for differential equations with coefficients in $m(\mathscr{D})$ (cf. §§3.2 and 4.1). The following lemma explains the situation more clearly.

LEMMA 4.2.1. *Set*

$$\mathscr{D} = \{ z ; 0 < |z - a| < r_0 \}, \tag{4.2.1}$$

and assume that the matrix $A(z)$ of a differential equation

$$\frac{dW}{dz} = A(z)W \tag{4.2.2}$$

on $\mathscr{D}$ belongs to $gl(n, \mathscr{O}(\mathscr{D}))$. Then, if there exist a matrix $P(z) \in GL(n, m(\mathscr{D}))$ and a matrix $A_0 \in gl(n, \mathbf{C})$ such that the transformation

$$W = P(z)V \tag{4.2.3}$$

changes (4.2.2) *to*

$$\frac{dV}{dz} = \frac{1}{z-a} A_0 V, \tag{4.2.4}$$

the matrix $A(z)$ is contained in $gl(n, m(\mathscr{D}))$.

PROOF. The three matrices P, A, and A_0 satisfy the equation

$$\frac{dP}{dz} = AP - \frac{1}{z-a} PA_0.$$

Hence

$$A(z) = \frac{1}{z-a} P(z)A_0P(z)^{-1} + \frac{dP(z)}{dz} P(z)^{-1}. \tag{4.2.5}$$

□

In particular, if $P(z)$ is in $GL(n, \mathscr{O}(\Delta(a, r_0)))$ the matrix $A(z)$ can be written in the form

$$A(z) = \frac{1}{z-a} \tilde{A}(z), \tag{4.2.6}$$

where

(i) $\tilde{A}(z) \in gl(n, \mathscr{O}(\Delta(a, r_0)))$,
(ii) $\tilde{A}(a) = P(a)A_0P(a)^{-1}$.

DEFINITION 4.2.2. *Assume that* $\mathscr{D}$ *is given by* (4.2.1) *and that a matrix* $A(z)$ *is in* $gl(n, \mathscr{O}(\Delta(a, r_0)))$. *In this case, if there exist a matrix* $P(z) \in GL(n, \mathscr{O}(\Delta(a, r_0)))$ *and a matrix* $A_0 \in gl(n, \mathbf{C})$ *such that the transformation*

$$W = P(z)V \tag{4.2.7}$$

changes the differential equation

$$\frac{dW}{dz} = \frac{1}{z-a}A(z)W \tag{4.2.8}$$

to

$$\frac{dV}{dz} = \frac{1}{z-a}A_0V, \tag{4.2.9}$$

then $z = a$ *is said to be* (*at worst*) **a normally regular singular point** *of differential equation* (4.2.8).

Generally speaking, by virtue of Theorem 3.9.5, $z = a$ is always at worst a regular singular point of (4.2.8), but not necessarily normal. The Hukuhara theorem (i.e., Theorem 3.9.4) shows this situation very clearly. In the case when $z = a$ is a normally regular singular point of (4.2.8), we have

$$A(a) = A_0 \tag{4.2.10}$$

if $P(a)$ is the identity matrix because of property (ii) of $\tilde{A}$. When we derived the Hukuhara theorem in §3.9, we remarked that, in general, the two matrices $A(a)$ and A_0 might not agree. Actually, condition (4.2.10) characterizes the normal regular singular points.

LEMMA 4.2.3. *Assume that a domain* $\mathscr{D}$ *is given by* (4.2.1) *and that a matrix* $A(z)$ *is in* $gl(n, \mathscr{O}(\Delta(a, r_0)))$. *In this case, if there exists a matrix* $Q(z) \in GL(n, \mathscr{M}(\mathscr{D}))$ *such that the transformation*

$$W = Q(z)U \tag{4.2.11}$$

changes differential equation (4.2.8) *to*

$$\frac{dU}{dz} = \frac{1}{z-a}A(a)U, \tag{4.2.12}$$

then $z = a$ *is* (*at worst*) *a normally regular singular point of* (4.2.8). (*We already explained that the converse is also true.*)

PROOF. Assume that the matrix $A(a)$ satisfies the conditions given in the theorem of Hukuhara (i.e., Theorem 3.9.4). Consequently, by virtue of the Hukuhara theorem, we can assume that the matrix $A(z)$ has the following form:

$$A(z) = \Lambda(z)^{-1}\{A(a) + C\}\Lambda(z), \tag{4.2.13}$$

where the three matrices $\Lambda(z)$, C, and L are the same as in Theorem 3.9.4. Now, if we apply the ideas given in Lemma 3.3.3, we can find a matrix $\Gamma \in GL(n, \mathbf{C})$ such that

$$\exp\{2\pi i[A(a) + L + C]\} = \exp\{2\pi i \Gamma^{-1} A(a) \Gamma\}. \tag{4.2.14}$$

This means that the Jordan canonical form of $A(a) + L + C$ and $A(a)$ have the same shapes. Therefore there exists a matrix $\tilde{\Gamma} \in GL(n, \mathbf{C})$ such that

$$\tilde{\Gamma}^{-1}\{A(a) + L + C\}\tilde{\Gamma} = A(a) + L.$$

Furthermore we can choose $\tilde{\Gamma}$ to be upper triangular. Thus, if the matrix $A(z)$ of (4.2.8) is given by (4.2.13), the transformation

$$W = \Lambda(z)^{-1}\tilde{\Gamma}\Lambda(z)V$$

changes (4.2.8) to

$$\frac{dV}{dz} = \frac{1}{z-a}A(a)V.$$

Since $\tilde{\Gamma}$ is upper triangular and $\ell_1, \cdots, \ell_n$ satisfy inequalities (3.9.22), the matrix $\Lambda(z)^{-1}\tilde{\Gamma}\Lambda(z)$ is contained in $GL(n, \mathscr{O}(\mathbf{C}))$. □

In the proof given above, we utilized the assumption of Lemma 4.2.3 to derive (4.2.14). Hereafter, we shall not use Lemma 4.2.3.

§4.3. Linear ordinary differential equations of Fuchsian type

Let us consider a domain

$$\mathscr{D} = \mathbf{C} - \{a_1, \cdots, a_p\}, \tag{4.3.1}$$

where $a_1, \cdots, a_p$ are a finite number of points fixed on the z-plane. If we choose positive numbers $r_1, \cdots, r_p$ sufficiently small, the closures of the disks $\Delta(a_1, r_1), \cdots, \Delta(a_p, r_p)$ are mutually disjoint. Set

$$\mathscr{D}_j = \{\, z\ ;\ 0 < |z - a_j| < r_j \,\} \qquad (j = 1, \cdots, p). \tag{4.3.2}$$

We shall denote by $\mathscr{m}(\mathscr{D})$ the set of all complex-valued functions $f(z)$ satisfying the conditions below:

(a) $f(z) \in \mathscr{O}(\mathscr{D})$,
(b) the restriction of $f(z)$ to $\mathscr{D}_j$ belongs to $\mathscr{m}(\mathscr{D}_j)$, $(j = 1, \cdots, p)$.

We also define $gl(n, \mathscr{m}(\mathscr{D}))$ and $GL(n, \mathscr{m}(\mathscr{D}))$ in the same manner as in §3.2.

DEFINITION 4.3.1. *A differential equation*

$$\frac{dW}{dz} = A(z)W \tag{4.3.3}$$

is said to be of **quasi-Fuchsian type** *on* $\mathscr{D}$ *if the conditions below are satisfied*:

(i) $A(z) \in gl(n, \mathscr{m}(\mathscr{D}))$,

(ii) *for every* j, $z = a_j$ *is at worst a regular singular point of* (4.3.3).

Differential equation (4.3.3) *is said to be of* **normally quasi-Fuchsian type** *on* $\mathscr{D}$ *if* $z = a_j$ *is at worst a normally regular singular point of* (4.3.3) *for every* j.

THEOREM 4.3.2. *Assume that two differential equations*

$$\frac{dW_1}{dz} = A_1(z)W_1 \tag{4.3.4}$$

and

$$\frac{dW_2}{dz} = A_2(z)W_2 \tag{4.3.5}$$

are both of quasi-Fuchsian type on $\mathscr{D}$. *Then, in order that there exists a matrix* $P(z) \in GL(n, \mathscr{m}(\mathscr{D}))$ *such that the transformation*

$$W_1 = P(z)W_2 \tag{4.3.6}$$

changes (4.3.4) *to* (4.3.5), *it is necessary and sufficient that the structure of analytic continuation defined on* $\mathscr{D}$ *by* (3.4.4) *and the structure defined by* (3.4.5) *are identical.*

PROOF. The necessity of the condition follows from Theorem 1.11.1. The sufficiency of the condition can be verified if we first find a matrix $P(z) \in GL(n, \mathscr{O}(\mathscr{D}))$ by means of Theorem 1.11.1 and then apply Lemma 4.1.2 to $P(z)$. □

LEMMA 4.3.3. *Assume that a matrix* $P(z)$ *belongs to* $GL(n, \mathscr{O}(\mathbf{C}))$ *and that the differential equation*

$$\frac{dW}{dz} = A(z)W \tag{4.3.7}$$

is of normally quasi-Fuchsian type on $\mathscr{D}$. *If the transformation*

$$W = P(z)V \tag{4.3.8}$$

changes (4.3.7) *to*

$$\frac{dV}{dz} = B(z)V, \tag{4.3.9}$$

then this differential equation is also of normally quasi-Fuchsian type on $\mathscr{D}$.

This lemma can be easily proved if we apply Definitions 4.2.2 and 4.3.1 in a proper way.

In the results given above, the situation at $z = \infty$ was totally obscured. To study the situation at $z = \infty$ more closely, it is necessary to assume some conditions there.

DEFINITION 4.3.4. Set

$$\mathscr{D}_\infty = \{ z \; ; \; |z| > R_0 \} \tag{4.3.10}$$

and assume that the matrix $A(z)$ of the differential equation

$$\frac{dW}{dz} = A(z)W \tag{4.3.11}$$

belongs to $gl(n, m(\mathscr{D}_\infty))$ (cf. §3.3). In this case, if there exist a matrix $P(z) \in GL(n, m(\mathscr{D}_\infty))$ and a matrix $A_0 \in gl(n, \mathbf{C})$ such that the transformation

$$W = P(z)V \tag{4.3.12}$$

changes (4.3.11) to

$$\frac{dV}{dz} = \frac{1}{z}A_0 V, \tag{4.3.13}$$

then $z = \infty$ is said to be (at worst) **a regular singular point** of (4.3.11). Furthermore if $P(\frac{1}{\zeta})$ belongs to $GL(n, \mathscr{O}(\Delta(0, \frac{1}{R_0})))$, then $z = \infty$ is said to be (at worst) **a normally regular singular point** of (4.3.11).

Taking those into account, we give the definition below.

DEFINITION 4.3.5. Assume that the matrix $A(z)$ of the differential equation

$$\frac{dW}{dz} = A(z)W \tag{4.3.14}$$

satisfies the following conditions:

(i) $A(z) \in gl(n, \mathscr{O}(\mathscr{D}))$, where $\mathscr{D}$ is given by (4.3.1),
(ii) every entry of $A(z)$ is a rational function of z,
(iii) for every j, $z = a_j$ is at worst a regular singular point of (4.3.14),
(iv) $z = \infty$ is at worst a regular singular point of (4.3.14).

In this case, differential equation (4.3.14) is said to be of **Fuchsian type** on $\mathscr{D}$. Furthermore, if $z = a_j$ is at worst a normally regular singular point of (4.3.14) for every j and if $z = \infty$ is also at worst a normally regular singular point of (4.3.14), then differential equation (4.3.14) is said to be of **normally Fuchsian type** on $\mathscr{D}$.

THEOREM 4.3.6. *Assume that two differential equations*

$$\frac{dW_1}{dz} = A_1(z)W_1 \tag{4.3.15}$$

and

$$\frac{dW_2}{dz} = A_2(z)W_2 \tag{4.3.16}$$

are both of Fuchsian type on $\mathcal{D}$. *Then, in order that there exists a matrix* $P(z) \in GL(n, \mathcal{O}(\mathcal{D}))$ *such that every entry of* $P(z)$ *is a rational function of* z *and the transfomation*

$$W_1 = P(z)W_2 \tag{4.3.17}$$

changes (4.3.15) *to* (4.3.16), *it is necessary and sufficient that the structure of analytic continuation defined on* $\mathcal{D}$ *by* (4.3.15) *and the structure defined by* (4.3.16) *are identical.*

PROOF. The necessity of the condition follows from Theorem 1.11.1. To prove the sufficiency of the condition, we first find a matrix $P(z) \in GL(n, \mathcal{m}(\mathcal{D}))$. Then, if we apply Lemma 4.1.2 to $P(\frac{1}{\zeta})$, we can show that the restriction of $P(z)$ to $\mathcal{D}_\infty$ belongs to $GL(n, \mathcal{m}(\mathcal{D}_\infty))$. This means that $P(z)$ admits at worst a finite number of poles on the Riemann sphere. Hence every entry of $P(z)$ is a rational function of z. □

We give two more lemmas which will be used later.

LEMMA 4.3.7. *If a differential equation*

$$\frac{dW}{dz} = A(z)W \tag{4.3.18}$$

is of Fuchsian type and of normally quasi-Fuchsian type on $\mathcal{D}$, *then* $A(z)$ *has the form*

$$A(z) = \sum_{k=1}^{p} \frac{1}{z - a_k} A_k + B(z), \tag{4.3.19}$$

where

(i) $A_k \in gl(n, \mathbf{C}) \qquad (k = 1, \cdots, p)$,
(ii) *all entries of* $B(z)$ *are polynomials in* z.

If (4.3.18) *is of normally Fuchsian type, then* $B(z)$ *is the zero matrix.*

This lemma can be easily proved by using the fact that every entry of $A(z)$ is a rational function of z which admits at worst a simple pole at $z = a_k$ for every k. Moreover, if (4.3.18) is of normally Fuchsian type, then $\lim_{z \to \infty} A(z) = 0$. Hence $B(z) = 0$.

LEMMA 4.3.8. *If differential equation* (4.3.18) *is of quasi-Fuchsian type on* $\mathcal{D}$ *and if* $z = \infty$ *is at worst a regular singular point of* (4.3.18), *then* (4.3.18) *is of Fuchsian type on* $\mathcal{D}$.

To prove this lemma, it suffices to show that every entry of $A(z)$ is a rational function of z (cf. Definition 4.3.5). The details are left to the reader.

§4.4. An existence theorem over the complex plane

Concerning the existence of a differential equation that defines a preassigned structure of analytic continuation, we proved Theorems 1.11.3 and 2.11.4 by using the existence theorem of Grauert (i.e., Theorem 2.0.1). In this section, we shall improve these results so that we can apply them to the case of quasi-Fuchsian type (cf. Definition 4.3.1). We shall use the notations such as $a_1, \cdots, a_p, \mathscr{D}, \mathscr{D}_1, \cdots, \mathscr{D}_p$ in the same sense as in §4.3. As the starting point of our discussion, we state here Theorem 2.11.4 again as a lemma.

LEMMA 4.4.1. *Let $\mathscr{D}$ be the domain given by* (4.3.1). *Assume that $\mathscr{C}$ is a structure of analytic continuation on $\mathscr{D}$ and that G is a linear group. Then, in order that we find a matrix $B(z)$ in $gl(n, \mathscr{O}(\mathscr{D}))$ such that*

(i) *$B(z)$ is contained in the Lie algebra $\mathbf{g}$ of G at every point z of $\mathscr{D}$,*
(ii) *the differential equation*

$$\frac{dU}{dz} = B(z)U \tag{4.4.1}$$

defines the structure $\mathscr{C}$ of analytic continuation,

it is necessary and sufficient that we find a system $\{\Delta(z_j, \rho_j), C_{jk}\}$ of connection matrices belonging to $\mathscr{C}$ such that every matrix C_{jk} is contained in G.

The main problem of this section is to improve (4.4.1) so that it becomes a differential equation of normally quasi-Fuchsian type. To do this, we need a preparation. Let $\mathscr{D}_j$ be the domain given by (4.3.2). We assume that the closures of disks $\Delta(a_1, r_1), \cdots, \Delta(a_p, r_p)$ are mutually disjoint. Then we have

$$\mathscr{D}_j \subset \mathscr{D} \qquad (j = 1,\ 2,\ \cdots,\ p),$$

and

$$\mathbf{C} = \mathscr{D} \cup \Delta(a_1, r_1) \cup \Delta(a_2, r_2) \cup \cdots \cup \Delta(a_p, r_p).$$

In order to apply the ideas of §3.1 on the domain $\mathscr{D}_j$, we restrict the linear group G by the condition below.

ASSUMPTION 4.4.2. *For every matrix $P \in G$ there is a one-parameter subgroup of G containing P.*

For example, $GL(n, \mathbf{C})$ satisfies this condition. The subgroup consisting of all upper (resp. lower) triangular matrices in $GL(n, \mathbf{C})$ also satisfies this assumption. However the special linear group $SL(n, \mathbf{C})$ does not satisfy Assumption 4.4.2. (cf. L. Markus [25], Example 1, p. 352). In the case to which Assumption 4.4.2 applies, we can improve Theorem 3.1.1 by utilizing the ideas given in §3.1 within the group G. We state such a result as a lemma.

LEMMA 4.4.3. *Set*

$$\mathscr{D}_* = \{ z \,;\, 0 < |z - a| < r \} \tag{4.4.2}$$

and assume that the matrix $B_*(z)$ *of the differential equation*

$$\frac{dU}{dz} = B_*(z)U \tag{4.4.3}$$

on $\mathscr{D}_*$ *is in* $gl(n, \mathscr{O}(\mathscr{D}_*))$ *and that* $B_*(z)$ *is contained in the Lie algebra* $\mathbf{g}$ *of a linear group* G. *We assume also that* G *satisfies Assumption* 4.4.2. *Then, there exist a matrix* $A_* \in \mathbf{g}$ *and a matrix* $P_*(z)$ *defining a holomorphic map* $: \mathscr{D}_* \to G$ *such that the transformation*

$$U = P_*(z)V \tag{4.4.4}$$

changes differential equation (4.4.3) *to*

$$\frac{dV}{dz} = \frac{1}{z-a} A_* V. \tag{4.4.5}$$

This lemma can be proved by utilizing ideas given in §3.1 and Lemma 2.11.3.

Using what has been explained so far as a preparation, we shall prove the following theorem.

THEOREM 4.4.4. *Let* $\mathscr{D}$ *be the domain given by* (4.3.1) *and let* $\mathscr{C}$ *be a structure of analytic continuation on* $\mathscr{D}$. *Assume that* G *is a linear group satisfying Assumption* 4.4.2. *Then, in order that there exists a matrix* $A(z) \in gl(n, \mathscr{M}(\mathscr{D}))$ *satisfying the conditions*

(1) $A(z)$ *is contained in the Lie algebra* $\mathbf{g}$ *of* G *at every point of* $\mathscr{D}$,

(2) *the differential equation*

$$\frac{dW}{dz} = A(z)W \tag{4.4.6}$$

defines the structure $\mathscr{C}$ *of analytic continuation,*

(3) (4.4.6) *is of normally quasi-Fuchsian type on* $\mathscr{D}$,

it is necessary and sufficient that there exist a system $\{\Delta(z_j, \rho_j), C_{jk}\}$ *of connection matrices belonging to the structure* $\mathscr{C}$ *such that every matrix* C_{jk} *is in* G.

PROOF. It suffices to show the sufficiency of the condition of the theorem. To do this, we assume that every C_{jk} is in G. First we apply Lemma

4.4.1 to construct differential equation (4.4.1) that satisfies conditions (i) and (ii) of Lemma 4.4.1. Next, we apply Lemma 4.4.3 to (4.4.1) on the domain $\mathscr{D}_j$. This means that we select matrices $P_1(z), \cdots, P_p(z)$ and $A_1, \cdots, A_p$ so that

(α) $P_j(z)$ is a holomorphic map : $\mathscr{D}_j \to G$,
(β) $A_j \in \mathbf{g}$,
(γ) the transformation

$$U = P_j(z)V_j \tag{4.4.7.j}$$

changes (4.4.1) to

$$\frac{dV_j}{dz} = \frac{1}{z - a_j} A_j V_j. \tag{4.4.8.j}$$

As we mentioned before, $\{ \mathscr{D}, \Delta(a_1, r_1), \cdots, \Delta(a_p, r_p) \}$ is an open covering of $\mathbf{C}$, and

$$\begin{cases} \mathscr{D} \cap \Delta(a_j, r_j) = \mathscr{D}_j, \\ \Delta(a_j, r_j) \cap \Delta(a_k, r_k) = \varnothing \qquad (j \neq k). \end{cases}$$

Then, if we apply the existence theorem of Grauert (i.e., Theorem 2.0.1), we can find matrices $F(z), F_1(z), \cdots, F_p(z)$ satisfying the following conditions:

(a) $F(z)$ is a holomorphic map : $\mathscr{D} \to G$,
(b) $F_j(z)$ is a holomorphic map : $\Delta(a_j, r_j) \to G$,
(c) we have

$$P_j(z) = F(z)F_j(z). \tag{4.4.9.j}$$

Assume that the transformation

$$U = F(z)W \tag{4.4.10}$$

changes (4.4.1) to (4.4.6). Thus we can prove that conditions (1) and (2) are satisfied by utilizing the relation

$$A(z) = F(z)^{-1}B(z)F(z) - F(z)^{-1}\frac{dF(z)}{dz}$$

and Theorem 1.11.1. Furthermore, if we compare transformations (4.4.10), (4.4.7.j) and (4.4.9.j), it is clear that the transformation

$$W = F_j(z)V_j \tag{4.4.11.j}$$

changes (4.4.6) to (4.4.8.j). Consequently, $z = a_j$ is at worst a normally regular singular point of (4.4.6). □

This theorem is a slight modification of a result due to H. Röhrl [27]. The following result is a refinement of Theorem 2.11.5.

THEOREM 4.4.5. *Let $\mathscr{D}$ be the domain given by* (4.3.1). *Assume that a matrix $A(z)$ belongs to $gl(n, \mathscr{m}(\mathscr{D}))$ and that the differential equation*

$$\frac{dW}{dz} = A(z)W \tag{4.4.12}$$

is of quasi-Fuchsian type on $\mathscr{D}$. We also assume that G is a linear group satisfying Assumption 4.4.2. *Then, in order that there exists a matrix $P(z) \in GL(n, \mathscr{m}(\mathscr{D}))$ such that the transformation*

$$W = P(z)V \tag{4.4.13}$$

changes (4.4.12) *to a differential equation*

$$\frac{dV}{dz} = B(z)V \tag{4.4.14}$$

of normally quasi-Fuchsian type on $\mathscr{D}$ with a matrix $B(z)$ in the Lie algebra $\mathbf{g}$ of G at every point z of $\mathscr{D}$, it is necessary and sufficient that the structure of analytic continuation defined on $\mathscr{D}$ by (4.4.12) *satisfies the requirement given in Theorem* 4.4.4.

REMARK 4.4.6. In the proof of Theorem 4.4.4, we determined the matrices $F(z), F_1(z), \cdots, F_p(z)$ by conditions (4.4.9.j). However, we can determine these matrices by more general conditions $P_j(z)Q_j(z)^{-1} = F(z)F_j(z)$, where the matrices $Q_j(z)$ are holomorphic maps : $\mathscr{D}_j \to G$ respectively. Since there are ample choices of the matrices $Q_j(z)$, we can assign at each $z = a_j$ a singularity of various kinds.

REMARK 4.4.7. In this section, we used Assumption 4.4.2. However, a result similar to Theorem 4.4.4 still holds in the case when G is $SL(n, \mathbf{C})$. We required in (3) of Theorem 4.4.4 that the differential equation be of normally quasi-Fuchsian type. If G is $SL(n, \mathbf{C})$, we must relax this requirement so that (4.4.6) should be merely of quasi-Fuchsian type. The author does not know to what extent such a generalization holds in general.

§4.5. An existence theorem of T. Kimura over the Riemann sphere

In this section, we shall explain some results concerning differential equations of Fuchsian type. The main result is the theorem below. The starting point of our discussion is an application of Theorem 4.4.4 to the particular case where the linear group G is $GL(n, \mathbf{C})$.

THEOREM 4.5.1 (T. Kimura). *Assume that*

(i) *$\mathscr{D}$ is the domain given by* (4.3.1),
(ii) *$\mathscr{C}$ is a structure of analytic continuation on $\mathscr{D}$,*
(iii) *$\mathscr{M}(\mathscr{C})$ is the structure of monodromy defined on $\mathscr{D}$ by $\mathscr{C}$ (cf. Theorem* 1.7.5),

(iv) *z_0 is a point of $\mathscr{D}$ and $\rho(\sigma)$ is a representation of $\pi_1(\mathscr{D}, z_0)$ in $\mathbf{C}^n$ which belongs to $\mathscr{M}(\mathscr{C})$,*

(v) *ℓ_0 is a simple closed curve which starts from and terminates at z_0 such that $a_1, \cdots, a_p$ are in the interior of ℓ_0 and that $\rho([\ell_0]) = I$* (the identity matrix).

Then, there exist matrices $A_1, \cdots, A_p$ in $gl(n, \mathbf{C})$ such that

(I) *the differential equation*

$$\frac{dW}{dz} = \left\{ \sum_{k=1}^{p} \frac{1}{z - a_k} A_k \right\} W \tag{4.5.1}$$

is of normally Fuchsian type on $\mathscr{D}$,

(II) *differential equation* (4.5.1) *defines the structure $\mathscr{C}$ on $\mathscr{D}$.*

PROOF. First, utilizing Theorem 4.4.4, we construct a differential equation

$$\frac{dU}{dz} = B(z)U \tag{4.5.2}$$

in such a way that

(α) $B(z)$ is a matrix in $gl(n, \mathfrak{m}(\mathscr{D}))$ and differential equation (4.5.2) is of normally quasi-Fuchsian type on $\mathscr{D}$,

(β) differential equation (4.5.2) defines the structure $\mathscr{C}$ on $\mathscr{D}$.

Next, we study differential equation (4.5.2) on a domain

$$\mathscr{D}_\infty = \{ z \,;\, |z| > R_0 \}, \tag{4.5.3}$$

where R_0 is a sufficiently large positive number. By virtue of assumption (v) of the theorem, we can apply the ideas given in §3.1 to find a matrix $P(z) \in GL(n, \mathscr{O}(\mathscr{D}_\infty))$ such that the transformation

$$U = P(z)V \tag{4.5.4}$$

changes (4.5.2) to

$$\frac{dV}{dz} = 0 \tag{4.5.5}$$

on $\mathscr{D}_\infty$.

If we apply the Kimura lemma (i.e., Lemma 3.7.1) to the matrix $P(z)^{-1}$, we can write the matrix $P(z)$ in the form

$$P(z) = E(z)\tilde{P}\left(\frac{1}{z}\right)\Lambda(z), \tag{4.5.6}$$

where

(1) $\tilde{P}(\zeta) \in GL(n, \mathscr{O}(\Delta(0, \frac{1}{R_0})))$,

(2) $E(z) \in GL(n, \mathscr{O}(\mathbf{C}))$,
(3) $\Lambda(z)$ is a diagonal matrix

$$\Lambda(z) = \begin{bmatrix} z^{\ell_1} & \cdots & 0 \\ \cdots & \cdots & \cdots \\ 0 & \cdots & z^{\ell_n} \end{bmatrix} \tag{4.5.7}$$

and $\ell_1, \cdots, \ell_n$ are integers.

If the transformation

$$U = E(z)W \tag{4.5.8}$$

changes (4.5.2) to the differential equation

$$\frac{dW}{dz} = A(z)W, \tag{4.5.9}$$

Lemma 4.3.3 and Theorem 1.11.1 imply that

(α') $A(z)$ is a matrix in $gl(n, m(\mathscr{D}))$ and differential equation (4.5.9) is of normally quasi-Fuchsian type on $\mathscr{D}$,
(β') differential equation (4.5.9) defines on $\mathscr{D}$ the structure $\mathscr{C}$ of analytic continuation.

Furthermore, by utilizing (4.5.4), (4.5.6), and (4.5.8), we can show that the transformation

$$W = \tilde{P}\left(\frac{1}{z}\right)\Lambda(z)V$$

changes (4.5.9) to (4.5.5) on $\mathscr{D}_\infty$. This means that

$$\frac{d}{dz}\left\{\tilde{P}\left(\frac{1}{z}\right)\Lambda(z)\right\} = A(z)\tilde{P}\left(\frac{1}{z}\right)\Lambda(z)$$

holds on $\mathscr{D}_\infty$. It follows from this that

$$\frac{d}{dz}\tilde{P}\left(\frac{1}{z}\right) = A(z)\tilde{P}\left(\frac{1}{z}\right) - \tilde{P}\left(\frac{1}{z}\right)\frac{1}{z}L, \tag{4.5.10}$$

where

$$L = \begin{bmatrix} \ell_1 & \cdots & o \\ \cdots & \cdots & \cdots \\ 0 & \cdots & \ell_n \end{bmatrix}. \tag{4.5.11}$$

Relation (4.5.10) shows that the transformation

$$W = \tilde{P}\left(\frac{1}{z}\right)\tilde{W}$$

changes (4.5.9) to the differential equation

$$\frac{d\tilde{W}}{dz} = \frac{1}{z}L\tilde{W},$$

where L is the matrix given by (4.5.11). Therefore $z = \infty$ is at worst a normally regular singular point of (4.5.9) (cf. Definition 4.3.4).

Thus, it was shown that differential equation (4.5.9) is of normally Fuchsian type on $\mathscr{D}$ (cf. Lemma 4.3.8). Therefore (4.5.9) has the form (4.5.1) by virtue of Lemma 4.3.7. □

The following result is a refinement of Theorem 2.11.5.

THEOREM 4.5.2. *Assume that*

(i) *$\mathscr{D}$ is the domain given by* (4.3.1),
(ii) *the differential equation*

$$\frac{dW}{dz} = A(z)W \tag{4.5.12}$$

is of Fuchsian type on $\mathscr{D}$,
(iii) *the structure $\mathscr{C}$ of analytic continuation defined on $\mathscr{D}$ by* (4.5.12) *satisfies condition* (v) *of Theorem* 4.5.1.

Then there exists a matrix $P(z) \in GL(n,\mathscr{O}(\mathscr{D}))$ such that every entry of $P(z)$ is a rational function of z and the transformation $W = P(z)V$ changes (4.5.12) *to a differential equation of normally Fuchsian type on $\mathscr{D}$.*

This theorem follows from Theorems 4.5.1 and 4.3.6.

In the case when the structure $\mathscr{C}$ of analytic continuation defined on $\mathscr{D}$ by differential equation (4.5.12) does not satisfy condition (v) of Theorem 4.5.1, let us change the independent variable z to a new variable u by a suitable transformation $z = \dfrac{\alpha u + \beta}{\gamma u + \delta}$ (α, β, γ, and δ being complex numbers) so that the singular points at ∞, $a_1, \cdots, a_p$ are moved to new locations at $b_0, b_1, \cdots, b_p$ in the u-plane respectively. Then condition (v) of Theorem 4.5.1 is satisfied. If we use this technique, $u = \infty$ becomes an additional singular point. This singular point is introduced technically through a change of the independent variable. We call such a singular point **an apparent singular point**. In general, it becomes necessary to introduce some apparent singular points if we consider the Riemann problem over a closed Riemann surface. [The translator omitted some remarks here. They will be restored in Appendix 5, §A.5.2, (3) and (4) of Remark A.5.2.14.]

§4.6. Triangulation of a differential equation of Fuchsian type

In this section, we shall consider the problem of triangulation of a differential equation of Fuchsian type by means of a transformation with rational coefficients. Since the subgroup of $GL(n, \mathbf{C})$ consisting of all invertible upper triangular matrices satisfies Assumption 4.4.2, we can apply Theorem 4.4.4.

THEOREM 4.6.1. *Assume that*

(i) $\mathscr{D}$ *is the domain given by* (4.3.1),
(ii) $\mathscr{C}$ *is a structure of analytic continuation on* $\mathscr{D}$,
(iii) $\{\Delta(z_j, \rho_j), C_{jk}\}$ *is a system of connection matrices belonging to* $\mathscr{C}$ *such that every matrix* C_{jk} *is upper triangular.*

Then, there exist an $n \times n$ *matrix* $\tilde{A}(z)$ *with polynomial entries and matrices* $A_1, \cdots, A_p$ *in* $gl(n, \mathbf{C})$ *such that*

(1) $A_1, \cdots, A_p$ *are upper triangular,*
(2) $\tilde{A}(z)$ *is upper triangular at every point of* $\mathscr{D}$,
(3) *the differential equation*

$$\frac{dW}{dz} = \left\{\sum_{k=1}^{p} \frac{1}{z-a_k} A_k + \tilde{A}(z)\right\} W \tag{4.6.1}$$

is of Fuchsian type on $\mathscr{D}$ *and of normally quasi-Fuchsian type on* $\mathscr{D}$,
(4) *differential equation* (4.6.1) *defines on* $\mathscr{D}$ *the structure* $\mathscr{C}$ *of analytic continuation.*

PROOF. By virtue of Theorem 4.4.4, we can construct a differential equation

$$\frac{dU}{dz} = B(z)U \tag{4.6.2}$$

in such a way that

(α) $B(z)$ belongs to $gl(n, \mathscr{m}(\mathscr{D}))$ and is upper triangular at every point of $\mathscr{D}$, and differential equation (4.6.2) is of normally quasi-Fuchsian type on $\mathscr{D}$,
(β) differential equation (4.6.2) defines on $\mathscr{D}$ the structure $\mathscr{C}$ of analytic continuation.

We consider differential equation (4.6.2) on a domain

$$\mathscr{D}_\infty = \{\, z\ ;\ |z| > R_0 \,\}, \tag{4.6.3}$$

where R_0 is a sufficiently large positive number. The subgroup of $GL(n, \mathbf{C})$ consisting of all invertible upper triangular matrices satisfies Assumption 4.4.2. Hence by utilizing the ideas of §3.1, we can find a matrix $P(z) \in GL(n, \mathscr{O}(\mathscr{D}_\infty))$ so that

(a) $P(z)$ is upper triangular at every point of $\mathscr{D}_\infty$,

(b) the transformation

$$U = P(z)V \tag{4.6.4}$$

changes (4.6.2) to

$$\frac{dV}{dz} = \frac{1}{z}A_0V, \tag{4.6.5}$$

where A_0 is an upper triangular matrix in $gl(n, \mathbf{C})$.

If we apply Lemma 3.4.3 to the matrix $P(z)^{-1}$, we can write the matrix $P(z)$ in the form

$$P(z) = E(z)Q(z), \tag{4.6.6}$$

where

(I) $Q(z) \in GL(n, \mathscr{M}(\mathscr{D}_\infty))$,
(II) $E(z) \in GL(n, \mathscr{O}(\mathbf{C}))$,
(III) Q and E are upper triangular.

If the transformation

$$U = E(z)W \tag{4.6.7}$$

changes (4.6.2) to

$$\frac{dW}{dz} = A(z)W, \tag{4.6.8}$$

then

(α') $A(z)$ is a matrix in $gl(n, \mathscr{M}(\mathscr{D}))$ which is upper triangular at every point of $\mathscr{D}$, and differential equation (4.6.8) is of normally quasi-Fuchsian type on $\mathscr{D}$,
(β') differential equation (4.6.8) defines on $\mathscr{D}$ the structure $\mathscr{C}$ of analytic continuation.

Furthermore, the transformation $W = Q(z)V$ changes (4.6.8) to (4.6.5) on $\mathscr{D}_\infty$. Therefore, $z = \infty$ is at worst a regular singular point of (4.6.8). Hence, (4.6.8) is of Fuchsian type on $\mathscr{D}$ (cf. Lemma 4.3.8). Lemma 4.3.7 implies that differential equation (4.6.8) has the form (4.6.1). □

The theorem below follows from Theorems 4.6.1 and 4.3.6 (cf. Example 2.11.6).

THEOREM 4.6.2. *Assume that*

(i) *$\mathscr{D}$ is the domain given by* (4.3.1),
(ii) *the differential equation*

$$\frac{dW}{dz} = A(z)W \tag{4.6.9}$$

is of Fuchsian type on $\mathscr{D}$ and defines on $\mathscr{D}$ a structure $\mathscr{C}$ of analytic continuation.

Then, in order that there exists a matrix $P(z) \in GL(n, \mathcal{O}(\mathcal{D}))$ *such that every entry of* $P(z)$ *is a rational function of* z *and that the transformation*

$$W = P(z)U \tag{4.6.10}$$

changes (4.6.9) *to the differential equation*

$$\frac{dU}{dz} = B(z)U \tag{4.6.11}$$

with a triangular matrix $B(z)$ *at every point of* $\mathcal{D}$, *it is necessary and sufficient that we find a system* $\{\Delta(z_j, \rho_j), C_{jk}\}$ *of connection matrices belonging to* $\mathcal{C}$ *such that every matrix* C_{jk} *is upper triangular.*

REMARK 4.6.3. By virtue of Theorem 4.6.1, we can assert that differential equation (4.6.11) is of Fuchsian type and of normally quasi-Fuchsian type on $\mathcal{D}$.

REMARK 4.6.4. As we explained it in Example 2.11.6, a structure $\mathcal{C}$ of analytic continuation satisfies the requirements given above if and only if, no matter what representation $\rho(\sigma)$ we may choose from the structure $\mathcal{M}(\mathcal{C})$ of monodromy, there exists a matrix $\Gamma \in GL(n, \mathbf{C})$ such that $\Gamma^{-1}\rho(\sigma)\Gamma$ is upper triangular for every $\sigma \in \pi_1(\mathcal{D}, z_0)$. This situation seems to be deeply related to the Galois theory of linear ordinary differential equations (cf. R. Kolchin [8] and J.-P. Ramis [126, 128]).

REMARK 4.6.5. The most essential assumptions in Theorems 4.6.1 and 4.6.2 are that the linear group G satisfies Assumption 4.4.2 and that Lemma 3.4.3 applies to G.

CHAPTER V

Linear Differential Equations on a Sectorial Domain

§5.0. Preliminaries

So far it has been sufficient to use systems of fundamental solutions defined on disks and systems of connection matrices defined by them as long as we dealt with regular singular points. However, to study the problem of "irregular singular points", we must investigate behavior of solutions on sectorial domains. The concept of structures of analytic continuation (or of monodromy) is not enough to understand the essential characteristic of irregular singular points (cf. Chapter 6). As a preliminary to the explanation of such situations and also a supplement to what we have written about regular singular points, we shall present here a reasonably complete picture of solutions of linear ordinary differential equations on a sectorial domain. We shall utilize **the characterization of regular singular points due to W. Jurkat and D. A. Lutz** as the central theme. Developing the entire story in such a way, we shall present most of the main ingredients. (Papers such as J. Moser [26] and Gérard–Levelt [16] are also interesting works concerning the problem of this chapter.)

§5.1. Asymptotic expansions

For a positive number R_0 and two real numbers a and b such that $a < b$, we denote by $\mathscr{D}(R_0, a, b)$ the sectorial domain $\{\ z;\ |z| > R_0,\ a < \arg\ z < b\ \}$. If $b - a$ is larger than 2π, $\mathscr{D}(R_0, a, b)$ is a domain on the Riemann surface of log z. We denote by $\mathscr{O}(R_0, a, b)$ the set of all complex-valued holomorphic functions on $\mathscr{D}(R_0, a, b)$. Also we denote by $C[[z^{-1}]]$ the set of all formal power series in z^{-1} with complex coefficients. This set $C[[z^{-1}]]$ admits properties quite similar to $C[[z-a]]$ (cf. §1.2). However, the derivative $\frac{dp}{dz}$ of $p = \sum_{m\geq 0} a_m z^m \in C[[z^{-1}]]$ is defined by

$$\frac{dp}{dz} = \sum_{m\geq 0} (-m)a_m z^{-m-1}\ ,$$

and this definition distinguishes $C[[z^{-1}]]$ from $C[[z-a]]$. We can regard C as a subset of $C[[z^{-1}]]$. Then, if $p \notin C$, we cannot annihilate p by differentiating it any number of times.

Now, we define a map $A_{sp} : \mathscr{A}(R_0, a, b) \to C[[z^{-1}]]$ as follows, i.e.,

$$A_{sp}(f) = p \tag{5.1.1}$$

means that

(i) $f \in \mathscr{O}(R_0, a, b)$,

(ii) $p = \sum_{m\geq 0} a_m z^{-m} \in C[[z^{-1}]]$,

(iii) an inequality

$$|f(z) - \sum_{m=0}^{N-1} a_m z^{-m}| \leq K_{N,R,\alpha,\beta}\ |z|^{-N} \tag{5.1.2}$$

holds on $\mathscr{D}(R, \alpha, \beta)$ for every positive integer N and every (R, α, β) satisfying the inequalities $0 < R_0 < R$ and $a < \alpha < \beta < b$, where $K_{N,R,\alpha,\beta}$ is a positive number determined by (N, R, α, β).

Map (5.1.1) is not defined for all f in $\mathscr{O}(R_0, a, b)$. The domain of definition of the map A_{sp} is denoted by $\mathscr{A}(R_0, a, b)$. It follows successively from inequalities (5.1.2) that

$$\lim_{z\to\infty} f(z) = a_0.$$

$$\lim_{z\to\infty} z^N \left\{ f(z) - \sum_{m=0}^{N-1} a_m z^{-m} \right\} = a_N \qquad (N \geq 0).$$

Hence $A_{sp}(f)$ is well defined for $f \in \mathscr{A}(R_0, a, b)$. Moreover

(1) if f and $g \in \mathscr{A}(R_0, a, b)$, then $f + g$ and $fg \in \mathscr{A}(R_0, a, b)$, and $A_{sp}(f + g) = A_{sp}(f) + A_{sp}(g)$ and $A_{sp}(fg) = A_{sp}(f)A_{sp}(g)$,

(2) if $f \in \mathscr{A}(R_0, a, b)$, then $\frac{df}{dz} \in \mathscr{A}(R_0, a, b)$ and

$$A_{sp}(\left(\frac{df}{dz}\right) = \frac{dA_{sp}(f)}{dz},$$

(3) if $A_{sp}(f) = \sum_{m\geq 2} a_m z^{-m}$, then

$$A_{sp}\left(\int_z^{\infty} f(s)ds\right) = \sum_{m\geq 1} \frac{a_{m+1}}{m} z^{-m},$$

where the path of integration is contained in $\mathscr{D}(R_0, a, b)$,

(4) $A_{sp}(f)$ is a surjective map : $\mathscr{A}(R_o, a, b) \to C[[z^{-1}]])$, but not injective

For these facts, see W. Wasow [11; Chap. III] and Y. Sibuya [9; Chap.1].

DEFINITION 5.1.1. For $f \in \mathscr{A}(R_0, a, b)$, the formal power series $A_{sp}(f)$ is called the **asymptotic expansion** of f as $z \to \infty$ in $\mathscr{D}(R_0, a, b)$.

§5.2. Linear differential equations on a sectorial domain

Let us fix a sectorial domain $\mathscr{D}(R_0, a, b) = \{ \zeta \,;\, |\zeta| > R_0,\ a < \arg \zeta < b \}$ on the ζ-plane. We denote by $gl(n, \mathscr{A}(R_0, a, b))$ the set of all $n \times n$ matrices with entries $a_{jk}(\zeta) \in \mathscr{A}(R_0, a, b)$. For a matrix

$$A(\zeta) \in gl(n, \mathscr{A}(R_0, a, b)),$$

we write the asymptotic expansion of the entry $a_{jk}(\zeta)$ in the form

$$A_{sp}(a_{jk}) = \sum_{m \geq 0} a_{jkm} \zeta^{-m}.$$

For every m, we denote by A_m the $n \times n$ matrix whose entries are the a_{jkm}. Then

$$A_{sp}(A) = \sum_{m \geq 0} A_m \zeta^{-m}.$$

This means that

$$\left| A(\zeta) - \sum_{m=0}^{N-1} A_m \zeta^{-m} \right| \leq K_{N,R,\alpha,\beta} |\zeta|^{-N}$$

holds on $\mathscr{D}(R, \alpha, \beta)$ for every positive integer N and every (R, α, β) satisfying the inequalities $0 < R_0 < R$ and $a < \alpha < \beta < b$, where $K_{N,R,\alpha,\beta}$ is a positive number determined by (N, R, α, β). Here $|A|$ denotes the sum of the absolute values of the entries of a matrix A.

Let s be a fixed positive integer. We denote by $\mathscr{D}_0$ the sectorial domain $\mathscr{D}(R_0^s, sa, sb)$ on the z-plane, i.e.,

$$\mathscr{D}_0 = \mathscr{D}(R_0^s, sa, sb).$$

If z is in $\mathscr{D}_0$, then $\zeta = z^{\frac{1}{s}}$ is in $\mathscr{D}(R_0, a, b)$. We consider a differential equation

$$\frac{dW}{dz} = z^{\frac{\kappa}{s}} A(z^{\frac{1}{s}}) W \tag{5.2.1}$$

on $\mathscr{D}_0$, where κ is a fixed integer and $A(\zeta)$ is in $gl(n, \mathscr{A}(R_0, a, b))$. If we change the independent variable by $\zeta = z^{\frac{1}{s}}$, differential equation (5.2.1) becomes

$$\frac{dW}{d\zeta} = s\zeta^{\kappa+s-1} A(\zeta) W \tag{5.2.2}$$

on $\mathscr{D}(R_0, a, b)$.

Since the sectorial domain $\mathcal{D}_0$ is connected and simply connected, we can apply Theorem 1.11.2 (cf. §1.11). This means that there exists a matrix $P(z) \in GL(n, \mathcal{O}(\mathcal{D}_0))$ such that the transformation

$$W = P(z)V \tag{5.2.3}$$

changes differential equation (5.2.1) to the differential equation

$$\frac{dV}{dz} = 0. \tag{5.2.4}$$

Consequently a solution to (5.2.1) is given by

$$W = P(z)C, \tag{5.2.5}$$

where $C \in gl(n, C)$.

DEFINITION 5.2.1. We say that $z = \infty$ is a **regular singular point** of differential equation (5.2.1) if there exists a nonnegative integer $N(\ell)$ for each direction ℓ contained in the sectorial domain $\mathcal{D}_0$ such that

$$\lim |z|^{-N(\ell)}|P(z)| = 0 \tag{5.2.6}$$

and

$$\lim |z|^{-N(\ell)}|P(z)^{-1}| = 0 \tag{5.2.7}$$

as $z \to \infty$ in the direction ℓ.

In Chapter 3 we also introduced a concept of regular singular points (cf. §3.2). The relation between these two definitions of regular singular points will be explained in detail later (cf. §5.8, Lemma 5.8.1).

LEMMA 5.2.2. *Let us denote* $\frac{\kappa}{s}$ *by* q *(i.e.,* $q = \frac{\kappa}{s}$*). If the rational number* q *satisfies the inequality*

$$q \leq -1, \tag{5.2.8}$$

then $z = \infty$ *is a regular singular point of* (5.2.1).

The proof of this lemma is left to the reader as an exercise.

§5.3. The order of solutions at $z = \infty$

As in §5.2, we consider differential equation (5.2.1) on the sectorial domain

$$\mathcal{D}_0 = \mathcal{D}(R_0^s, sa, sb). \tag{5.3.1}$$

This equation is changed to differential equation (5.2.4) by transformation (5.2.3). Therefore a solution to (5.2.1) is given by (5.2.5). Since every entry of the matrix $A(\zeta)$ admits an asymptotic expansion on $\mathcal{D}(R_0, a, b)$, $|A(\zeta)|$ is bounded on $\mathcal{D}(R, \alpha, \beta)$ if (R, α, β) satisfies the inequalities

$$0 < R_0 < R, \quad a < \alpha < \beta < b. \tag{5.3.2}$$

Fixing $\arg\ z = \theta$, set $t = |z|^{q+1}$, i.e.,

$$t = e^{-i(q+1)\theta} z^{q+1}.$$

In particular, if we assume that

$$q = \frac{\kappa}{s} > -1, \tag{5.3.3}$$

we can derive

$$\frac{dP(z)}{dt} = \frac{e^{i(q+1)\theta}}{q+1} A(z^{\frac{1}{s}})P(z) \tag{5.3.4}$$

from differential equation (5.2.1). Consequently, if (R, α, β) satisfies inequalities (5.3.2), the inequalities

$$|P(z)| \le K \exp\left\{L|z|^{q+1}\right\} \tag{5.3.5}$$

and

$$|P(z)^{-1}| \le K \exp\left\{L|z|^{q+1}\right\} \tag{5.3.6}$$

hold on $\mathscr{D}(R^s, s\alpha, s\beta)$ for some positive numbers K and L. Although the two positive numbers K and L depend on the choice of (R, α, β), the quantity q is independent of (R, α, β). Note that q was given by (5.3.3).

DEFINITION 5.3.1. The **order of solutions of differential equation (5.2.1) at** $z = \infty$ is defined to be the greatest lower bound of those positive numbers ρ for which two inequalities

$$|P(z)| \le K_{R,\alpha,\beta}\ \exp\left\{L_{R,\alpha,\beta}\ |z|^{\rho}\right\} \tag{5.3.7}$$

and

$$|P(z)^{-1}| \le K_{R,\alpha,\beta}\ \exp\left\{L_{R,\alpha,\beta}\ |z|^{\rho}\right\} \tag{5.3.8}$$

hold on $\mathscr{D}(R^s, s\alpha, s\beta)$ for some positive numbers $K_{R,\alpha,\beta}$ and $L_{R,\alpha,\beta}$ whenever (R, α, β) satisfies inequalities (5.3.2).
Inequalities (5.3.7) and (5.3.8) must hold for all (R, α, β) with ρ fixed. In the definition given above we used a matrix $P(z)$ which gives a solution (5.2.5). The order of solutions at $z = \infty$ is independent of the choice of such a matrix $P(z)$.

LEMMA 5.3.2. *Let ρ_0 be the order of solutions of differential equation (5.2.1) at $z = \infty$. Set also $q = \frac{\kappa}{s}$. Then $\rho_0 = 0$ if $q \le -1$. If $q > -1$, the inequality*

$$\rho_0 \le q + 1 \tag{5.3.9}$$

holds.

If $q > -1$, (5.3.9) follows from (5.3.5) and (5.3.6). The case when $q \le -1$ was left to the reader at Lemma 5.2.2.

§5.4. Improvement of estimates of order of solutions

As in the previous section, we consider differential equation (5.2.1) on sectorial domain (5.3.1). Since every entry of the matrix $A(\zeta)$ admits an asymptotic expansion on $\mathscr{D}(R_0, a, b)$, we set

$$A_{sp}(A) = \sum_{m \ge 0} A_m \zeta^{-m}, \tag{5.4.1}$$

where the A_m are $n \times n$ matrices in $gl(n, C)$ (cf. §5.2). In this section, assuming that the matrix A_0 is nilpotent, we shall estimate the order of solutions at $z = \infty$. By virtue of Lemma 5.3.2, the order is 0 if $q = \frac{\kappa}{s} \le -1$. Therefore the case when $q > -1$ is the main concern. We state the result as a lemma.

LEMMA 5.4.1. *Let ρ_0 be the order of solutions of differential equation (5.2.1) at $z = \infty$. Also set $q = \frac{\kappa}{s}$. If the matrix A_0 is nilpotent, then the inequality*

$$\rho_0 \le \max\left\{ 0,\ q + 1 - \frac{1}{ns} \right\} \tag{5.4.2}$$

holds.

PROOF. To prove this lemma, we assume that A_0 is an upper triangular matrix whose entries on the main diagonal are all zero. We leave it to the reader to verify that such an assumption does not cause any loss of generality. Next, we define a diagonal matrix $\Lambda(z)$ by

$$\Lambda(z) = \begin{bmatrix} 1 & 0 & 0 & \cdots & 0 \\ 0 & z^p & 0 & \cdots & 0 \\ 0 & 0 & z^{2p} & \cdots & 0 \\ \cdots & \cdots & \cdots & \cdots & \cdots \\ 0 & 0 & 0 & \cdots & z^{(n-1)p} \end{bmatrix}, \tag{5.4.3}$$

where

$$p = -\frac{1}{ns}. \tag{5.4.4}$$

Then the inequalities

$$|\Lambda(z)^{-1} A_0 \Lambda(z)| \le K|z|^p \tag{5.4.5}$$

and

$$|\Lambda(z)^{-1}\{A(z^{\frac{1}{s}}) - A_0\}\Lambda(z)| \le K|z|^p \tag{5.4.6}$$

hold on $\mathscr{D}(R^s, s\alpha, s\beta)$ provided that (R, α, β) satisfies inequalities (5.3.2), where K is a positive number determined by (R, α, β).

The transformation

$$W = \Lambda(z)V \tag{5.4.7}$$

changes differential equation (5.2.1) to

$$\frac{dV}{dz} = z^q \left\{\Lambda(z)^{-1}A(z^{\frac{1}{s}})\Lambda(z) - z^{-1-q}L\right\} V, \tag{5.4.8}$$

where

$$L = \begin{bmatrix} 0 & 0 & 0 & \cdots & 0 \\ 0 & p & 0 & \cdots & 0 \\ 0 & 0 & 2p & \cdots & 0 \\ \cdots & \cdots & \cdots & \cdots & \cdots \\ 0 & 0 & 0 & \cdots & (n-1)p \end{bmatrix}.$$

Consequently, by virtue of inequalities (5.4.5) and (5.4.6) and Lemma 5.3.2, we can show that the order of solutions of differential equation (5.4.8) at $z = \infty$ satisfies inequality (5.4.2). On the other hand, the order ρ_0 of solutions of differential equation (5.2.1) at $z = \infty$ is not changed by transformation (5.4.7). Hence ρ_0 also satisfies inequality (5.4.2). Thus we complete the proof of Lemma 5.4.1. □

§5.5. The case when A_0 is not a nilpotent matrix

In this section we shall still consider differential equation (5.2.1) on sectorial domain (5.3.1). In §5.4 we studied the case when A_0 of the asymptotic expansion (5.4.1) of the matrix $A(\zeta)$ is a nilpotent matrix. In this section, assuming that A_0 is not a nilpotent matrix, we shall find the order of solutions at $z = \infty$ exactly.

LEMMA 5.5.1. *Let ρ_0 be the order of solutions of differential equation (5.2.1) at $z = \infty$. Also set $q = \frac{\kappa}{s}$. If there exists a nonzero complex number λ such that $A_0 - \lambda I$ is nilpotent, then*

$$\rho_0 = \max\{0,\ q+1\}. \tag{5.5.1}$$

Here I is the identity matrix.

PROOF. If we set $\tilde{A}(\zeta) = A(\zeta) - \lambda I$, then

$$A_{sp}(\tilde{A}) = (A_0 - \lambda I) + \sum_{m\geq 1} A_m \zeta^{-m}. \tag{5.5.2}$$

Therefore, by virtue of Lemma 5.4.1, the order $\tilde{\rho}_0$ of solutions of the differential equation

$$\frac{d\tilde{W}}{dz} = z^q \tilde{A}(z^{\frac{1}{s}})\tilde{W} \tag{5.5.3}$$

at $z = \infty$ satisfies the inequality

$$\tilde{\rho}_0 \leq \max\left\{0,\ q+1-\frac{1}{ns}\right\}. \tag{5.5.4}$$

Furthermore, if $q > -1$, the transformation

$$W = \exp\left\{\frac{\lambda}{q+1} z^{q+1}\right\} \tilde{W} \tag{5.5.5}$$

changes differential equation (5.2.1) to (5.5.3). □

If we utilize Lemma 5.5.2 which will be given below, we can prove that conclusion (5.5.1) of Lemma 5.5.1 is also true in the general case when A_0 is not nilpotent. To state Lemma 5.5.2 precisely, we assume that the matrix A_0 has the following form:

$$A_0 = \begin{bmatrix} E_1 & 0 & \cdots & 0 \\ 0 & E_2 & \cdots & 0 \\ \cdots & \cdots & \cdots & \cdots \\ 0 & 0 & \cdots & E_\ell \end{bmatrix}, \tag{5.5.6}$$

where

(i) for each j, E_j is an $n_j \times n_j$ matrix and $\sum_{\ell \geq j \geq 1} n_j = n$,
(ii) for each j, there exists a complex number λ_j such that $E_j - \lambda_j I_j$ is nilpotent; here I_j is the $n_j \times n_j$ identity matrix,
(iii) $\ell \geq 2$,
(iv) $\lambda_j \neq \lambda_k$ if j and k are distinct.

LEMMA 5.5.2. *We can determine a positive number δ_0 by the matrix A_0 in such a way that, for any fixed two real numbers α and β satisfying the inequalities $a < \alpha < \beta < b$ and $\beta - \alpha \leq \delta_0$, there exists an $n \times n$ matrix $T(\zeta)$ such that*

(1) *the entries of $T(\zeta)$ belong to $\mathcal{A}(R, \alpha, \beta)$, where R is a positive number not less than R_0 and determined by $A(\zeta)$, α and β,*
(2) $A_{sp}(T) = I + \sum_{m\geq 1} T_m \zeta^{-m} \qquad (T_m \in gl(n, C))$,
(3) *the transformation*

$$W = T(z^{\frac{1}{s}})V \tag{5.5.7}$$

changes differential equation (5.2.1) *to a differential equation*

$$\frac{dV}{dz} = z^q B(z^{\frac{1}{s}})V \tag{5.5.8}$$

with a matrix $B(\zeta)$ of the form

$$B(\zeta) = \begin{bmatrix} F_1(\zeta) & 0 & \cdots & 0 \\ 0 & F_2(\zeta) & \cdots & 0 \\ \cdots & \cdots & \cdots & \cdots \\ 0 & 0 & \cdots & F_\ell(\zeta) \end{bmatrix}, \tag{5.5.9}$$

where, for each j, $F_j(\zeta)$ is an $n_j \times n_j$ matrix whose entries belong to $\mathscr{A}(R,\alpha,\beta)$ and

$$A_{sp}(F_j) = E_j + \sum_{m\geq 1} F_{jm}\zeta^{-m} \qquad (F_{jm} \in gl(n_j, C)).$$

We can derive the following theorem from Lemmas 5.2.1 and 5.2.2.

THEOREM 5.5.3. *Let ρ_0 be the order of solutions of differential equation (5.2.1) at $z = \infty$. Also set $q = \frac{\kappa}{s}$. If A_0 is not a nilpotent matrix, then*

$$\rho_0 = \max\{\ 0,\ q + 1\ \}. \tag{5.5.10}$$

We shall explain an outline of the proof of Lemma 5.5.2. The proof of Theorem 5.5.3 is left to the reader. The relation

$$\frac{dT(z^{\frac{1}{s}})}{dz} = z^q\{A(z^{\frac{1}{s}})T(z^{\frac{1}{s}}) - T(z^{\frac{1}{s}})B(z^{\frac{1}{s}})\} \tag{5.5.11}$$

is satisfied by the matrices A, B, and T of differential equations (5.2.1), (5.5.8), and transformation (5.5.7). Conversely, if we determine B and T by (5.5.11), then transformation (5.5.7) changes (5.2.1) to (5.5.8). If we set $\zeta = z^{\frac{1}{s}}$, then (5.5.11) becomes

$$\frac{dT(\zeta)}{d\zeta} = s\zeta^{\kappa+s-1}\{A(\zeta)T(\zeta) - T(\zeta)B(\zeta)\}. \tag{5.5.12}$$

To determine $B(\zeta)$ and $T(\zeta)$ by equation (5.5.12), we write $T(\zeta)$ and $A(\zeta)$ respectively in the forms below:

$$T = I + \begin{bmatrix} 0 & T_{12} & T_{13} & \cdots & T_{1\,\ell-1} & T_{1\,\ell} \\ T_{21} & 0 & T_{23} & \cdots & T_{2\,\ell-1} & T_{2\,\ell} \\ \cdots & \cdots & \cdots & \cdots & \cdots & \cdots \\ T_{\ell 1} & T_{\ell 2} & T_{\ell 3} & \cdots & T_{\ell\,\ell-1} & 0 \end{bmatrix}$$

and

$$A = \begin{bmatrix} A_{11} & A_{12} & \cdots & A_{1\,\ell} \\ A_{21} & A_{22} & \cdots & A_{2\,\ell} \\ \cdots & \cdots & \cdots & \cdots \\ A_{\ell 1} & A_{\ell 2} & \cdots & A_{\ell\,\ell} \end{bmatrix},$$

where T_{jk} and A_{jk} are $n_j \times n_k$ matrices and I is the $n \times n$ identity matrix. We also write $B(\zeta)$ as (5.5.9). Then (5.5.12) becomes a system of equations of the following form:

$$\begin{cases} 0 = A_{kk}(\zeta) + \sum_{h\neq k} A_{kh}(\zeta)T_{hk}(\zeta) - F_k(\zeta) \qquad (k = 1, \cdots, \ell), \\[2ex] \dfrac{dT_{jk}(\zeta)}{d\zeta} = s\zeta^{\kappa+s-1}\left\{A_{jk}(\zeta) + \sum_{h\neq k} A_{jh}(\zeta)T_{hk}(\zeta) - T_{jk}(\zeta)F_k(\zeta)\right\} \\ \hfill (j \neq k). \end{cases}$$

We can rewrite this system as

$$F_k(\zeta) = A_{kk}(\zeta) + \sum_{h \neq k} A_{kh}(\zeta) T_{hk}(\zeta) \qquad (k = 1, \cdots, \ell) \tag{5.5.13}$$

and

$$\frac{dT_{jk}(\zeta)}{d\zeta} = s\zeta^{\kappa+s-1} \Big\{ A_{jk}(\zeta) + A_{jj}(\zeta)T_{jk}(\zeta) - T_{jk}(\zeta)A_{kk}(\zeta) + \sum_{h \neq j,k} A_{jh}(\zeta)T_{hk}(\zeta) - T_{jk}(\zeta) \sum_{h \neq k} A_{kh}(\zeta)T_{hk}(\zeta) \Big\} \qquad (j \neq k). \tag{5.5.14}$$

If we determine the T_{jk} $(j \neq k)$ by (5.5.14), then the F_k are given by (5.5.13). To determine the T_{jk} by (5.5.14), we utilize the asymptotic method of solving nonlinear ordinary differential equations. For carrying out this plan, we need four conditions (i), (ii), (iii), and (iv) concerning form (5.5.6) of A_0. In particular, conditions (ii) and (iv) are important. For more information in detail (in particular, how to determine δ_0), see Y. Sibuya [29]. [We shall come back to this problem in Chapter 6; cf. §§6.1 and 6.8.]

If we examine the proof of Theorem 5.5.3 carefully, the following fact will be well understood.

THEOREM 5.5.4. *In the case when A_0 is not a nilpotent matrix, $z = \infty$ is a regular singular point of differential equation* (5.2.1) *if and only if q satisfies the inequality*

$$q \leq -1. \tag{5.5.15}$$

Furthermore if $z = \infty$ is not a regular singular point, the order ρ_0 of solutions at $z = \infty$ satisfies the inequality

$$\rho_0 \geq \frac{1}{s}. \tag{5.5.16}$$

This means that "$\rho_0 = 0$" is a necessary and sufficient condition that $z = \infty$ be a regular singular point. Furthermore ρ_0 satisfies inequality (5.5.16) if $\rho_0 > 0$. In the succeeding sections, we shall explain how to derive similar results for the general case.

§5.6. The case of a single equation of higher order

In this section we consider a single differential equation

$$\frac{d^n w}{dz^n} = \sum_{k=1}^{n} z^{\frac{\nu_k}{s}} a_k(z^{\frac{1}{s}}) \frac{d^{n-k} w}{dz^{n-k}} \tag{5.6.1}$$

on sectorial domain (5.3.1), where $p_1, \cdots, p_n$ are integers and $a_1(\zeta), \cdots, a_n(\zeta)$ belong to $\mathscr{A}(R_o, a, b)$ as functions of ζ. We write the aysmptotic expansion of $a_k(\zeta)$ in the form

$$A_{sp}(a_k) = \sum_{m\geq 0} a_{km}\zeta^{-m}, \tag{5.6.2}$$

where $a_{km} \in C$. Also we assume that

$$a_{k0} \neq 0 \tag{5.6.3}$$

if $A_{sp}(a_k) \neq 0$. Let us set

$$q_k = \begin{cases} -\infty & (\text{if } A_{sp}(a_k) = 0), \\ \dfrac{p_k}{s} & (\text{if } A_{sp}(a_k) \neq 0), \end{cases} \qquad (k = 1, \cdots, n), \tag{5.6.4}$$

and

$$q = \max_{n\geq k\geq 1} \frac{q_k}{k}. \tag{5.6.5}$$

[The rational number q is closely related with the *Newton polygon* of differential equation (5.6.1). We shall explain the Newton polygon in Appendix 1, §A.1.4.]

If we set

$$w_k = z^{-(k-1)q} \frac{d^{k-1}w}{dz^{k-1}} \quad (k = 1, \cdots, n), \tag{5.6.6}$$

differential equation (5.6.1) becomes the system below:

$$\begin{cases} \dfrac{dw_k}{dz} = z^q\left\{-(k-1)qz^{-q-1}w_k + w_{k+1}\right\}, \qquad (k = 1, \cdots, n-1), \\ \dfrac{dw_n}{dz} = z^q\left\{-(n-1)qz^{-q-1}\, w_n + \displaystyle\sum_{h=1}^{n} z^{q_h - hq} a_h(z^{\frac{1}{s}})\, w_{n-h+1}\right\}. \end{cases} \tag{5.6.7}$$

(Although the notations above are slightly ambiguous, the author believes that nothing is grossly improper.)

Definition (5.6.5) of the rational number q implies that

$$q_h - hq \leq 0 \qquad (h = 1, \cdots, n), \tag{5.6.8}$$

where the equality holds for at least an h. If $q \leq -1$, the order ρ_0 of solutions of (5.6.7) at $z = \infty$ is 0 because of Lemma 5.3.2. (To prove this precisely, we must treat z^{-q-1} carefully.) In the case when $q > -1$, since we can use Theorem 5.5.3 by virtue of assumption (5.6.3), we obtain $\rho_0 = q + 1$. To state this result more precisely, we write the matrix on the right-hand side of (5.6.7) in the form $z^q A(z)$ and consider the differential equation

$$\frac{dW}{dz} = z^q A(z) W, \tag{5.6.9}$$

where W is an $n \times n$ matrix. Then as a conclusion, we can derive the following theorem.

THEOREM 5.6.1. *The order ρ_0 of solutions of differential equation* (5.6.9) (*corresponding to the* nth *order equation* (5.6.1)) *at $z = \infty$ is given by*

$$\rho_0 = \max\{0, \ q + 1\}, \tag{5.6.10}$$

where q is the rational number determined by (5.6.5).

From Theorem 5.5.4 we can derive the theorem below.

THEOREM 5.6.2. *In order that $z = \infty$ be a regular singular point of differential equation* (5.6.9) *which corresponds to n-th order equation* (5.6.1), *it is necessary and sufficient that the rational number q determined by* (5.6.5) *satisfies the inequality*

$$q \leq -1. \tag{5.6.11}$$

Furthermore, if $z = \infty$ is not a regular singular point, the order ρ_0 of solutions at $z = \infty$ satisfies the inequality

$$\rho_0 \geq \frac{1}{ns}. \tag{5.6.12}$$

This means again that "$\rho = 0$ " is a necessary and sufficient condition that $z = \infty$ be a regular singular point. Furthermore ρ_0 satisfies inequality (5.6.12) if $\rho_0 > 0$. Actually this result equally applies to the general case.

§5.7. Existence of a cyclic vector

In this section we consider differential equation (5.2.1) on sectorial domain (5.3.1). To apply the results concerning nth order equation (5.6.1) to (5.2.1) (in particular those comments given at the end of §5.6), we utilize a **cyclic vector**. The existence of such a vector was proved by P. Deligne [1; Lemma 1.3 on p. 42].

We introduce some notations. Let $C[[\zeta^{-1}]]$ be the set of all formal power series in ζ^{-1}, and denote by $\mathscr{K}$ the quotient field of $C[[\zeta^{-1}]]$. Then $\mathscr{K}$ is given by

$$\mathscr{K} = \{ \zeta^k p \,; k \in Z, \quad p \in C[[\zeta^{-1}]] \},$$

where Z is the ring of all integers. Next, we denote by $\hat{A}$ the asymptotic expansion of the matrix $A(\zeta)$ on the right-hand side of differential equation

(5.2.2), i.e.,

$$\hat{A} = \sum_{m \geq 0} A_m \zeta^{-m}.$$

The expansion $\hat{A}$ is a matrix in $gl(n, C[[\zeta^{-1}]])$. Also we denote by $\mathscr{K}^n$ the set of all n-dimensional row vectors $(\eta_1, \cdots, \eta_n)$ with entries $\eta_j \in \mathscr{K}$, i.e.,

$$\mathscr{K}^n = \{(\eta_1, \eta_2, \cdots, \eta_n);\ \eta_j \in \mathscr{K}, \quad j = 1, \cdots, n\}.$$

The set $\mathscr{K}^n$ is an n-dimensional vector space over $\mathscr{K}$.

The inverse W^{-1} of a solution W to differential equation (5.2.2) satisfies the differential equation

$$\frac{dU}{d\zeta} + s\zeta^{\kappa+s-1}\, UA(\zeta) = 0.$$

Set

$$L(y) = \zeta \frac{dy}{d\zeta} + s\zeta^{\kappa+s} y\hat{A} \tag{5.7.1}$$

for $y \in \mathscr{K}^n$. Since $\frac{d}{d\zeta}$ is a differential operator and $\mathscr{K}^n$ is a vector space over $\mathscr{K}$, L is not a linear operator, although L is additive. Because of this situation, the manipulation with L is a little complicated [cf. Appendix 1]. In spite of this, the following identity holds:

$$L(\zeta^\ell y) = \zeta^\ell \{L(y) + \ell y\},$$

where $\ell \in Z$ and $y \in \mathscr{K}^n$. If we write this identity in the form

$$L(\zeta^\ell y) = \zeta^\ell (L + \ell) y,$$

we have

$$L^k(\zeta^\ell y) = \zeta^\ell (L + \ell)^k y \qquad (\ell \in Z,\ \ y \in \mathscr{K}^n,\ \ k = 0, 1, 2, \cdots). \tag{5.7.2}$$

The main concern of this section is to prove the lemma below.

LEMMA 5.7.1. *There exists a vector $y \in \mathscr{K}^n$ such that the n vectors*

$$y,\ L(y),\ L^2(y),\ \cdots,\ L^{n-1}(y) \tag{5.7.3}$$

are linearly independent over $\mathscr{K}$.

PROOF. For a vector $y \in \mathscr{K}^n$, set

$$N = \mathscr{N}(y) \tag{5.7.4}$$

if $\{y,\ L(y),\ \cdots,\ L^{N-1}(y)\}$ is linearly independent but $\{y,\ L(y),\ \cdots,\ L^N(y)\}$ is not. We shall derive a contradiction from the inequality

$$\max_{y \in \mathscr{K}^n} \mathscr{N}(y) < n. \tag{5.7.5}$$

To do this, we choose a z from $\mathscr{K}^n$ so that

$$\mathscr{N}(z) = \max_{y \in \mathscr{K}^n} \mathscr{N}(y) < n. \tag{5.7.6}$$

We also set

$$N_0 = \mathscr{N}(z). \tag{5.6.7}$$

Now if we set

$$u = z + \lambda \zeta^{\ell} y \tag{5.7.8}$$

for an arbitrary vector $y \in \mathscr{K}^n$, an arbitrary complex number λ, and an arbitrary integer ℓ, then

$$L^k(u) = L^k(z) + \lambda \zeta^{\ell} (L + \ell)^k y \qquad (k = 0, , 1, 2 \cdots). \tag{5.7.9}$$

Choosing $N_0 + 1$ vectors $y_1, \cdots, y_{N_0+1}$ from $\mathscr{K}^n$ in any manner, we construct an $(N_0 + 1) \times n$ matrix by using them as row vectors. Then, choosing $N_0 + 1$ column vectors from this matrix in any manner, we form an $(N_0 + 1) \times (N_0 + 1)$ determinant which is written in the form

$$\det(y_1, \cdots, y_{N_0+1}).$$

Since $\{u, L(u), \cdots, L^{N_0}(u)\}$ is not linearly independent, we must have

$$\det(u, L(u), \cdots, L^{N_0}(u)) = 0. \tag{5.7.10}$$

Then if we utilize (5.7.9), the left-hand side of (5.7.10) becomes a polynomial in λ. Computing the coefficient of the linear term in λ of this polynomial, we derive

$$\sum_{k=0}^{N_0} \det(z, L(z), \cdots, L^{k-1}(z), (L + \ell)^k y, L^{k+1}(z), \cdots, L^{N_0}(z)) = 0. \tag{5.7.11}$$

The left-hand side of (5.7.11) is a polynomial of degree N_0 in ℓ. Since ℓ is an arbitrary integer, every coefficient of this polynomial must be zero. In particular, computing the coefficient of the term of the highest degree in ℓ, we derive

$$\det(z, L(z), \cdots, L^{N_0-1}(z), y) = 0. \tag{5.7.12}$$

Thus we conclude that $\{z, L(z), \cdots, L^{N_0-1}(z), y\}$ is not linearly independent. Since y is an arbitrary vector in $\mathscr{K}^n$ and $N_0 < n$, this contradicts the fact that the dimension of $\mathscr{K}^n$ over $\mathscr{K}$ is n. □

Lemma 5.7.1 was given in terms of formal power series. However the same result applies to more general cases as we can see it clearly if we examine the proof given above very closely. In particular we can apply the lemma to functions in $\mathscr{A}(R_0, a, b)$ in the following way. First write the entries $\eta_1, \cdots, \eta_n$ of the cyclic vector $y \in \mathscr{K}^n$ given in Lemma 5.7.1 in the form

$$\eta_j = \zeta^{\ell_j} p_j, \quad \ell_j \in Z, \quad p_j \in C[[\zeta^{-1}]] \qquad (j = 1, \cdots, n).$$

Since the map $A_{sp} : \mathscr{A}(R_0, a, b) \rightarrow C[[\zeta^{-1}]]$ is surjective, there exist n functions $f_1, \cdots, f_n \in \mathscr{A}(R_0, a, b)$ such that

$$A_{sp}(f_j) = p_j \qquad (j = 1, \cdots, n).$$

Denote by $g(\zeta)$ the vector whose entries are $g_j(\zeta) = \zeta^{\ell_j} f_j(\zeta)$ $(j = 1, \cdots, n)$, i.e.,

$$g(\zeta) = (g_1(\zeta), \ \cdots, \ g_n(\zeta)).$$

Setting also

$$\mathscr{L}(g) = \zeta \frac{dg}{d\zeta} + s\zeta^{\kappa+s} g A(\zeta),$$

we construct $\mathscr{L}^2(g), \mathscr{L}^3(g), \cdots, \mathscr{L}^{n-1}(g)$ successively. Then, denote by $G(\zeta)$ the matrix whose row vectors are $g, \mathscr{L}(g), \cdots, \mathscr{L}^{n-1}(g)$. The determinant of $G(\zeta)$ is written in the form

$$\det G(\zeta) = \zeta^{\ell} \gamma(\zeta) \tag{5.7.13}$$

where ℓ is an integer and γ is a function in $\mathscr{A}(R_0, a, b)$. Thus we can derive the result below from Lemma 5.7.1.

LEMMA 5.7.2. *The function γ on the right-hand side of* (5.7.13) *satisfies the condition*

$$A_{sp}(\gamma) \neq 0. \tag{5.7.14}$$

If we choose a positive number R sufficiently large, we can express each entry of $G(\zeta)^{-1}$ as a product of a power of ζ and a function in $\mathscr{A}(R, a, b)$. The transformation

$$U = G(\zeta) W \tag{5.7.15}$$

changes (5.2.2) to

$$\zeta \frac{dU}{d\zeta} = \left\{ \zeta \frac{dG(\zeta)}{d\zeta} + s\zeta^{\kappa+s} G(\zeta) A(\zeta) \right\} G(\zeta)^{-1} U. \tag{5.7.16}$$

To study the matrix on the right-hand side of (5.7.16) more closely, we set

$$H(\zeta) = \zeta \frac{dG(\zeta)}{d\zeta} + s\zeta^{\kappa+s} G(\zeta) A(\zeta)$$

(i.e., $H = \mathscr{L}(G)$). The row vectors of this matrix are $\mathscr{L}(g), \mathscr{L}^2(g), \cdots, \mathscr{L}^n(g)$. Therefore $H(\zeta)G(\zeta)^{-1}$ has the following form:

$$HG^{-1} = \begin{bmatrix} 0 & 1 & 0 & 0 & \cdots & 0 & 0 \\ 0 & 0 & 1 & 0 & \cdots & 0 & 0 \\ 0 & 0 & 0 & 1 & \cdots & 0 & 0 \\ \cdots & \cdots & \cdots & \cdots & \cdots & \cdots & \cdots \\ 0 & 0 & 0 & 0 & \cdots & 1 & 0 \\ 0 & 0 & 0 & 0 & \cdots & 0 & 1 \\ b_1 & b_2 & b_3 & b_4 & \cdots & b_{n-1} & b_n \end{bmatrix},$$

where the b_j are the product of powers of ζ and functions in $\mathscr{A}(R, a, b)$. Transformation (5.7.14) does not change the order of solutions at $z = \infty$. Moreover, the results of §5.6 apply to differential equation (5.7.16). Thus we derive the theorem below.

THEOREM 5.7.3. *If we denote by ρ_0 the order of solutions of differential equation* (5.2.1) *at $z = \infty$, then ρ_0 is a rational number such that*

(i) *$z = \infty$ is a regular singular point of* (5.2.1) *if and only if*

$$\rho_0 = 0, \tag{5.7.17}$$

(ii) *if $\rho_0 > 0$, the inequality*

$$\rho_0 \geq \frac{1}{ns} \tag{5.7.18}$$

holds.

§5.8. Singularity of a transformation

In this section we shall still consider differential equation (5.2.1) on sectorial domain (5.3.1). The lemma below follows from the results given in §§5.6 and 5.7.

LEMMA 5.8.1. *If we denote by ρ_0 the order of solutions of differential equation* (5.2.1) *at $z = \infty$, we can choose a positive integer p, a positive number R, and an $n \times n$ matrix $T(\eta)$ as follows:*

(i) *the entries of $T(\eta)$ are rational functions of η,*
(ii) *the determinant of $T(\eta)$ does not vanish identically in η,*
(iii) *the transformation*

$$W = T(z^{\frac{1}{ps}})U \tag{5.8.1}$$

changes differential equation (5.2.1) *to*

$$\frac{dU}{dz} = z^{\rho_0 - 1} B(z^{\frac{1}{ps}})U \tag{5.8.2}$$

with a matrix $B(\eta)$ *whose entries belong to* $\mathscr{A}(R, \frac{a}{p}, \frac{b}{p})$ *as functions of* η.

If the asymptotic expansion of the matrix $B(\eta)$ on the right-hand side of differential equation (5.8.2) is given by

$$A_{sp}(B) = \sum_{m \geq 0} B_m \eta^{-m}, \tag{5.8.3}$$

B_0 is not a nilpotent matrix if $\rho_0 > 0$ because of Lemma 5.4.1 (and Theorem 5.5.3). If we examine the proof of Lemma 5.8.1 very carefully, we will understand clearly the relation between the two definitions of regular singular points given in §3.2 and §5.2 respectively. We leave it to the reader to look into these situations in detail.

The main concern of this section is to study the behavior of the matrix $T(\eta)$ at $\eta = \infty$. To do this, we shall utilize the following lemma.

LEMMA 5.8.2. *If a matrix* $T(\eta)$ *satisfies conditions* (i) *and* (ii) *of lemma* 5.8.1, *there exist n integers* $\lambda_1, \cdots, \lambda_n$ *and two matrices* $P(\eta)$ *and* $Q(\eta)$ *such that*

(1) *the entries of* $P(\eta)$ *and* $Q(\eta)$ *are rational functions of* η,

(2) *if the expansions of* $P(\eta)$ *and* $Q(\eta)$ *at* $\eta = \infty$ *are*

$$P(\eta) = \eta^{\alpha} \sum_{m \geq 0} P_m \eta^{-m} \quad \textit{and} \quad Q(\eta) = \eta^{\beta} \sum_{m \geq 0} Q_m \eta^{-m}$$

respectively, then $\alpha = 0$, $\beta = 0$, *and the determinants of* P_0 *and* Q_0 *are not zero,*

(3) $\lambda_1 \leq \lambda_2 \leq \cdots \leq \lambda_n$,

(4) *if we denote by* Λ *the diagonal matrix whose entries on the main diagonal are* $\lambda_1, \cdots, \lambda_n$, *then*

$$T(\eta) = P(\eta)\eta^{\Lambda}Q(\eta). \tag{5.8.4}$$

Furthermore, $\lambda_1, \cdots, \lambda_n$ *are uniquely determined by* $T(\eta)$.

We can prove Lemma 5.8.2 by means of the echelon form of $T(\eta)$. We leave the proof of this lemma also to the reader.

Next, we give the definition below.

DEFINITION 5.8.3. Assuming that the matrix $T(\eta)$ of Lemma 5.8.1 is written in form (5.8.4) of Lemma 5.8.2, we set

$$\sigma(T) = \lambda_n - \lambda_1 \tag{5.8.5}$$

and call $\sigma(T)$ the **span** of T (with respect to η). Furthermore $\sigma(T)/ps$ is called the **order of singularity** of transformation (5.8.1) at $z = \infty$.

Hereafter we shall consider the problem of minimizing the order of singularity of transformation (5.8.1) at $z = \infty$ under the assumption that (5.8.1) satisfies the conditions of Lemma 5.8.1. We denote by $\varsigma(T)$ the order of singularity of (5.8.1) at $z = \infty$. Then we can summarize the main result as follows.

THEOREM 5.8.4. (D.Lutz) *There exists a transformation* (5.8.1) *satisfying the conditions of Lemma* 5.8.1 *such that the inequality*

$$\varsigma(T) \leq (n-1)(q_0 + 1 - \rho_0) \tag{5.8.6}$$

holds, where

$$q_0 = \max\{q, \ -1\} \qquad (q = \frac{\kappa}{s}). \tag{5.8.7}$$

PROOF. Let us fix a transformation (5.8.1) satisfying the conditions of Lemma 5.8.1. We write its matrix $T(\eta)$ in form (5.8.4). If we set

$$\begin{cases} W = P(z^{\frac{1}{ps}})\hat{W}, \qquad \hat{U} = Q(z^{\frac{1}{ps}})U, \\ \hat{A}(z) = P(z^{\frac{1}{ps}})^{-1}\left\{ z^q A(z^{\frac{1}{s}})P(z^{\frac{1}{ps}}) - \dfrac{dP(z^{\frac{1}{ps}})}{dz}\right\}, \\ \hat{B}(z) = \left\{ z^{\rho_0 - 1}Q(z^{\frac{1}{ps}})B(z^{\frac{1}{ps}}) + \dfrac{dQ(z^{\frac{1}{ps}})}{dz}\right\} Q(z^{\frac{1}{ps}})^{-1}, \end{cases} \tag{5.8.8}$$

we obtain

$$\frac{d\hat{W}}{dz} = \hat{A}(z)\hat{W} \tag{5.8.9}$$

and

$$\frac{d\hat{U}}{dz} = \hat{B}(z)\hat{U}. \tag{5.8.10}$$

Furthermore, differential equation (5.8.9) is changed to (5.8.10) by the transformation

$$\hat{W} = z^{\frac{1}{ps}\Lambda}\ \hat{U}. \tag{5.8.11}$$

Consequently,

$$\hat{B}(z) = z^{-\frac{1}{ps}\Lambda}\ \hat{A}(z)\ z^{\frac{1}{ps}\Lambda} - \frac{1}{psz}\Lambda. \tag{5.8.12}$$

If we denote by $\hat{a}_{jk}(z)$ (resp. $\hat{b}_{jk}(z)$) the entry on the jth row and the kth column of $\hat{A}(z)$ (resp. $\hat{B}(z)$), it follows from (5.8.12) that

$$\begin{cases} \hat{b}_{jj}(z) = \hat{a}_{jj}(z) - \dfrac{\lambda_j}{psz} \qquad (j = 1, \cdots, n), \\ \hat{b}_{jk}(z) = z^{\frac{1}{ps}(\lambda_k - \lambda_j)}\ \hat{a}_{jk}(z) \qquad (j \neq k). \end{cases} \tag{5.8.12'}$$

We can find a positive number $\hat{R}$, functions $\alpha_{jk}(\eta) \in \mathscr{A}(\hat{R}, \frac{a}{p}, \frac{b}{p})$, and integers p_{jk} satisfying the inequalities

$$\frac{p_{jk}}{ps} \leq q_0$$

so that we can write $\hat{a}_{jk}(z)$ in the form

$$\hat{a}_{jk}(z) = z^{\frac{1}{ps}p_{jk}}\ \alpha_{jk}(z^{\frac{1}{ps}})) \tag{5.8.13}$$

The choice of the p_{jk} is not unique. We choose the smallest p_{jk} satisfying the requirements given above. Hence we put $p_{jk} = -\infty$ if $A_{sp}(\alpha_{jk}) = 0$. In this way, we see that

$$N_0 = ps\rho_0 \tag{5.8.14}$$

is an integer, and that the inequalities

$$\frac{1}{ps}(\lambda_k - \lambda_j + p_{jk}) \leq \rho_0 - 1 \qquad (j, k = 1, \cdots, n) \tag{5.8.15}$$

hold.

By utilizing the results above conversely, we shall prove Theorem 5.8.4. To do this, we set up the following plan. First, we observe that the matrix

$$\hat{T}(\eta) = P(\eta)\eta^{\Lambda}$$

satisfies the conditions of Lemma 5.8.1 if we define $\hat{B}(z)$ by (5.8.12), provided that n integers $\lambda_1 \leq \lambda_2 \leq \cdots \leq \lambda_n$ satisfy inequalities (5.8.15). This follows from the fact that the transformation

$$W = \hat{T}(z^{\frac{1}{ps}})\hat{U}$$

changes (5.2.1) to (5.8.10). Accordingly, starting with the given $\lambda_1, \cdots, \lambda_n$, we shall try to modify $\lambda_1, \cdots, \lambda_n$ without changing inequalities (5.8.15) so that the inequality

$$\frac{1}{ps}(\lambda_n - \lambda_1) \leq (n-1)(q_0 + 1 - \rho_0) \tag{5.8.16}$$

will be eventually satisfied.

To accomplish the plan above, we assume first that (5.8.15) holds. Then we modify $\lambda_1, \cdots, \lambda_n$ as follows:

$$\begin{cases} \lambda_1^* = \lambda_1, \\ \lambda_2^* = \min\{\lambda_2 - \lambda_1,\ ps(q_0 + 1 - \rho_0)\} + \lambda_1, \\ \lambda_j^* = \lambda_j - (\lambda_2 - \lambda_2^*) \qquad (j = 3, \cdots, n). \end{cases} \tag{5.8.17}$$

Lemma 5.3.2 and definition (5.8.7) of q_0 imply that $q_0 + 1 - \rho_0 \geq 0$, and hence $0 \leq \lambda_2^* - \lambda_1^* \leq \lambda_2 - \lambda_1$. Moreover

$$\lambda_j^* - \lambda_k^* = \lambda_j - \lambda_k \qquad (j, k = 2, \cdots, n).$$

Therefore it follows that $\lambda_1^* \le \lambda_2^* \le \cdots \le \lambda_n^*$ and

$$\frac{1}{ps}(\lambda_k^* - \lambda_j^* + p_{jk}) \le \rho_0 - 1 \qquad (j,k = 2, \cdots, n).$$

Further, since $\lambda_j^* - \lambda_1^* = \lambda_j - \lambda_1 - (\lambda_2 - \lambda_2^*) \le \lambda_j - \lambda_1$, we have

$$\frac{1}{ps}(\lambda_j^* - \lambda_1^* + p_{1j}) \le \rho_0 - 1 \qquad (j = 1, \cdots, n).$$

If $\lambda_2^* \ne \lambda_2$ and $j \ge 2$, then

$$\begin{aligned}\lambda_1^* - \lambda_j^* &= \lambda_1 - \lambda_j + (\lambda_2 - \lambda_2^*)\\ &= \lambda_1 - \lambda_j + \lambda_2 - ps(q_0 + 1 - \rho_0) - \lambda_1\\ &\le -ps(q_0 + 1 - \rho_0)\\ &\le ps(\rho_0 - 1) - p_{j1}.\end{aligned}$$

Thus we see that λ_1^*, λ_2^*, $\cdots$, λ_n^* satisfy inequalities (5.8.15). Moreover, (5.8.17) implies that

$$\lambda_2^* - \lambda_1^* \le ps(q_0 + 1 - \rho_0).$$

Accordingly we can assume now that λ_1, $\cdots$, λ_n satisfy inequalities (5.8.15) and that λ_1 and λ_2 satisfy the inequality

$$\frac{1}{ps}(\lambda_2 - \lambda_1) \le q_0 + 1 - \rho_0. \tag{5.8.18}$$

Therefore, under this assumption, we set

$$\begin{cases}\lambda_1^* = \lambda_1,\\ \lambda_2^* = \lambda_2,\\ \lambda_3^* = \min\{\lambda_3 - \lambda_2,\ ps(q_0 + 1 - \rho_0)\} + \lambda_2,\\ \lambda_j^* = \lambda_j - (\lambda_3 - \lambda_3^*) \qquad (j = 4, \cdots, n).\end{cases} \tag{5.8.19}$$

Then we can again verify that λ_1^*, λ_2^*, $\cdots$, λ_n^* satisfy inequalities (5.8.15) and $\lambda_1^* \le \cdots \le \lambda_n^*$. Further, the inequality

$$\frac{1}{ps}(\lambda_3^* - \lambda_1^*) \le 2(q_0 + 1 - \rho_0)$$

follows from (5.8.18) and (5.8.19). In this way, we arrive eventually at inequality (5.8.16). □

§5.9. Characterization of singular points

We defined the order of solutions at $z = \infty$ by Definition 5.3.1 for differential equation (5.2.1) on sectorial domain (5.3.1). In this section and the succeeding section, we shall discuss an algebraic characterization of the order of solutions at $z = \infty$. The characterization [of the case $\rho_0 > 0$] given below is the same as Théorème 1.9 on p. 45 of P. Deligne [1]. Deligne gave this theorem as a result due to N. Katz. Actually this characterization of singular points of solutions is one of the highlights of the works due to D. Lutz [22, 23, 24] and W. B. Jurkat and D. Lutz [21]. We need preparation for explaining this result.

DEFINITION 5.9.1. We set

$$p(B) = k \tag{5.9.1}$$

if an $n \times n$ matrix $B(\eta)$ has the following form:

$$B(\eta) = \eta^k \tilde{B}(\eta), \tag{5.9.2}$$

where

(i) k is an integer,
(ii) the entries of $\tilde{B}(\eta)$ belong to $\mathscr{A}(R, \alpha, \beta)$ as functions of η for a suitable choice of (R, α, β),
(iii) if $A_{sp}(\tilde{B}) = \sum_{m \geq 0} \tilde{B}_m \eta^{-m}$, then $\tilde{B}_0 \neq 0$.

Furthermore we set

$$p(B) = -\infty \tag{5.9.3}$$

if $A_{sp}(\tilde{B}) = 0$. Also we set

$$\wp(B) = \frac{1}{\tau} p(B) \tag{5.9.4}$$

when we deal with the matrix $B(z^{\frac{1}{\tau}})$ for a fixed positive integer τ.

LEMMA 5.9.2. *Assume that $\sigma(T)$ (resp. $p(B)$) is determined by Definition 5.8.3 (resp. Definition 5.9.1) applied to a matrix $T(\eta)$ (resp. $B(\eta)$). Then*

$$p(T^{(m)} B T^{-1}) \leq \sigma(T) + p(B) - m, \tag{5.9.5}$$

where $T^{(m)}(\eta)$ is the mth derivative of $T(\eta)$ with respect to η.

We can prove this lemma by writing $T(\eta)$ as (5.8.4) and computing $T^{(m)}(\eta)$. Similarly, we can prove the lemma below.

LEMMA 5.9.3. *Let $T(\eta)$ and $B(\eta)$ be the same as in Lemma 5.9.2. Fixing a positive integer τ, we consider $T(z^{\frac{1}{\tau}})$ and $B(z^{\frac{1}{\tau}})$. If we denote by $T^{[m]}(z^{\frac{1}{\tau}})$ the mth derivative of $T(z^{\frac{1}{\tau}})$ with respect to z, then*

$$\wp(T^{[m]} B T^{-1}) \leq \varsigma(T) + \wp(B) - m, \tag{5.9.6}$$

where

$$\varsigma(T) = \frac{1}{\tau} \sigma(T). \tag{5.9.7}$$

The proofs of these two lemmas above are left to the reader.

We consider the differential operator

$$\mathscr{L}(X) = \frac{dX}{dz} + z^{\frac{\kappa}{s}} XA(z^{\frac{1}{s}}) \tag{5.9.8}$$

which is similar to the operator given in §5.7, where κ and $A(\zeta)$ are the same as on the right-hand side of (5.2.1) and X is an $n \times n$ matrix. Utilizing this operator, we set

$$\mathscr{A}_0(z^{\frac{1}{s}}) = I, \quad \mathscr{A}_m(z^{\frac{1}{s}}) = \mathscr{L}(\mathscr{A}_{m-1}), \qquad (m = 1, 2, \cdots) \tag{5.9.9}$$

where I is the $n \times n$ identity matrix. Now, the main result of this section is given by the theorem below.

THEOREM 5.9.4. (D. Lutz) *Let ρ_0 be the order of solutions of differential equation* (5.2.1) *at $z = \infty$ and set*

$$p_m = | \wp(\mathscr{A}_m) - m(\rho_0 - 1) |, \qquad (m = 1, 2, \cdots). \tag{5.9.10}$$

If

$$\rho_0 > 0, \tag{5.9.11}$$

then $\{ p_m \}$ is a bounded sequence of rational numbers.

PROOF. Utilizing Lemma 5.8.1, we change (5.2.1) to (5.8.2) by transformation (5.8.1). Because of (5.9.11), the matrix B_0 of asymptotic expansion (5.8.3) of the matrix B on the right-hand side of (5.8.2) satisfies the condition

$$B_0^m \neq 0 \qquad (m = 1, 2, \cdots), \tag{5.9.12}$$

i.e., B_0 is not a nilpotent matrix. Now, introducing the differential operator

$$\tilde{\mathscr{L}}(X) = \frac{dX}{dz} + z^{\rho_0 - 1} XB(z^{\frac{1}{ps}}), \tag{5.9.13}$$

we set

$$\mathscr{B}_0(z^{\frac{1}{ps}}) = I, \quad \mathscr{B}_m(z^{\frac{1}{ps}}) = \tilde{\mathscr{L}}(\mathscr{B}_{m-1}), \qquad (m = 1, 2, \cdots). \tag{5.9.14}$$

Then

$$\wp(\mathscr{B}_m) = m(\rho_0 - 1) \qquad m = 1, 2, \cdots. \tag{5.9.15}$$

Since differential equation (5.2.1) changes to (5.8.2) by transformation (5.8.1), we have

$$\tilde{\mathscr{L}}(T) = \mathscr{A}_1 T$$

and hence

$$\tilde{\mathscr{L}}^m(T) = \mathscr{A}_m T, \qquad (m = 1, 2, \cdots) \tag{5.9.16}$$

successively. Similarly we derive

$$\mathscr{L}^m(T^{-1}) = \mathscr{B}_m T^{-1}, \qquad (m = 1, 2, \cdots). \tag{5.9.17}$$

If we denote by $T^{[m]}$ (resp. $(T^{-1})^{[m]}$) the mth derivative of T (resp. T^{-1}) with respect to z, we can write (5.9.16) and (5.9.17) in the following forms:

$$\sum_{\ell=0}^{m} \frac{m!}{\ell!(m-\ell)!} T^{[\ell]} \mathscr{B}_{m-\ell} T^{-1} = \mathscr{A}_m, \tag{5.9.16$'$}$$

and

$$\sum_{\ell=0}^{m} \frac{m!}{\ell!(m-\ell)!} (T^{-1})^{[\ell]} \mathscr{A}_{m-\ell} T = \mathscr{B}_m, \tag{5.9.17$'$}$$

respectively. Utilizing Lemma 5.9.3, we can derive

$$\begin{cases} \wp(\mathscr{A}_m) \leq \max_{m \geq \ell \geq 0} \{ \varsigma(T) + \wp(\mathscr{B}_{m-\ell}) - \ell \}, \\ \wp(\mathscr{B}_m) \leq \max_{m \geq \ell \geq 0} \{ \varsigma(T^{-1}) + \wp(\mathscr{A}_{m-\ell}) - \ell \} \end{cases}$$

from (5.9.16$'$) and (5.9.17$'$) respectively. By virtue of Theorem 5.8.4, we can assume that

$$\varsigma(T) = \varsigma(T^{-1}) \leq (n-1)(q_0 + 1 - \rho_0). \tag{5.9.18}$$

Utilizing (5.9.15) and (5.9.18) we derive

$$\wp(\mathscr{A}_m) \leq (n-1)(q_0 + 1 - \rho_0) + m(\rho_0 - 1) \qquad (m = 1, 2, \cdots) \tag{5.9.19}$$

and

$$m(\rho_0 - 1) \leq \max_{m \geq \ell \geq 0} \{ \wp(\mathscr{A}_{m-\ell}) - \ell \} + (n-1)(q_0 + 1 - \rho_0) \qquad (m = 1, 2, \cdots). \tag{5.9.20}$$

Inequalities (5.9.19) give an upper bound for the quantities $\wp(\mathscr{A}_m) - m(\rho_0 - 1)$ $(m = 1, 2, \cdots)$. To obtain a lower bound from (5.9.20), we shall derive a contradiction from the assumption that there exists a nonnegative integer m_0 such that the conditions

$$\wp(\mathscr{A}_m) < \max_{m \geq \ell \geq 0} \{ \wp(\mathscr{A}_{m-\ell}) - \ell \} \qquad (m > m_0) \tag{5.9.21}$$

and

$$\wp(\mathscr{A}_{m_0}) = \max_{m_0 \geq \ell \geq 0} \{ \wp(\mathscr{A}_{m_0-\ell}) - \ell \} \tag{5.9.22}$$

are satisfied.

First we shall deduce from this assumption the inequalities below:

$$\wp(\mathscr{A}_m) < \wp(\mathscr{A}_{m_0}) - (m - m_0) \tag{5.9.23}$$

which must hold for all m satisfying $m > m_0$. In fact, because of (5.9.21), there exists an m' less than m such that

$$\wp(\mathscr{A}_m) < \wp(\mathscr{A}_{m'}) - (m - m').$$

If $m' \leq m_0$, then (5.9.23) follows from

$$\wp(\mathscr{A}_{m'}) - (m_0 - m') \leq \wp(\mathscr{A}_{m_0})$$

which is a consequence of (5.9.22). If $m' > m_0$, assuming that

$$\wp(\mathscr{A}_{m'}) < \wp(\mathscr{A}_{m_0}) - (m' - m_0),$$

we can derive (5.9.23) by induction. Note that $m' \leq m_0$ if $m \leq m_0 + 1$. Thus (5.9.23) follows from (5.9.21) and (5.9.22).

Now, if we use (5.9.23) to rewrite (5.9.20), we derive

$$m(\rho_0 - 1) \leq \wp(\mathscr{A}_{m_0}) - (m - m_0) + (n-1)(q_0 + 1 - \rho_0) \qquad (m = 1, 2, \cdots)$$

or

$$m \leq \frac{1}{\rho_0}\{\, \wp(\mathscr{A}_{m_0}) + m_0 + (n-1)(q_0 + 1 - \rho_0)\,\} \qquad (m = 1, 2, \cdots)$$

which is a contradiction. Consequently, there must be an increasing sequence $\{\, m_j \,\}$ of positive integers such that

$$\begin{cases} \wp(\mathscr{A}_{m_j}) = \max_{m_j \geq \ell \geq 0} \{\wp(\mathscr{A}_{m_j - \ell}) - \ell\,\} & (j = 1, 2, \cdots), \\ \wp(\mathscr{A}_m) < \max_{m \geq \ell \geq 0} \{\wp(\mathscr{A}_{m-\ell}) - \ell\,\} & (m \neq m_j), \\ \lim_{j \to \infty} m_j = \infty. \end{cases} \tag{5.9.24}$$

Assuming (5.9.24) we shall estimate the differences $m_{j+1} - m_j$. To do this, we prove for m satisfying the inequality

$$m_j < m < m_{j+1} \tag{5.9.25}$$

that we have

$$\max_{m \geq \ell \geq 0} \{\, \wp(\mathscr{A}_{m-\ell}) - \ell \,\} = \wp(\mathscr{A}_{m_j}) - (m - m_j). \tag{5.9.26}$$

In fact, if there exists an m that satisfies (5.9.25) but does not satisfy (5.9.26), let m be the smallest of such m. Then we must have the inequalities below:

$$\begin{cases} \wp(\mathscr{A}_{m_j}) - (m - m_j) \geq \wp(\mathscr{A}_{m-\ell}) - \ell & (\ell = 1, 2, \cdots, m), \\ \wp(\mathscr{A}_{m_j}) - (m - m_j) < \wp(\mathscr{A}_m). \end{cases}$$

Hence

$$\wp(\mathscr{A}_m) = \max_{m \geq \ell \geq 0} \{\, \wp(\mathscr{A}_{m-\ell}) - \ell \,\}$$

which contradicts the second inequality of (5.9.24). Thus we see that (5.9.26) follows from (5.9.25).

Now, if we apply (5.9.20) to the left-hand side of (5.9.26) and apply (5.9.19) to the right-hand side of (5.9.26), the inequality

$$m(\rho_0-1)-(n-1)(q_0+1-\rho_0) \leq (n-1)(q_0+1-\rho_0)+m_j(\rho_0-1)-(m-m_j)$$

is derived. Hence

$$m - m_j \leq \frac{2}{\rho_0}(n-1)(q_0+1-\rho_0).$$

Thus we obtain the estimate

$$m_{j+1} - m_j \leq \frac{2}{\rho_0}(n-1)(q_0+1-\rho_0)+1. \tag{5.9.27}$$

Hereafter we shall explain the last stage of the proof of Theorem 5.9.4. First, because of definition (5.9.9) of the $\mathscr{A}_m$ and definition (5.8.7) of q_0, the inequality

$$\wp(\mathscr{A}_{m'}) \leq \wp(\mathscr{A}_m) + q_0(m'-m) \tag{5.9.28}$$

holds if $m < m'$. Next, it follows from (5.9.20) and (5.9.24) that

$$m_{j+1}(\rho_0-1) \leq \wp(\mathscr{A}_{m_{j+1}}) + (n-1)(q_0+1-\rho_0).$$

For m satisfying inequality (5.9.25) we have

$$\wp(\mathscr{A}_{m_{j+1}}) \leq \wp(\mathscr{A}_m) + |q_0|(m_{j+1}-m_j)$$

because of (5.9.28). Therefore

$$m(\rho_0-1) \leq \wp(\mathscr{A}_m)+|q_0|(m_{j+1}-m_j)+(n-1)(q_0+1-\rho_0)+(m_{j+1}-m_j)|\rho_0-1|$$

for m satisfying (5.9.25). □

§5.10. Necessary and sufficient conditions

In §5.9, we gave a characterization of the case when $z=\infty$ is not a regular singular point (i.e., the case when $\rho_0 > 0$). In this section we consider the case when $z=\infty$ is a regular singular point. We shall use the same notations as in §5.9. Our starting point is the theorem below.

THEOREM 5.10.1. (D. Lutz) *A necessary and sufficient condition that $z=\infty$ be a regular singular point of differential equation* (5.2.1) *is that the inequalities*

$$\wp(\mathscr{A}_m) \leq -m+(n-1)(q_0+1) \qquad (m=1,2,\cdots) \tag{5.10.1}$$

hold.

PROOF. To prove the sufficiency of the condition, it suffices to derive $\rho_0 = 0$ from (5.10.1) (cf. Theorem 5.7.3). If $\rho_0 > 0$, it would follow from Theorem 5.9.4 that

$$\lim_{m\to\infty} \frac{1}{m}\wp(\mathscr{A}_m) = \rho_0 - 1 > -1. \tag{5.10.2}$$

This is impossible because of (5.10.1).

To prove the necessity, it suffices to follow the proof of Theorem 5.9.4 up to inequality (5.9.19), assuming however that $\rho_0 = 0$ and replacing (5.9.15) by the inequalities

$$\wp(\mathscr{B}_m) \le -m, \qquad (m = 0, 1, 2, \cdots). \quad \square$$

Criterion (5.10.1) of Theorem 5.10.1 consists of infinitely many conditions. If we can replace them by a finite number of conditions, we will have a definitely better criterion. Such a result is given in the theorem below. This result is well known as **the criterion of W. B. Jurkat and D. A. Lutz.**

THEOREM 5.10.2. (Jurkat–Lutz)*A necessary and sufficient condition that* $z = \infty$ *be a regular singular point of differential equation* (5.2.1) *is that the inequalities*

$$\wp(\mathscr{A}_m) \le -m + (n-1)(q_0 + 1) \tag{5.10.3}$$

hold for $m = n,\ n+1,\ \cdots,\ (n-1)\{2ns(q_0+1) - 1\} + 1$.

PROOF. It suffices to prove the sufficiency of the condition. In the proof of Theorem 5.10.1, we used (5.10.2). We look at this part of the proof of Theorem 5.10.1 more closely. If $\rho_0 > 0$, inequalities (5.9.20) should hold. Then applying (5.10.3) to the right-hand side of (5.9.20), we derive

$$m(\rho_0 - 1) \le -m + (n-1)(q_0+1) + (n-1)(q_0 + 1 - \rho_0). \tag{5.10.4}$$

Utilizing, at this point, (ii) of Theorem 5.7.3, i.e.,

$$\rho_0 \ge \frac{1}{ns}, \tag{5.7.18}$$

we derive

$$m\left(\frac{1}{ns} - 1\right) \le -m + (n-1)(q_0+1) + (n-1)\left(q_0 + 1 - \frac{1}{ns}\right). \tag{5.10.5}$$

Hence we obtain

$$m \le (n-1)\{2ns(q_0+1) - 1\}$$

which contradicts the condition of the theorem.

We did not assume (5.10.3) for $m = 0, 1, \cdots, (n-1)$, since the inequalities

$$\wp(\mathscr{A}_m) \leq q_0 m \qquad (m = 0, 1, \cdots)$$

hold and hence inequalities (5.10.3) hold for $m = 0, 1, \cdots, (n-1)$ no matter what cases we may consider. □

Thus we have finished an introduction to work due to W. B. Jurkat and D. A. Lutz [cf., also W. B. Jurkat [77]].

CHAPTER VI

The Stokes Phenomena

§6.0. Preliminaries

In this chapter, treating the problem of analytic continuation of solutions from one sectorial domain to another sectorial domain as the Stokes phenomena (cf. Definitions 6.2.1, 6.10.3, 6.2.2, 6.10.5, and 6.10.7), we shall classify differential equations by means of the Stokes phenomena (cf. §§6.3 and 6.11) and construct differential equations which define preassigned Stokes phenomena (cf. §§6.4–6.7, 6.9, and 6.10). We shall construct such differential equations in a way similar to the existence theorem of Grauert (Theorem 2.9.4) by refining the decomposition theorem of Cartan (Theorem 2.2.1). We remark that this problem was also treated with different methods, for example, by G. D. Birkhoff [12, 14] and W. J. Trjitzinsky [33].

§6.1. Systems of fundamental solutions in the neighborhood of $z = \infty$

In this chapter we shall study a differential equation

$$\frac{dW}{dz} = z^{\kappa} A(z) W \tag{6.1.1}$$

on a full neighborhood of $z = \infty$. Here, κ is a nonnegative integer, $A(\frac{1}{\zeta})$ is a matrix in $gl(n, \mathscr{O}(\Delta(0, \frac{1}{R_0})))$ and R_0 is a fixed positive number. Differential equation (6.1.1), in general, admits a singular point at $z = \infty$. (Note that, in Chapter 5, we considered differential equations on sectorial domains.)

Let

$$T(A) = \sum_{m \geq 0} A_m \zeta^m \tag{6.1.2}$$

be the Taylor series of $A(\frac{1}{\zeta})$ at $\zeta = 0$. If the matrix A_0 has n distinct eigenvalues $\lambda_{1,0}, \lambda_{2,0}, \cdots, \lambda_{n,0}$, the matrix $A(\frac{1}{\zeta})$ also has n distinct eigenvalues $\lambda_1(\zeta), \cdots, \lambda_n(\zeta)$ for $\zeta \in \Delta(0, \frac{1}{R_1})$ provided that R_1 is a sufficiently large positive number. Furthermore, the $\lambda_j(\zeta)$ belong to $\mathscr{O}(\Delta(0, \frac{1}{R_1}))$ and

$$\lambda_j(0) = \lambda_{j,0} \qquad (j = 1, \cdots, n). \tag{6.1.3}$$

Utilizing these eigenvalues, we define a diagonal matrix $\Lambda(\zeta)$ by

$$\Lambda(\zeta) = \begin{bmatrix} \lambda_1(\zeta) & 0 & \cdots & 0 \\ 0 & \lambda_2(\zeta) & \cdots & 0 \\ \cdots & \cdots & \cdots & \cdots \\ 0 & 0 & \cdots & \lambda_n(\zeta) \end{bmatrix}. \tag{6.1.4}$$

As in Chapter 5, we denote by $\mathscr{D}(R, a, b)$ the sectorial domain

$$|z| > R, \qquad a < \arg z < b. \tag{6.1.5}$$

If N sectorial domains

$$\mathscr{S}_\ell = \mathscr{D}(R, a_\ell, b_\ell) \qquad (\ell = 1, \cdots, N) \tag{6.1.6}$$

satisfy the condition

$$\bigcup_{\ell=1}^{N} \mathscr{S}_\ell = \{\, z \; ; \; |z| > R \,\}, \tag{6.1.7}$$

we call $\{ \mathscr{S}_1, \cdots, \mathscr{S}_N \}$ **a covering at** $z = \infty$. The result below follows from Lemma 5.5.2.

THEOREM 6.1.1. *There exist a positive number R, a positive integer N, a covering* (6.1.6) *at $z = \infty$, and N matrices $P_1(z), \cdots, P_N(z)$ such that*

(i) *$P_\ell(z)^{-1}$ exists, and $P_\ell(z)$ and $P_\ell(z)^{-1}$ belong to $gl(n, \mathscr{A}(R, a_\ell, b_\ell))$,*

(ii) *if we write the asymptotic expansion of P_ℓ in the form*

$$A_{sp}(P_\ell) = \sum_{m \geq 0} T_m z^{-m}, \tag{6.1.8}$$

the right-hand side of (6.1.8) *is independent of ℓ,*

(iii) *the transformation*

$$W = P_\ell(z) U \tag{6.1.9}$$

changes differential equation (6.1.1) *to the differential equation*

$$\frac{dU}{dz} = z^{\kappa} \Lambda\left(\frac{1}{z}\right) U \tag{6.1.10}$$

on $\mathscr{S}_\ell$.

For a choice of R, N, and covering (6.1.6), see G. D. Birkhoff [12; §6, pp. 463–469], W. Wasow [11; §15.3, pp. 82–87] and Y. Sibuya [9; §56. pp. 225–232]. [See also J. -P. Ramis and Y. Sibuya [129].] It follows from condition (i) of the theorem that the matrix T_0 on the right-hand side of asymptotic expansion (6.1.8) of P_ℓ belongs to $GL(n, C)$, i.e., the determinant of T_0 is not zero.

Let us solve differential equation (6.1.10) on the sectorial domain $\mathscr{S}_\ell$. To do this, writing

$$z^\kappa \lambda_j\left(\frac{1}{z}\right) = \sum_{m=0}^{\kappa+1} \lambda_{j,m} z^{\kappa-m} + z^{-2}\mu_j\left(\frac{1}{z}\right) \qquad (j = 1, \cdots, n),$$

we set

$$\gamma_j(z) = \sum_{m=0}^{\kappa} \frac{\lambda_{j,m}}{\kappa - m + 1} z^{\kappa-m+1} + \lambda_{j,\kappa+1} \log z + \int_\infty^z t^{-2}\mu_j\left(\frac{1}{t}\right) dt$$
$$(j = 1, \cdots, n), \tag{6.1.11}$$

where the $\mu_j(\zeta)$ belong to $\mathscr{O}(\Delta(0, \frac{1}{R_1}))$. Moreover if we set

$$\Gamma(z) = \begin{bmatrix} \gamma_1(z) & 0 & \cdots & 0 \\ 0 & \gamma_2(z) & \cdots & 0 \\ \cdots & \cdots & \cdots & \cdots \\ 0 & 0 & \cdots & \gamma_n(z) \end{bmatrix}, \tag{6.1.12}$$

the matrix

$$U = \exp\{\ \Gamma(z)\ \} \tag{6.1.13}$$

satisfies differential equation (6.1.10).

In this way, we can construct a fundamental solution

$$W = W_\ell(z) = P_\ell(z) \exp\{\ \Gamma(z)\ \} \tag{6.1.14}$$

of differential equation (6.1.1) on the sectorial domain $\mathscr{S}_\ell$.

DEFINITION 6.1.2. We call $\{\ \mathscr{S}_\ell, W_\ell\ ;\ \ell = 1, \cdots, N\ \}$ a system of fundamental solutions of differential equation (6.1.1) in the neighborhood of $z = \infty$.

In the argument above, we assumed that the matrix A_0 has n distinct eigenvalues. However, detailed information is available for more general cases (cf. W. J. Trjitzinsky [33] and K. Takahasi [32]). [The translator omitted some remarks here. We shall come back to this problem for the general case in §6.8.]

§6.2. The Stokes phenomena

In §6.1 a fundamental solution on $\mathcal{S}_\ell$ was given by (6.1.14). If we set

$$\begin{aligned} \nu_j(z) &= \gamma_j(z) - \int_\infty^z t^{-2}\mu_j(\frac{1}{t})dt \\ &= \sum_{m=0}^{\kappa} \frac{\lambda_{j,m}}{\kappa - m + 1} z^{\kappa-m+1} + \lambda_{j,\kappa+1} \log z , \end{aligned} \tag{6.2.1}$$

$$\Gamma_0(z) = \begin{bmatrix} \nu_1(z) & 0 & \cdots & 0 \\ 0 & \nu_2(z) & \cdots & 0 \\ \cdots & \cdots & \cdots & \cdots \\ 0 & 0 & \cdots & \nu_n(z) \end{bmatrix} \tag{6.2.2}$$

and

$$Q_\ell(z) = P_\ell(z)\exp\{ \Gamma(z) - \Gamma_0(z) \} \qquad (\ell = 1, \cdots, N); \tag{6.2.3}$$

fundamental solution (6.1.14) is written in the form

$$W = W_\ell(z) = Q_\ell(z)\exp\{ \Gamma_0(z) \}. \tag{6.2.4}$$

Furthermore

(i′) $Q_\ell(z)$ and $Q_\ell(z)^{-1}$ belong to $gl(n, \mathcal{A}(R, a_\ell, b_\ell))$,
(ii′) the asymptotic expansion of Q_ℓ:

$$A_{sp}(Q_\ell) = \sum_{m\geq 0} \tilde{T}_m z^{-m} \tag{6.2.5}$$

is independent of ℓ; in particular $\tilde{T}_0 = T_0$.

By means of the behavior of Γ_0 we can see fairly well how $W_\ell(z)$ behaves as $z \to \infty$ on $\mathcal{S}_\ell$. Also, if the condition

$$\mathcal{S}_k \cap \mathcal{S}_\ell \neq \emptyset \tag{6.2.6}$$

is satisfied, there exists a matrix $C_{k\ell}$ in $GL(n, C)$ such that

$$W_\ell(z) = W_k(z)C_{k\ell} \qquad (z \in \mathcal{S}_k \cap \mathcal{S}_\ell). \tag{6.2.7}$$

More precisely speaking, we must require that $\mathcal{S}_k \cap \mathcal{S}_\ell$ be connected. Thus we assume that the $\mathcal{S}_\ell$ satisfy the condition

$$b_\ell - a_\ell < \pi \qquad (\ell = 1, \cdots, N) \tag{6.2.8}$$

respectively. By means of relation (6.2.7) we can see fairly well how $W_\ell(z)$ behaves as $z \to \infty$ on $\mathcal{S}_k$. Since $\{\mathcal{S}_1, \cdots, \mathcal{S}_N\}$ is a covering at $z = \infty$, the behavior of solutions of (6.1.1) near $z = \infty$ is understood fairly well in this way.

Let us study some properties of the matrices $C_{k\ell}$. First of all, if

$$\mathscr{S}_h \cap \mathscr{S}_k \cap \mathscr{S}_\ell \neq \varnothing, \tag{6.2.9}$$

the relation

$$C_{hk}C_{k\ell} = C_{h\ell} \tag{6.2.10}$$

holds. Also, if condition (6.2.6) is satisfied, (6.2.7) implies that

$$\exp\{\ \Gamma_0(z)\ \}\ C_{k\ell}\ \exp\{\ -\Gamma_0(z)\ \} = Q_k(z)^{-1}Q_\ell(z). \tag{6.2.11}$$

Therefore, if we denote by $F_{k\ell}(z)$ the left-hand side of (6.2.11), the matrix $F_{k\ell}(z)$ belongs to $gl(n, \mathscr{A}(\mathscr{S}_k \cap \mathscr{S}_\ell))$ and

$$A_{sp}(F_{k\ell}) = I. \tag{6.2.12}$$

Here, I is the $n \times n$ identity matrix and $\mathscr{A}(\mathscr{S}_k \cap \mathscr{S}_\ell)$ is the set of all functions in $\mathscr{O}(\mathscr{S}_k \cap \mathscr{S}_\ell)$ which admit asymptotic expansions as $z \to \infty$ on $\mathscr{S}_k \cap \mathscr{S}_\ell$. We can derive (6.2.12) from the fact that $Q_k(z)$ and $Q_\ell(z)$ admit the same asymptotic expansion on $\mathscr{S}_k \cap \mathscr{S}_\ell$.

DEFINITION 6.2.1. We call $\{\mathscr{S}_\ell, C_{k\ell}, \Gamma_0(z)\}$ a **Stokes phenomenon** at $z = \infty$ if the following conditions are satisfied:

(1) $\{\mathscr{S}_1, \cdots, \mathscr{S}_N\}$ is a covering at $z = \infty$,

(2) the sectorial domains $\mathscr{S}_\ell = \mathscr{D}(R, a_\ell, b_\ell)$ satisfy the conditions $b_\ell - a_\ell < \pi$,

(3) $\Gamma_0(z)$ is an $n \times n$ diagonal matrix whose entries $\nu_1(z), \cdots, \nu_n(z)$ on the main diagonal are given by

$$\nu_j(z) = \sum_{m=0}^{\kappa} \nu_{j,m} z^{\kappa-m+1} + \alpha_j \log z, \tag{6.2.13}$$

where the $\nu_{j,m}$ and the α_j are complex numbers and κ is a non-negative integer,

(4) the $C_{k\ell}$ are matrices in $GL(n, C)$ defined for all (k, ℓ) satisfying the condition $\mathscr{S}_k \cap \mathscr{S}_\ell \neq \varnothing$,

(5) $C_{hk}C_{k\ell} = C_{h\ell}$ if $\mathscr{S}_h \cap \mathscr{S}_k \cap \mathscr{S}_\ell \neq \varnothing$,

(6) if $\mathscr{S}_k \cap \mathscr{S}_\ell \neq \varnothing$, the matrix $F_{k\ell}(z)$ defined by

$$F_{k\ell}(z) = \exp\{\ \Gamma_0(z)\ \}\ C_{k\ell}\ \exp\{\ -\Gamma_0(z)\ \} \tag{6.2.14}$$

belongs to $gl(n, \mathscr{A}(\mathscr{S}_k \cap \mathscr{S}_\ell))$ and $A_{sp}(F_{k\ell}) = I$.

DEFINITION 6.2.2. Given a system $\{\mathscr{S}_\ell, W_\ell; \ell = 1, \cdots, N\}$ of fundamental solutions of differential equation (6.1.1) near $z = \infty$, assume that a Stokes phenomenon $\{\mathscr{S}_\ell, C_{k\ell}, \Gamma_0(z)\}$ at $z = \infty$ is determined as explained above. Then this Stokes phenomenon is called the Stokes phenomenon defined at $z = \infty$ by the system $\{\mathscr{S}_\ell, W_\ell; \ell = 1, \cdots, N\}$ of fundamental solutions.

In Chapter 1, we explained the one-to-one correspondence between structures of analytic continuation and structures of monodromy. If $\{\mathscr{S}_\ell, C_{k\ell}, \Gamma_0(z)\}$ is a Stokes phenomenon at $z = \infty$, then $\{C_{k\ell}\}$ is "a system of connection matrices" which defines a structure of monodromy. However, there is no one-to-one correspondence between these two.

To understand the problem given by G. D. Birkhoff [14], it suffices to formulate the Stokes phenomena as in Definition 6.2.1. However, to consider such general cases as studied by W. J. Trjitzinsky, we must use matrices more general than $\exp\{\ \Gamma_0(z)\ \}$ [cf. §6.8]. Nevertheless, we can still handle quite general cases by means of the ideas which will be explained hereafter. [§§6.8–6.11 were added in this translation to explain the general case.]

To define $\Gamma_0(z)$ precisely, the choice of $\log\ z$ should not be ambiguous. Actually, it suffices to fix a branch of $\log\ z$ on each sectorial domain $\mathscr{S}_\ell$ in any manner. To understand Definition 6.2.1, we must suppose that $\Gamma_0(z)$ has been given already. However, even if we change the choice of $\Gamma_0(z)$, we may still change the choice of $C_{k\ell}$ so that the conditions of Definition 6.2.1 remain valid. To explain those situations in detail, we must consider an equivalence relation among Stokes phenomena. [We shall come back to these problems in §§6.9 and 6.10.]

§6.3. Differential equations defining the same Stokes phenomenon

Let $\{\mathscr{S}_\ell, C_{k\ell}, \Gamma_0(z)\}$ be the Stokes phenomenon defined at $z = \infty$ by a system $\{\mathscr{S}_\ell, W_\ell; \ell = 1, \cdots, N\}$ of fundamental solutions of differential equation (6.1.1) near $z = \infty$. Let $P(\zeta)$ be a matrix in $GL(n, \mathscr{O}(\Delta(0, \frac{1}{R})))$. Assume that the transformation

$$W = P\left(\frac{1}{z}\right) V \tag{6.3.1}$$

changes (6.1.1) to

$$\frac{dV}{dz} = z^\chi B(z) V. \tag{6.3.2}$$

Since $W_\ell(z)$ is a fundamental solution of (6.1.1) on $\mathscr{S}_\ell$, the matrix

$$V_\ell(z) = P\left(\frac{1}{z}\right)^{-1} W_\ell(z) \tag{6.3.3}$$

is a fundamental solution of (6.3.2) on $\mathscr{S}_\ell$. Thus we can construct a system $\{\mathscr{S}_\ell, V_\ell; \ell = 1, \cdots, N\}$ of fundamental solutions of (6.3.2) near $z = \infty$.

Furthermore, this system of fundamental solutions also defines the Stokes phenomenon $\{\mathcal{S}_\ell, C_{k\ell}, \Gamma_0(z)\}$ at $z = \infty$. This means that transformation (6.3.1) does not change the Stokes phenomena. In this section we shall explain that the converse of the result above is also true.

THEOREM 6.3.1. *Assume that a system* $\{\mathcal{S}_\ell, W_\ell; \ell = 1, \cdots, N\}$ *of fundamental solutions of differential equation* (6.1.1) *near* $z = \infty$ *defines a Stokes phenomenon* $\{\mathcal{S}_\ell, C_{k\ell}, \Gamma_0(z)\}$ *at* $z = \infty$. *Assume also that differential equation* (6.3.2) *has a system* $\{\mathcal{S}_\ell, V_\ell; \ell = 1, \cdots, N\}$ *of fundamental solutions near* $z = \infty$ *which also defines the same Stokes phenomenon* $\{\mathcal{S}_\ell, C_{k\ell}, \Gamma_0(z)\}$ *at* $z = \infty$. *Here* χ *is a nonnegative integer and* $B(\frac{1}{\zeta})$ *is a matrix in* $gl(n, \mathcal{O}(\Delta(0, \frac{1}{R})))$. *Let*

$$\mathcal{S}_\ell = \mathcal{D}(R, a_\ell, b_\ell) \qquad (\ell = 1, \cdots, N). \tag{6.3.4}$$

Then, there exists a matrix $P(\zeta)$ *in* $GL(n, \mathcal{O}(\Delta(0, \frac{1}{R})))$ *such that transformation* (6.3.1) *changes* (6.1.1) *to* (6.3.2).

PROOF. If we write

$$W_\ell(z) = Q_\ell(z) \exp\{\, \Gamma_0(z)\, \}$$

and

$$V_\ell(z) = \tilde{Q}_\ell(z) \exp\{\, \Gamma_0(z)\, \},$$

the matrices $Q_\ell(z), Q_\ell(z)^{-1}, \tilde{Q}_\ell(z)$ and $\tilde{Q}_\ell(z)^{-1}$ belong to $gl(n, \mathcal{A}(R, a_\ell, b_\ell))$ because of the assumptions of the theorem. In addition to this, we have

$$W_\ell(z) = W_k(z) C_{k\ell}, \qquad V_\ell(z) = V_k(z) C_{k\ell}$$

for $z \in \mathcal{S}_k \cap \mathcal{S}_\ell$ if $\mathcal{S}_k \cap \mathcal{S}_\ell \neq \emptyset$. Therefore, if we set

$$P\left(\frac{1}{z}\right) = W_\ell(z) V_\ell(z)^{-1} \qquad (z \in \mathcal{S}_\ell), \tag{6.3.5}$$

$P(\frac{1}{z})$ is single valued and holomorphic for $|z| > R$. Furthermore, $P(\zeta)$ is also holomorphic at $\zeta = 0$ since

$$P\left(\frac{1}{z}\right) = Q_\ell(z) \tilde{Q}_\ell(z)^{-1} \qquad (z \in \mathcal{S}_\ell). \quad \square$$

The integer χ (resp. κ) was used on the right-hand side of differential equation (6.3.2) (resp. (6.1.1)). However, in the case above, we may assume that $\chi = \kappa$ since

$$z^\chi B(z) = P\left(\frac{1}{z}\right)^{-1} \left\{ z^\kappa A(z) P\left(\frac{1}{z}\right) - \frac{d}{dz} P\left(\frac{1}{z}\right) \right\}.$$

We can generalize Theorem 6.3.1 also in various manners [cf. §6.11].

§6.4. Differential equations defining a given Stokes phenomenon

Hereafter, we shall explain how to construct a differential equation that defines a preassigned Stokes phenomenon $\{\mathscr{S}_\ell, C_{k\ell}, \Gamma_0(z)\}$ at $z = \infty$. Of three numbers R, a_ℓ, b_ℓ which determine the sectorial domain $\mathscr{S}_\ell$, we will change R if necessary, but a_ℓ and b_ℓ will not be changed. After all, by virtue of Theorem 3.3.1, we do not have to change R except for deriving some estimates as $z \to \infty$. Therefore, whenever it is necessary to choose R sufficiently large, we will assume implicitly that R is large enough without mentioning it explicitly. The details will be left to the reader.

The matrix

$$F_{k\ell}(z) = \exp\{\Gamma_0(z)\}\, C_{k\ell}\, \exp\{-\Gamma_0(z)\} \tag{6.4.1}$$

(given in Definition 6.2.1) is defined for all (k, ℓ) such that $\mathscr{S}_k \cap \mathscr{S}_\ell \neq \varnothing$. On the right-hand side of (6.4.1), we assume without any ambiguity that $\exp\{\Gamma_0(z)\}$ (resp. $\exp\{-\Gamma_0(z)\}$) has been given on the sectorial domain $\mathscr{S}_k$ (resp. $\mathscr{S}_\ell$). Then, because of condition (5) of Definition 6.2.1, the relation

$$F_{hk}(z)F_{k\ell}(z) = F_{h\ell}(z) \qquad (z \in \mathscr{S}_h \cap \mathscr{S}_k \cap \mathscr{S}_\ell) \tag{6.4.2}$$

holds if $\mathscr{S}_h \cap \mathscr{S}_k \cap \mathscr{S}_\ell \neq \varnothing$. Consequently we can apply Theorem 2.0.1 to find matrices $Q_\ell(z) \in GL(n, \mathscr{O}(\mathscr{S}_\ell))$ $(\ell = 1, \cdots, N)$ such that

$$F_{k\ell}(z) = Q_k(z)^{-1}Q_\ell(z) \qquad (z \in \mathscr{S}_k \cap \mathscr{S}_\ell). \tag{6.4.3}$$

Hence if we set

$$W_\ell(z) = Q_\ell(z)\, \exp\{\Gamma_0(z)\} \qquad (\ell = 1, \cdots, N), \tag{6.4.4}$$

a differential equation is determined for $|z| > R$ by means of $\{\mathscr{S}_\ell, W_\ell; \ell = 1, \cdots, N\}$ so that the system $\{\mathscr{S}_\ell, C_{k\ell}\}$ of "connection matrices" is defined by this differential equation. However, if we leave the Q_ℓ as they were defined in this manner, we would never know how they behave as $z \to \infty$. The main concern hereafter is to find a more specific Q_ℓ by utilizing condition (6) of Definition 6.2.1. The final result is summarized in the theorem below.

THEOREM 6.4.1. *Assume that N sectorial domains*

$$\mathscr{S}_\ell = \mathscr{D}(R, a_\ell, b_\ell) \qquad (\ell = 1, \cdots, N) \tag{6.4.5}$$

form a covering at $z = \infty$ and that $n \times n$ matrices $F_{k\ell}(z) \in gl(n, \mathscr{A}(\mathscr{S}_k \cap \mathscr{S}_\ell))$ are defined for all (k, ℓ) satisfying the condition

$$\mathscr{S}_k \cap \mathscr{S}_\ell \neq \varnothing. \tag{6.4.6}$$

Also assume that, if

$$\mathcal{S}_h \cap \mathcal{S}_k \cap \mathcal{S}_\ell \neq \emptyset, \tag{6.4.7}$$

the relation

$$F_{hk}(z)F_{k\ell}(z) = F_{h\ell}(z) \qquad (z \in \mathcal{S}_h \cap \mathcal{S}_k \cap \mathcal{S}_\ell) \tag{6.4.8}$$

holds and that

$$A_{sp}(F_{k\ell}) = I, \tag{6.4.9}$$

where I is the $n \times n$ identity matrix. Then there exist a positive number R' ($\geq R$) and $n \times n$ matrices $Q_\ell(z)$ ($\ell = 1, \cdots, N$) such that

(i) *$Q_\ell(z)$ and $Q_\ell(z)^{-1}$ belong to $gl(n, \mathcal{A}(R', a_\ell, b_\ell))$,*
(ii) $\lim_{z\to\infty} Q_\ell(z) = I$,
(iii) *the relation*

$$F_{k\ell}(z) = Q_k(z)^{-1}Q_\ell(z) \qquad (z \in \mathcal{S}'_k \cap \mathcal{S}'_\ell) \tag{6.4.10}$$

holds whenever condition (6.4.6) is satisfied, where $\mathcal{S}'_\ell = \mathcal{D}(R', a_\ell, b_\ell)$.

If we apply this theorem to a preassigned Stokes phenomenon $\{\mathcal{S}_\ell, C_{k\ell}, \Gamma_0(z)\}$, we can choose matrices satisfying conditions (i) and (ii) of the theorem as the Q_ℓ on the right-hand side of (6.4.3). If we define the W_ℓ of (6.4.4) by means of these Q_ℓ, we have

$$\frac{dW_\ell}{dz} = \frac{dQ_\ell(z)}{dz} \exp\{\, \Gamma_0(z) \,\} + Q_\ell(z) \frac{d\Gamma_0(z)}{dz} \exp\{\, \Gamma_0(z) \,\}$$

and hence

$$\frac{dW_\ell}{dz} W_\ell(z)^{-1} = \left\{ \frac{dQ_\ell(z)}{dz} + Q_\ell(z) \frac{d\Gamma_0(z)}{dz} \right\} Q_\ell(z)^{-1}. \tag{6.4.11}$$

which results in

$$z^{-\kappa} \frac{dW_\ell(z)}{dz} W_\ell(z)^{-1} \in gl(n, \mathcal{A}(\mathcal{S}'_\ell)) \tag{6.4.12}$$

Now, if we set

$$A(z) = z^{-\kappa} \frac{dW_\ell(z)}{dz} W_\ell(z)^{-1} \qquad (z \in \mathcal{S}'_\ell), \tag{6.4.13}$$

then $A(z)$ is single valued and holomorphic for $|z| > R'$. Hence

$$A\left(\frac{1}{\zeta}\right) \in gl(\left(n, \mathcal{O}\left(\Delta\left(0, \frac{1}{R'}\right)\right)\right),$$

and the differential equation

$$\frac{dW}{dz} = z^{\kappa} A(z) W \tag{6)4.14}$$

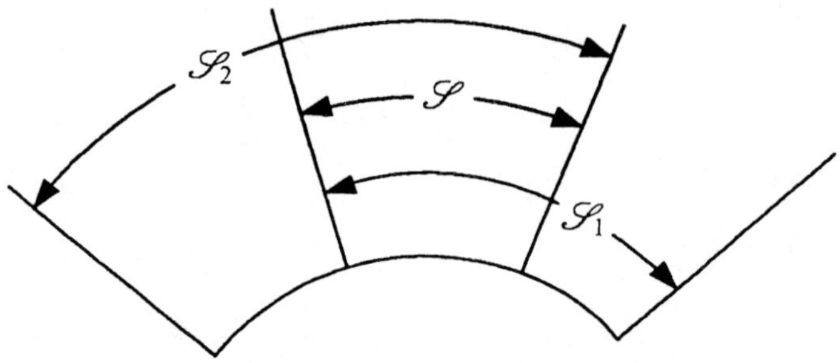

FIGURE 4

defines the Stokes phenomenon $\{\mathcal{S}'_\ell, C_{k\ell}, \Gamma_0(z)\}$ at $z = \infty$.

Hereafter we shall focus our attention on the proof of Theorem 6.4.1.

§6.5. An improvement of the decomposition theorem of Cartan

To prove Theorem 6.4.1, we shall improve Theorem 2.2.1 (i.e., the decomposition theorem of Cartan). The rough idea is similar to that of §2.3, but we must look at the details more carefully. First of all, we shall consider the following case.

If two sectorial domains

$$\mathcal{S}_1 = \mathcal{D}(R, a, b), \qquad \mathcal{S}_2 = \mathcal{D}(R, c, d)$$

satisfy the conditions

$$0 < b - a < \pi, \qquad 0 < d - c < \pi, \qquad a < c < b < d,$$

we have

$$\mathcal{S}_1 \cap \mathcal{S}_2 = \mathcal{D}(R, c, b) \qquad \text{(cf. Figure 4)}.$$

Set $\mathcal{S} = \mathcal{S}_1 \cap \mathcal{S}_2$ and denote the δ-neighborhood of $\mathcal{S}_1$, $\mathcal{S}_2$, and $\mathcal{S}$ by $\mathcal{S}_1(\delta)$, $\mathcal{S}_2(\delta)$, and $\mathcal{S}(\delta)$, respectively. Assume that an $n \times n$ matrix $F(z) \in gl(n, \mathcal{O}(\mathcal{S}(\delta_0)))$ satisfies the inequalities

$$|F(z)| \leq K_N\, |z|^{-N} \qquad (z \in \mathcal{S}(\delta_0)) \tag{6.5.1}$$

for every nonnegative integer N, where δ_0 and K_N are suitable positive numbers. Let us fix positive numbers δ and p so that the inequalities $0 < \delta < \delta_0$ and $0 < p < 1$ are satisfied. The two numbers δ and p are arbitrary otherwise. We also divide the boundary of $\mathcal{S}(\delta)$ into two parts γ_1 and γ_2 as in Figure 5.

Then

$$F(z) = \frac{1}{2\pi i}\int_{\gamma_2} \frac{F(\zeta)}{\zeta - z}\, d\zeta - \frac{1}{2\pi i}\int_{\gamma_1} \frac{F(\zeta)}{\zeta - z}\, d\zeta$$

holds at every point z of $\mathcal{S}(\delta)$.

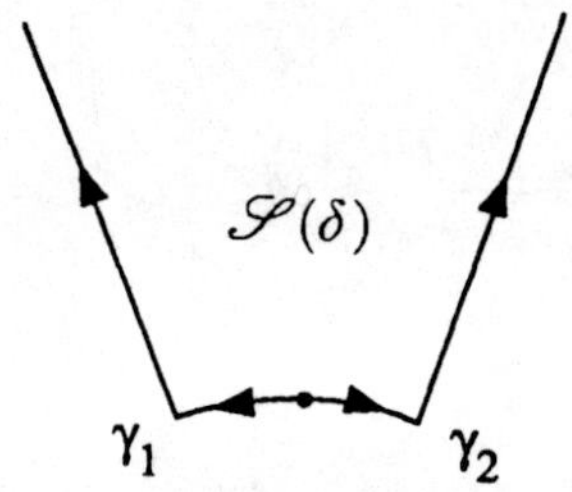

FIGURE 5

Now, if we set

$$\begin{cases} F_1(z) = \dfrac{1}{2\pi i} \displaystyle\int_{\gamma_1} \dfrac{F(\zeta)}{\zeta - z}\, d\zeta, \\ F_2(z) = \dfrac{1}{2\pi i} \displaystyle\int_{\gamma_2} \dfrac{F(\zeta)}{\zeta - z}\, d\zeta, \end{cases} \tag{6.5.2}$$

we have

$$F_1(z) \in gl(n, \mathscr{O}(\mathscr{S}_1(\delta))), \qquad F_2(z) \in gl(n, \mathscr{O}(\mathscr{S}_2(\delta))) \tag{6.5.3}$$

and

$$F(z) = F_2(z) - F_1(z) \qquad (z \in \mathscr{S}(\delta)). \tag{6.5.4}$$

LEMMA 6.5.1. *The two matrices $F_1(z)$ and $F_2(z)$ satisfy the inequalities*

$$|F_1(z)| \leq \frac{L\, K_2}{2\pi(1-p)\delta} \tag{6.5.5}$$

and

$$|F_2(z)| \leq \frac{L\, K_2}{2\pi(1-p)\delta} \tag{6.5.6}$$

respectively on $\mathscr{S}_1(p\delta)$ and $\mathscr{S}_2(p\delta)$, where L is a positive number determined by $\mathscr{S}(\delta_0)$, while K_2 is one of the K_N ($N = 2$) on the right-hand side of condition (6.5.1).

It is not difficult to prove this lemma. The number L is chosen so that the two integrals

$$\int_{\gamma_1} |\zeta|^{-2}\, |d\zeta| \qquad \text{and} \qquad \int_{\gamma_2} |\zeta|^{-2}\, |d\zeta|$$

are bounded above by L.

Let us look at $F_1(z)$ and $F_2(z)$ more closely. To do this, we utilize the identity below:

$$\frac{1}{\zeta - z} = -\sum_{m=0}^{N} \frac{\zeta^m}{z^{m+1}} + \frac{\zeta^{N+1}}{\zeta - z}\,\frac{1}{z^{N+1}}, \tag{6.5.7}$$

where N is an arbitrary nonnegative integer. Because of this identity, if we set

$$T_{j,m} = \frac{-1}{2\pi i} \int_{\gamma_j} \zeta^{m-1}\, F(\zeta)\, d\zeta \qquad (j = 1, 2\, ;\ m \geq 1) \tag{6.5.8}$$

and

$$E_{j,N+1}(z) = \frac{1}{2\pi i}\int_{\gamma_j} \frac{\zeta^{N+1}\,F(\zeta)}{\zeta - z}\,d\zeta \qquad (j = 1,2\,;\ N \geq 0), \tag{6.5.9}$$

then

$$F_j(z) = \sum_{m=1}^{N+1} T_{j,m}\,z^{-m} + E_{j,N+1}(z)\,z^{-(N+1)} \qquad (j = 1,2\ :\ N \geq 0). \tag{6.5.10}$$

LEMMA 6.5.2. *The matrices $T_{j,m}$ and $E_{j,N+1}(z)$ satisfy the inequalities below:*

$$|T_{j,m}| \leq \frac{L\,K_{m+1}}{2\pi} \qquad (j = 1,2\,;\ m \geq 1\ ), \tag{6.5.11}$$

and

$$|E_{j,N+1}(z)| \leq \frac{L\,K_{N+3}}{2\pi(1-p)\delta} \qquad (z \in \mathcal{S}_j(p\delta)\,;\ j = 1,2\,;\ N \geq 0). \tag{6.5.12}$$

Here, the K_h are the positive numbers on the right-hand side of condition (6.5.1) *and L is the same positive number as in Lemma* 6.5.1.

It is also not difficult to prove this lemma.

Next, we define a matrix $G(z)$ by the equation

$$(I + F_1(z))(I + F(z)) = (I + G(z))(I + F_2(z)). \tag{6.5.13}$$

Utilizing (6.5.4), we derive

$$G(z) = F_1(z)F(z)(I + F_2(z))^{-1}. \tag{6.5.14}$$

LEMMA 6.5.3. *The matrix $G(z)$ satisfies the inequality*

$$|G(z)| \leq \left(\frac{L\,K_2}{2\pi(1-p)\delta}\right)\left\{\frac{1}{1 - \dfrac{L\,K_2}{2\pi(1-p)\delta}}\right\} K_N\,|z|^{-N} \qquad (N = 1,2,\cdots) \tag{6.5.15}$$

on $\mathcal{S}(p\delta)$, where we assume that the inequality

$$\frac{L\,K_2}{2\pi(1-p)\delta} < 1 \tag{6.5.16}$$

is satisfied.

This lemma follows from (6.5.1), (6.5.5), and (6.5.6). Note that R, a, b, c, and d determine the sectorial domains $\mathcal{S}_1$ and $\mathcal{S}_2$ and that if R is sufficiently large, the positive number L becomes very small. Hence assumption (6.5.16) does not cause any damage.

LEMMA 6.5.4. *Assume that*

$$p = \frac{1}{2}, \qquad \delta = \left(\frac{1}{2}\right)^k \delta_0 \tag{6.5.17}$$

for some nonnegative number k,

$$0 < K_m \leq \left(\frac{1}{2}\right)^{2k} C_m \qquad (m = 1, 2, \cdots), \tag{6.5.18}$$

$$\frac{L}{2\pi} \leq 1, \qquad \frac{L}{\pi\delta_0} \leq 1, \tag{6.5.19}$$

and

$$\frac{L\, C_2}{\pi\delta_0} \leq \left(\frac{1}{2}\right)^3. \tag{6.5.20}$$

Then the following inequalities hold:

$$\frac{L\, K_2}{2\pi(1-p)\delta} \leq \left(\frac{1}{2}\right)^{k+3} \leq \left(\frac{1}{2}\right)^k, \tag{6.5.21}$$

$$\frac{L\, K_{m+1}}{2\pi} \leq \left(\frac{1}{2}\right)^{2k} C_{m+1} \leq \left(\frac{1}{2}\right)^k C_{m+1}, \tag{6.5.22}$$

$$\frac{L\, K_{N+3}}{2\pi(1-p)\delta} \leq \left(\frac{1}{2}\right)^k C_{N+3}, \tag{6.5.23}$$

and

$$\left(\frac{L\, K_2}{2\pi(1-p)\delta}\right)\left\{\frac{1}{1 - \dfrac{L\, K_2}{2\pi(1-p)\delta}}\right\} K_m \leq \left(\frac{1}{2}\right)^{3k+2} C_m \leq \left(\frac{1}{2}\right)^{2(k+1)} C_m \qquad (m = 1, 2, \cdots). \tag{6.5.24}$$

It is also not difficult to prove this lemma. Using what we have explained above as preparation, we shall prove the lemma below.

LEMMA 6.5.5. *Assume that two sectorial domains $\mathcal{S}_1 = \mathcal{D}(R, a, b)$ and $\mathcal{S}_2 = \mathcal{D}(R, c, d)$ satisfy the conditions*

$$0 < b - a < \pi, \qquad 0 < d - c < \pi, \qquad a < c < b < d. \tag{6.5.25}$$

(Then we have $\mathcal{S}_1 \cap \mathcal{S}_2 = \mathcal{D}(R, c, b)$. We denote $\mathcal{S}_1 \cap \mathcal{S}_2$ by $\mathcal{S}$, and the δ-neighborhoods of $\mathcal{S}_1$, $\mathcal{S}_2$, and $\mathcal{S}$ are denoted by $\mathcal{S}_1(\delta)$, $\mathcal{S}_2(\delta)$, and $\mathcal{S}(\delta)$, respectively.) We assume also that an $n \times n$ matrix $F(z) \in gl(n, \mathcal{O}(\mathcal{S}(\delta_0)))$ satisfies the inequalities

$$|F(z)| \leq K_N\, |z|^{-N} \qquad (z \in \mathcal{S}(\delta_0)) \tag{6.5.1}$$

for all nonnegative integers N. Then there exist a positive number R' not less than R and two $n \times n$ matrices $H_1(z)$ and $H_2(z)$ such that

(i) $H_1(z) \in gl(n, \mathscr{A}(R', a, b))$, $\quad H_2(z) \in gl(n, \mathscr{A}(R', c, d))$,
(ii) $\lim_{z \to \infty} H_j(z) = 0$,
(iii) *we have*

$$I + F(z) = (I + H_1(z))^{-1}(I + H_2(z)) \tag{6.5.26}$$

at every point z of the sectorial domain $\mathscr{D}(R', c, b)$.

PROOF. We define sectorial domains $\mathscr{S}_1'$, $\mathscr{S}_2'$, and $\mathscr{S}'$, respectively by

$$\mathscr{S}_1' = \mathscr{D}(R', a, b), \qquad \mathscr{S}_2' = \mathscr{D}(R', c, d), \qquad \mathscr{S}' = \mathscr{D}(R', c, b).$$

Then $\mathscr{S}' = \mathscr{S}_1' \cap \mathscr{S}_2'$. Let k be a nonnegative integer. We divide the boundary of $\mathscr{S}'((\frac{1}{2})^k \delta_0)$ into two parts $\gamma_{1,k}$ and $\gamma_{2,k}$ as we did above. If R' is sufficiently large, we can find a positive number L such that

(1) the inequalities

$$\int_{\gamma_{j,k}} |\zeta|^{-2}\, |d\zeta| \leq L \qquad (j = 1,\ 2)$$

are valid for every k,

(ii) the inequalities

$$\frac{L}{2\pi} \leq 1, \qquad \frac{L}{\pi\delta_0} \leq \frac{1}{2} \tag{6.5.19$'$}$$

and

$$\frac{LK_2}{\pi\delta_0} \leq \left(\frac{1}{2}\right)^4 \tag{6.5.20$'$}$$

hold.

We shall define matrices $F_{1,k}(z)$, $F_{2,k}(z)$, and $G_k(z)$ successively in $\mathscr{S}_1'((\frac{1}{2})^k \delta_0)$, $\mathscr{S}_2'((\frac{1}{2})^k \delta_0)$, and $\mathscr{S}'((\frac{1}{2})^k \delta_0)$, respectively as follows. First, we set

$$G_1(z) = F(z) \qquad \left(z \in \mathscr{S}'(\frac{1}{2}\delta_0)\right)$$

and

$$\begin{cases} F_{1,2}(z) = \dfrac{1}{2\pi i} \displaystyle\int_{\gamma_{1,1}} \dfrac{F(\zeta)}{\zeta - z}\, d\zeta & \left(z \in \mathscr{S}_1'\left(\left(\dfrac{1}{2}\right)^2 \delta_0\right)\right), \\ F_{2,2}(z) = \dfrac{1}{2\pi i} \displaystyle\int_{\gamma_{2,1}} \dfrac{F(\zeta)}{\zeta - z}\, d\zeta & \left(z \in \mathscr{S}_2'\left(\left(\dfrac{1}{2}\right)^2 \delta_0\right)\right). \end{cases}$$

Then

$$G_1(z) = F_{2,2}(z) - F_{1,2}(z) \qquad \left(z \in \mathcal{S}'(\left(\frac{1}{2}\right)^2 \delta_0)\right),$$

and

$$|F_{j,2}(z)| \leq \left(\frac{1}{2}\right)^3 \qquad \left(z \in \mathcal{S}_j'(\left(\frac{1}{2}\right)^2 \delta_0)\right)$$

are derived from Lemmas 6.5.1–6.5.4.

Next if we set

$$G_2(z) = F_{1,2}(z) G_1(z) (I + F_{2,2}(z))^{-1} \qquad \left(z \in \mathcal{S}'(\left(\frac{1}{2}\right)^2 \delta_0)\right),$$

we have

$$|G_2(z)| \leq \left(\frac{1}{2}\right)^2 K_N\, |z|^{-N} \qquad \left(z \in \mathcal{S}'(\left(\frac{1}{2}\right)^2 \delta_0)\right), \qquad (N = 1, 2, \cdots).$$

Furthermore, if we set

$$T_{j,2,m} = \frac{-1}{2\pi i} \int_{\gamma_{j,1}} \zeta^{m-1}\, F(\zeta)\, d\zeta,$$

and

$$E_{j,2,N+1} = \frac{1}{2\pi i} \int_{\gamma_{j,1}} \frac{\zeta^{N+1}\, F(\zeta)}{\zeta - z}\, d\zeta \qquad \left(z \in \mathcal{S}_j'(\left(\frac{1}{2}\right)^2 \delta_0)\right),$$

we have

$$|T_{j,2,m}| \leq K_{m+1},$$

$$|E_{j,2,N+1}(z)| \leq K_{N+3} \qquad \left(z \in \mathcal{S}_j'(\left(\frac{1}{2}\right)^2 \delta_0)\right)$$

and

$$F_{j,2}(z) = \sum_{m=1}^{N+1} T_{j,2,m}\, z^{-m} + E_{j,2,N+1}(z)\, z^{-(N+1)} \qquad (N = 1, 2, \cdots).$$

In general, if $G_k(z)$ is defined on $\mathcal{S}'((\frac{1}{2})^k \delta_0)$ so that

$$|G_k(z)| \leq \left(\frac{1}{2}\right)^{2(k-1)} K_N\, |z|^{-N} \qquad \left(z \in \mathcal{S}'(\left(\frac{1}{2}\right)^k \delta_0)\right)$$

holds, the matrix

$$F_{j,k+1}(z) = \frac{1}{2\pi i} \int_{\gamma_{j,k}} \frac{G_k(\zeta)}{\zeta - z} d\zeta \qquad \left(z \in \mathcal{S}_j'(\left(\frac{1}{2}\right)^{k+1} \delta_0)\right)$$

satisfies the inequality

$$|F_{j,k+1}(z)| \leq \left(\frac{1}{2}\right)^{k+2} \qquad \left(z \in \mathcal{S}_j'(\left(\frac{1}{2}\right)^{k+1} \delta_0)\right).$$

Next, if we set

$$G_{k+1}(z) = F_{1,k+1}(z) G_k(z)(I + F_{2,k+1}(z))^{-1} \qquad \left(z \in \mathcal{S}'(\left(\frac{1}{2}\right)^{k+1} \delta_0)\right), \tag{6.5.27}$$

we obtain

$$|G_{k+1}(z)| \leq \left(\frac{1}{2}\right)^{2k} K_N \, |z|^{-N} \qquad \left(z \in \mathcal{S}'(\left(\frac{1}{2}\right)^{k+1} \delta_0)\right).$$

Also, if we set

$$T_{j,k+1,m} = \frac{-1}{2\pi i} \int_{\gamma_{j,k}} \zeta^{m+1} \, G_k(\zeta) \, d\zeta$$

and

$$E_{j,k+1,N+1}(z) = \frac{1}{2\pi i} \int_{\gamma_{j,k}} \frac{\zeta^{N+1} G_k(\zeta)}{\zeta - z} \, d\zeta,$$

we obtain

$$|T_{j,k+1,m}| \leq \left(\frac{1}{2}\right)^{2(k-1)} K_{m+1},$$

$$|E_{j,k+1,N+1}(z)| \leq \left(\frac{1}{2}\right)^{k-1} K_{N+3}$$

and

$$F_{j,k+1}(z) = \sum_{m=1}^{N+1} T_{j,k+1,m} \, z^{-m} \; + \; E_{j,k+1,N+1}(z) \, z^{-(N+1)}.$$

It follows from (6.5.27) and the relation $G_k(z) = F_{2,k+1}(z) - F_{1,k+1}(z)$ that

$$(I + F_{1,k+1}(z))(I + G_k(z)) = (I + G_{k+1}(z))(I + F_{2,k+1}(z)). \tag{6.5.28}$$

Hence

$$\begin{aligned}(I + F_{1,k+1}(z))(I + F_{1,k}(z)) \cdots (I + F_{1,2}(z))(I + F(z)) \\ = (I + G_{k+1}(z))(I + F_{2,k+1}(z))(I + F_{2,k}(z)) \cdots (I + F_{2,2}(z)).\end{aligned}$$

Therefore if we set

$$I + H_j(z) = \lim_{k \to \infty} (I + F_{j,k+1}(z)) \cdots (I + F_{j,2}(z)) \qquad (z \in \mathcal{S}_j'),$$

then (6.5.26) holds on $\mathcal{S}'$.

To prove that $H_j(z) \in gl(n, \mathscr{A}(\mathscr{S}_j'))$, we proceed as follows. First, if we set

$$\tilde{F}_{j,k}(z) = 2^{k-2} F_{j,k}(z) \qquad (k = 2, 3, \cdots),$$

the asymptotic expansions of the $\tilde{F}_{j,k}(z)$ are majorized by $\sum_{m \geq 1} K_{m+1} z^{-m}$. Furthermore, we have

$$|\tilde{F}_{j,k}(z) - \sum_{m=1}^{N+1} 2^{k-2} T_{j,k,m} z^{-m}| \leq K_{N+3} |z|^{-(N+1)}.$$

Now, letting u be a variable, we define a function

$$\lambda(u) = \prod_{k=2}^{\infty} \left(1 + \left(\frac{1}{2}\right)^{k-2} u\right).$$

Then $\lambda(u)$ is holomorphic for $|u| < \infty$. If we set

$$u = \sum_{m=1}^{N+1} K_{m+1} z^{-m} + K_{N+3} z^{-(N+1)},$$

$\lambda(u)$ becomes a convergent power series in z^{-1}. Utilizing this series as a majorant, we can prove that the $(I + H_j(z))$ admit asymptotic expansions. □

§6.6. An improvement of the decomposition theorem of Cartan (continued)

In Lemma 6.5.5 we assumed asymptotic property (6.5.1) of $F(z)$ on $\mathscr{S}(\delta_0)$. Actually, it is sufficient to assume that

$$\begin{cases} F(z) \in gl(n, \mathscr{A}(\mathscr{S})), \\ A_{sp}(F) = 0, \end{cases}$$

In such a case, if we set

$$\begin{cases} \tilde{\mathscr{S}_1} = \mathscr{D}(R + \varepsilon, a, b - \varepsilon), \\ \tilde{\mathscr{S}_2} = \mathscr{D}(R + \varepsilon, c + \varepsilon, d), \\ \tilde{\mathscr{S}} = \mathscr{D}(R + \varepsilon, c + \varepsilon, b - \varepsilon) = \tilde{\mathscr{S}_1} \cap \tilde{\mathscr{S}_2} \end{cases}$$

for a sufficiently small positive number ε, Lemma 6.5.5 applies to $\tilde{\mathscr{S}_1}$, $\tilde{\mathscr{S}_2}$, $\tilde{\mathscr{S}}$ and $F(z)$. This means that we can choose R', $H_1(z)$, and $H_2(z)$ so that

$$\begin{cases} H_1(z) \in gl(n, \mathscr{A}(R', a, b - \varepsilon)), \\ H_2(z) \in gl(n, \mathscr{A}(R', c + \varepsilon, d)), \\ \lim_{z \to \infty} H_j(z) = 0 \end{cases}$$

and that

$$I + F(z) = (I + H_1(z))^{-1}(I + H_2(z)) \tag{6.6.1}$$

at every point $z \in \mathscr{D}(R', c+\varepsilon, b-\varepsilon)$. Now, if we rewrite (6.6.1) as

$$(I + H_1(z))(I + F(z)) = I + H_2(z),$$

the left-hand side belongs to $gl(n, \mathscr{A}(R', c, b-\varepsilon))$. Hence $H_2(z)$ belongs to $gl(n, \mathscr{A}(R', c, d))$. Similarly, $H_1(z)$ belongs to $gl(n, \mathscr{A}(R', a, b))$. Thus we obtain the result below.

THEOREM 6.6.1. *Assume that two sectorial domains* $\mathscr{S}_1 = \mathscr{D}(R, a, b)$ *and* $\mathscr{S}_2 = \mathscr{D}(R, c, d)$ *satisfy the conditions*

$$0 < b - a < \pi, \qquad 0 < d - c < \pi, \qquad a < c < b < d. \tag{6.6.2}$$

We assume also that an $n \times n$ *matrix* $F(z) \in gl(n, \mathscr{A}(\mathscr{S}_1 \cap \mathscr{S}_2))$ *satisfies the condition*

$$A_{sp}(F) = I. \tag{6.6.3}$$

Then, there exist a positive number R' $(\geq R)$ *and two* $n \times n$ *matrices* $Q_1(z)$ *and* $Q_2(z)$ *such that*

(i) $Q_1(z)$ *and* $Q_1(z)^{-1}$ *belong to* $gl(n, \mathscr{A}(R', a, b))$,
(ii) $Q_2(z)$ *and* $Q_2(z)^{-1}$ *belong to* $gl(n, \mathscr{A}(R', c, d))$,
(iii) $\lim_{z\to\infty} Q_j(z) = I$,
(iv) *the relation*

$$F(z) = Q_1(z)^{-1} Q_2(z) \tag{6.6.4}$$

holds at every point z *of* $\mathscr{D}(R', a, b) \cap \mathscr{D}(R', c, d)$.

In Theorem 6.6.1 we assumed condition (6.6.2). However even in the case when $\mathscr{S}_1 \cap \mathscr{S}_2$ splits into two connected components as shown by Figure 6, we can prove a similar result, utilizing a method which is not much different from the method given above. We shall also use such a result in succeeding sections, although we do not state the result as a theorem. In particular, we can do without the assumption that $b - a < \pi$ and $d - c < \pi$. The details will be left to the reader.

§6.7. Proof of Theorem 6.4.1

To prove Theorem 6.4.1, we can assume without any loss of generality that

$$\begin{aligned}
&\mathscr{S}_h \cap \mathscr{S}_{h+1} \neq \varnothing && (h = 1, \cdots, N-1),\\
&\mathscr{S}_N \cap \mathscr{S}_1 \neq \varnothing, &&\\
&\mathscr{S}_h \cap \mathscr{S}_k = \varnothing && (k \neq h-1,\ h,\ h+1),\\
&\mathscr{S}_1 \cap \mathscr{S}_k = \varnothing && (k \neq 1,\ 2,\ N),\\
&\mathscr{S}_N \cap \mathscr{S}_k = \varnothing && (k \neq N-1,\ N, 1).
\end{aligned}$$

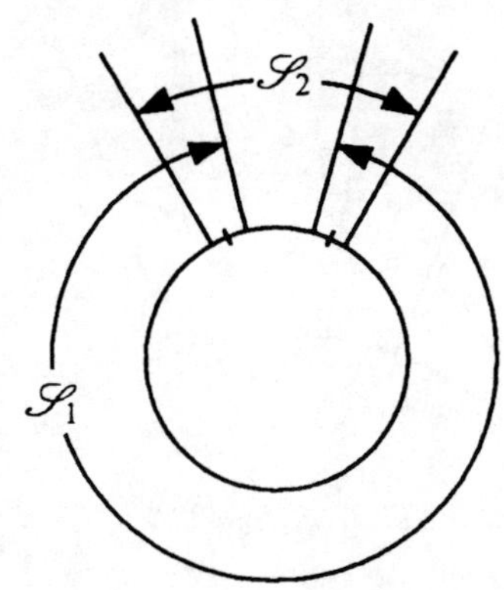

FIGURE 6

(If necessary, we can consider a much finer covering as in the proof of Theorem 2.9.4.)

First we consider the case shown by Figure 7 on page 162. Assume that two matrices Q_1 and Q_2 are well defined on $\mathscr{S}_1$and $\mathscr{S}_2$ respectively and that the relation

$$Q_1^{-1}Q_2 = F_{12}$$

holds on $\mathscr{S}_1 \cap \mathscr{S}_2$. Since $A_{sp}(F_{12}) = I$, we must have $A_{sp}(Q_1) = A_{sp}(Q_2)$. Let Q be a matrix such that $A_{sp}(Q) = A_{sp}(Q_1)$ on $\mathscr{S}_1 \cup \mathscr{S}_2$ and set

$$\tilde{Q}_j = Q^{-1}Q_j.$$

Then $\tilde{Q}_1^{-1}\tilde{Q}_2 = F_{12}$ and $A_{sp}(\tilde{Q}_1) = A_{sp}(\tilde{Q}_2) = I$. Thus we assume now that $A_{sp}(Q_1) = A_{sp}(Q_2) = I$.

Next, set

$$\tilde{F}_{23} = Q_2F_{23}.$$

Since $(\mathscr{S}_1 \cup \mathscr{S}_2) \cap \mathscr{S}_3 = \mathscr{S}_2 \cap \mathscr{S}_3$ and $A_{sp}(\tilde{F}_{23}) = I$, we can choose Q and Q_3 on $\mathscr{S}_1 \cup \mathscr{S}_2$ and $\mathscr{S}_3$ respectively so that

$$Q^{-1}Q_3 = Q_2F_{23}.$$

Then, set $\tilde{Q}_j = QQ_j$ $(j = 1, 2)$ to derive

$$\tilde{Q}_1^{-1}\tilde{Q}_2 = F_{12}, \qquad \tilde{Q}_2^{-1}Q_3 = F_{23}.$$

Proceeding in this way successively, we can complete the proof of Theorem 6.4.1.

In Theorem 6.4.1, we assumed that $A_{sp}(F_{kh}) = I$. However, if a problem is considered only on a sectorial domain, we do not need this assumption. Cases similar to Theorem 6.6.1 are such examples. We need this condition only when we consider the problem on a full neighborhood of $z = \infty$. As a

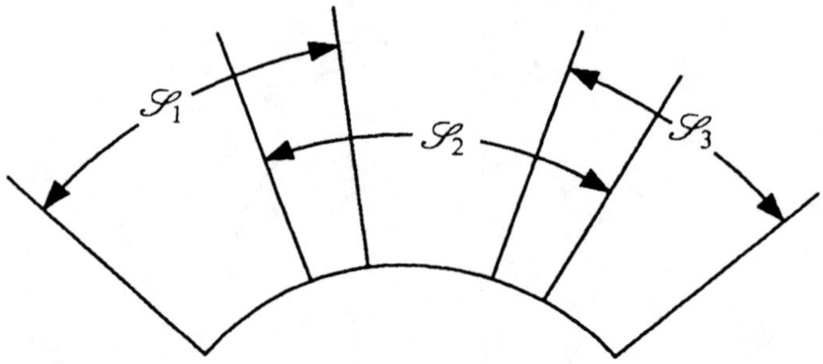

FIGURE 7

matter of fact, generally speaking, it is sufficient to assume that the problem is formally solvable [cf. Theorem 6.9.8].

We explained how to find a differential equation that defines a preassigned Stokes phenomenon. In the case when the C_{kj} and $\Gamma_0(z)$ depend on some auxiliary parameters, we must study behaviors of the matrix $A(z)$ of the corresponding differential equation with respect to these parameters. Such problems have not been solved completely [cf. Appendix 5; see also Y. Sibuya [9, 30, 31] and C. -H. Lin–Y. Sibuya [88, 89]]. A result similar to Theorem 6.4.1 was proved by G. D. Birkhoff [14]. His method is different from the method in this chapter. We note also that paper [15] of D. Drasin is closely related to the materials in this chapter. [The translator omitted some remarks here. They will be restored later in Appendix 5, §A.5.2, (5) of Remark A.5.2.14 .]

§6.8. Formal and asymptotic structures

In §6.1 we defined a system $\{\mathscr{S}_\ell, W_\ell; \ell = 1, \cdots, N\}$ of fundamental solutions of differential equation (6.1.1) in the neighborhood of $z = \infty$ (cf. Definition 6.1.2). In this definition we assumed that the matrix A_0 had n distinct eigenvalues. In this section we shall explain how to remove this restriction. The main problem of this section is to generalize Theorem 6.1.1. We shall accomplish this in two steps: *formal* and *asymptotic*.

Let us consider a differential equation

$$z\frac{dW}{dz} + \hat{A}(z)W = 0, \tag{6.8.1}$$

where W is an $n \times n$ unknown matrix and $\hat{A}(z) \in gl(n, \mathscr{K})$. Here $\mathscr{K}$ is the quotient field of $C[[z^{-1}]]$, i.e., $\mathscr{K} = C[[z^{-1}]][z]$ (cf. §5.7). The theorem below is a fundamental result due to M. Hukuhara [67] and H. L. Turrittin [147].

THEOREM 6.8.1. *There exist a positive integer* s *and an* $n \times n$ *matrix* $\hat{P}(\zeta)$ *such that*

(i) $\hat{P}(\zeta) \in gl(n, C[[\zeta^{-1}]])$,
(ii) $\det \hat{P}(\zeta) \neq 0$ *as an element of* $C[[\zeta^{-1}]]$,
(iii) *the transformation*

$$z = \zeta^s, \qquad W = \hat{P}(\zeta)U \tag{6.8.2}$$

changes differential equation (6.8.1) *to*

$$\zeta\frac{dU}{d\zeta} + B(\zeta)U = 0, \tag{6.8.3}$$

where $B(\zeta)$ *is an* $n \times n$ *matrix of the following form:*

$$B(\zeta) = \begin{bmatrix} B_1(\zeta) & 0 & \cdots & 0 \\ 0 & B_2(\zeta) & \cdots & 0 \\ \cdots & \cdots & \cdots & \cdots \\ 0 & 0 & \cdots & B_p(\zeta) \end{bmatrix},$$

$$B_j(\zeta) = \begin{bmatrix} B_{j1}(\zeta) & 0 & \cdots & 0 \\ 0 & B_{j2}(\zeta) & \cdots & 0 \\ \cdots & \cdots & \cdots & \cdots \\ 0 & 0 & \cdots & B_{jm_j}(\zeta) \end{bmatrix},$$

$$B_{jk}(\zeta) = \lambda_j(\zeta)I_{n_{jk}} + J_{n_{jk}}.$$

Here $I_{n_{jk}}$ *is the* $n_{jk} \times n_{jk}$ *identity matrix,* $J_{n_{jk}}$ *is an* $n_{jk} \times n_{jk}$ *nilpotent matrix of the form*

$$J_{n_{jk}} = \begin{bmatrix} 0 & 1 & 0 & \cdots & 0 \\ 0 & 0 & 1 & \cdots & 0 \\ \cdots & \cdots & \cdots & \cdots & \cdots \\ 0 & 0 & 0 & \cdots & 1 \\ 0 & 0 & 0 & \cdots & 0 \end{bmatrix},$$

$$\lambda_j(\zeta) = \sum_{\ell=0}^{d_j} \lambda_{j\ell}\,\zeta^\ell, \qquad \lambda_{j\ell} \in C$$

and

$$\lambda_j(\zeta) - \lambda_i(\zeta) \notin Z \qquad \text{if } i \neq j.$$

SKETCH OF PROOF. If $\hat{A}(z) = 0$ identically, choose $s = 1$ and $P = I$ (the $n \times n$ identity matrix). Then $B = 0$.

Assume that $\hat{A}$ is not identically equal to zero. Since $\hat{A} \in gl(n, \mathcal{K})$, we can write $\hat{A}$ in the form

$$\hat{A} = z^k \sum_{m \geq 0} A_m\, z^{-m},$$

where $k \in Z$, $A_m \in gl(n, C)$, and $A_0 \neq 0$.

Case 1. When $k > 0$ and A_0 has some distinct eigenvalues:
In this case block diagonalize (6.8.1) by using the method of §5.5. By doing this, we can apply induction on the size of the matrix $\hat{A}$.

Case 2. When $k > 0$ and $A_0 = \lambda I + J$ where $\lambda \in C$ and J is nilpotent:
In this case ignore λI and look at the differential equation

$$z\frac{dW}{dz} + (\hat{A}(z) - \lambda z^k I)W = 0, \tag{6.8.4}$$

where

$$\hat{A}(z) - \lambda z^k I = z^k \left(J + z^{-1} \sum_{m \geq 1} A_m z^{-m+1} \right). \tag{6.8.5}$$

Utilizing a cyclic vector (cf. §5.7), we can change (6.8.5) to "a single nth order equation". Then, apply the method of §5.6 to change the nilpotent matrix J to a nonnilpotent matrix. Through this process the integer s is naturally introduced [cf. the Newton polygon in Appendix 1, §A.1.4]. Furthermore either $k > 0$ becomes smaller or we go back to Case 1.

Case 3. When $k \leq 0$:
In this case apply the method of Hukuhara (cf. §3.9).

The main idea of the proof is that we arrive at Case 3 after repeating the processes of Case 1 and Case 2 *for a finite number of times.*

REMARK 6.8.2. The transformation

$$W = \hat{P}(\zeta)U \tag{6.8.6}$$

is a product of a finite number of transformations

$$W_j = \hat{P}_j(\zeta)U_j \qquad (j = 1, \cdots, N) \tag{6.8.7}$$

of Cases 1, 2, and 3, i.e.,

$$\hat{P}(\zeta) = \hat{P}_1(\zeta)\hat{P}_2(\zeta)\cdots\hat{P}_N(\zeta). \tag{6.8.8}$$

Even though the entries of $\hat{A}(z)$ are formal power series, the entries of $B(\zeta)$ are polynomials in ζ. If we set

$$\Phi_{jk}(\zeta) = \zeta^{-\lambda_{j0}} \exp\left\{ -\sum_{\ell=1}^{d_j} \frac{\lambda_{j\ell}}{\ell}\zeta^\ell \right\} \exp\{ -(\log\ \zeta)\ J_{n_{jk}} \} \tag{6.8.9}$$

and

$$\Phi_j(\zeta) = \begin{bmatrix} \Phi_{j1}(\zeta) & 0 & \cdots & 0 \\ 0 & \Phi_{j2}(\zeta) & \cdots & 0 \\ \cdots & \cdots & \cdots & \cdots \\ 0 & 0 & \cdots & \Phi_{jm_j}(\zeta) \end{bmatrix}, \tag{6.8.10}$$

then

$$\Phi(\zeta) = \begin{bmatrix} \Phi_1(\zeta) & 0 & \cdots & 0 \\ 0 & \Phi_2(\zeta) & \cdots & 0 \\ \cdots & \cdots & \cdots & \cdots \\ 0 & 0 & \cdots & \Phi_p(\zeta) \end{bmatrix} \tag{6.8.11}$$

satisfies differential equation (6.8.3) (cf. (6.1.13)). Therefore

$$W = \hat{P}(\zeta)\Phi(\zeta) \tag{6.8.12}$$

is a formal fundamental solution of (6.8.1).

REMARK 6.8.3. The uniqueness of $(\hat{P}, s, B)$ was carefully and systematically examined by A. H. M. Levelt [91] (cf. also W. Jurkat [77]).

Now, let us consider the asymptotic meaning of formal solution (6.8.12). To do this, we replace differential equation (6.8.1) by

$$z\frac{dW}{dz} + A(z)W = 0, \tag{6.8.13}$$

where $A(z) = z^k\tilde{A}(z)$ for some $k \in Z$ and $\tilde{A} \in gl(n, \mathscr{A}(R_0, a, b))$, and

$$z^k A_{sp}(\tilde{A}) = \hat{A}. \tag{6.8.14}$$

Then $X = \hat{P}$ is a formal solution of the differential equation

$$\zeta\frac{dX}{d\zeta} + sA(\zeta^s)X - XB(\zeta) = 0. \tag{6.8.15}$$

Note that $A(\zeta^s) \in gl(n, \mathscr{A}(R_0^{\frac{1}{s}}, \frac{a}{s}, \frac{b}{s})[\zeta])$ and $\zeta^{ks}A_{sp}(\tilde{A}(\zeta^s)) = \hat{A}(\zeta^s)$. It is known that for every direction θ in the sectorial domain $\mathscr{D}(R_0^{\frac{1}{s}}, \frac{a}{s}, \frac{b}{s})$ (i.e., $\frac{a}{s} < \theta < \frac{b}{s}$), there exist a sectorial subdomain $\mathscr{D}(R^{\frac{1}{s}}, \frac{\alpha}{s}, \frac{\beta}{s})$ of $\mathscr{D}(R_0^{\frac{1}{s}}, \frac{a}{s}, \frac{b}{s})$ and a matrix $P(\zeta) \in gl(n, \mathscr{A}(R^{\frac{1}{s}}, \frac{\alpha}{s}, \frac{\beta}{s}))$ such that

(i) $\frac{\alpha}{s} < \theta < \frac{\beta}{s}$,
(ii) $A_{sp}(P) = \hat{P}$,
(iii) $X = P(\zeta)$ satisfies differential equation (6.8.15).

(cf. M. Hukuhara [67], H. L. Turrittin [147] and J. -P. Ramis and Y. Sibuya [129]; see also Remark 6.8.5. below). Since P satisfies (6.8.15), the matrix

$$W = P(z^{\frac{1}{s}})\Phi(z^{\frac{1}{s}}) \tag{6.8.16}$$

is a solution of (6.8.13) on the sectorial domain $\mathscr{D}(R, \alpha, \beta)$.

If $A(z) \in gl(n, C\{z^{-1}\}[z])$, the result above applies to every direction θ in the z-plane. Thus we arrive at the following generalization of Theorem 6.1.1.

THEOREM 6.8.4. *If* $A(z) \in gl(n, C\{z^{-1}\}[z])$, *then*
(I) *there exists a formal transformation* (6.8.2) *which changes differential equation* (6.8.13) *to differential equation* (6.8.3) *of Theorem* 6.8.1,
(II) *there exist a positive number* R, *a positive integer* N, *a covering* (6.1.6) *at* $z = \infty$, *and* N *matrices* $P_1(z)$, $P_2(z)$, $\cdots$, $P_N(z)$ *such that*

(i) $P_\ell(\zeta^s) \in gl(n, \mathscr{A}(R^{\frac{1}{s}}, \frac{a_\ell}{s}, \frac{b_\ell}{s}))$,
(ii) $A_{sp}(P_\ell(\zeta^s)) = \hat{P}(\zeta) \qquad (\ell = 1, \cdots, N)$,
(iii) *the transformation*

$$W = P_\ell(z)U \tag{6.8.17}$$

changes differential equation (6.8.13) *to* (6.8.3) *in* $\mathscr{S}_\ell$.

REMARK 6.8.5. Generally speaking if $A \in gl(n, \mathscr{A}(R_0, a, b)[z])$ and $B \in gl(n, \mathscr{A}(R_0, a, b)[z])$ and if $\hat{P} \in gl(n, C[[z^{-1}]])$ is a formal solution of the differential equation

$$z\frac{dX}{dz} + AX - XB = 0, \tag{6.8.18}$$

then for every direction θ in the sectorial domain $\mathscr{D}(R_0, a, b)$, there exist a sectorial subdomain $\mathscr{D}(R, \alpha, \beta)$ of $\mathscr{D}(R_0, a, b)$ and a matrix $P \in gl(n, \mathscr{A}(R, \alpha, \beta))$ such that

(i) $\alpha < \theta < \beta$,
(ii) $A_{sp}(P) = \hat{P}$,
(iii) $X = P(z)$ satisfies differential equation (6.8.18).

In particular, if $\hat{P} \in GL(n, C[[z^{-1}]])$, let $Q(z)$ be any matrix in $gl(n, \mathscr{A}(R_0, a, b))$ such that $A_{sp}(Q) = \hat{P}$ and set

$$X = QY. \tag{6.8.19}$$

Then

$$z\frac{dY}{dz} + BY - YB = -Q^{-1}\left\{z\frac{dQ}{dz} + AQ - QB\right\}Y. \tag{6.8.20}$$

Since

$$A_{sp}\left(Q^{-1}\left\{z\frac{dQ}{dz} + AQ - QB\right\}\right) = 0,$$

the identity matrix I is a formal solution of (6.8.20). If we solve (6.8.20), then (6.8.19) gives a solution of (6.8.18). This shows that the sectorial domain $\mathscr{D}(R, \alpha, \beta)$ is determined only by B and independent of A.

Further discussions of the Hukuhara–Turrittin theorem will be continued in Appendix 1.

§6.9. Asymptotic systems at $z = \infty$

In this section and succeeding sections, we shall explain the materials of §§6.2–6.4 by using more general ideas.

Set

$$\mathscr{D}(R) = \{ z \; ; \; |z| > R \}, \tag{6.9.1}$$

where R is a nonnegative number. Let us consider two differential equations

$$\frac{dW}{dz} = A_1(z)W \tag{6.9.2}$$

and

$$\frac{dV}{dz} = A_2(z)V, \tag{6.9.3}$$

where W and V are $n \times n$ unknown matrices and $A_j(z) \in gl(n, m(\mathscr{D}(R_j)))$ $(j = 1, 2)$ for some nonnegative numbers R_1 and R_2 (cf. (3.3.2) of §3.3 for the definition of $m(\mathscr{D})$). Note that if A_1 and A_2 are replaced by their Laurent expansions at $z = \infty$, then (6.9.2) and (6.9.3) become two differential equations in $gl(n, C\{z^{-1}\}[z])$.

DEFINITION 6.9.1. Two differential equations (6.9.2) and (6.9.3) are said to be **formally equivalent** if there exists an $n \times n$ matrix $\hat{P}(z) \in GL(n, C[[z^{-1}]])$ such that the transformation

$$W = \hat{P}(z)V \tag{6.9.4}$$

changes (6.9.2) to (6.9.3).

DEFINITION 6.9.2. Two differential equations (6.9.2) and (6.9.3) are said to be **holomorphically equivalent** if there exists an $n \times n$ matrix $P(\zeta) \in GL(n, \mathscr{O}(\Delta(0, \frac{1}{R_3})))$ such that the transformation

$$W = P\left(\frac{1}{z}\right) V \tag{6.9.5}$$

changes (6.9.2) to (6.9.3) on $\mathscr{D}(R_3)$, where R_3 is a positive number not less than $\max(R_1, R_2)$.

REMARK 6.9.3. If (6.9.2) and (6.9.3) are holomorphically equivalent, then they are also formally equivalent.

Let $\mathscr{E}$ be an equivalence class of differential equations considered above with respect to the formal equivalence. In this section and succeeding sections, we shall explain the holomorphic equivalence of differential equations in $\mathscr{E}$ by means of Stokes phenomena.

Let us denote by $GL(n, \mathscr{A}(R, a, b))$ the set of all $n \times n$ matrices $Q \in gl(n, \mathscr{A}(R, a, b))$ such that $Q^{-1} \in gl(n, \mathscr{A}(R, a, b))$ (cf. the beginning of §5.2 for the definition of $gl(n, \mathscr{A}(R, a, b))$).

As in §6.1, we fix a covering $\{\mathscr{S}_1, \cdots, \mathscr{S}_N\}$ at $z = \infty$, where the $\mathscr{S}_\ell$ are defined by (6.1.6). We assume that

$$0 < b_\ell - a_\ell < \pi.$$

Set

$$\mathscr{S}_{jk} = \mathscr{S}_j \cap \mathscr{S}_k,$$

where (j,k) is an ordered pair.

DEFINITION 6.9.4. A finite set $\{F_{jk}(z)\}$ of $n \times n$ matrices is called an **asymptotic system** at $z = \infty$ if

(1) $F_{jk}(z) \in GL(n, \mathscr{A}(\mathscr{S}_{jk}))$ is given whenever $\mathscr{S}_{jk} \neq \emptyset$,
(2) $F_{jh}(z)F_{hk}(z) = F_{jk}(z)$ whenever $\mathscr{S}_j \cap \mathscr{S}_h \cap \mathscr{S}_k \neq \emptyset$

DEFINITION 6.9.5. An asymptotic system $\{F_{jk}(z)\}$ at $z = \infty$ is said to be **trivial** at $z = \infty$ if there exist $n \times n$ matrices $P_1(z), \cdots, P_N(z)$ such that

(i) $P_j(z) \in GL(n, \mathscr{A}(\mathscr{S}_j))$,
(ii) $F_{jk}(z) = P_j(z)^{-1}P_k(z)$ whenever $\mathscr{S}_{jk} \neq \emptyset$.

The set $\{P_j(z)\}$ is called a **cross section** of $\{F_{jk}(z)\}$ at $z = \infty$.

DEFINITION 6.9.6. An asymptotic system $\{F_{jk}(z)\}$ at $z = \infty$ is said to be **formally trivial** at $z = \infty$ if there exist $n \times n$ matrices $\hat{Q}_1(z), \cdots, \hat{Q}_N(z)$ such that

(i) $\hat{Q}_j \in GL(n, C[[z^{-1}]])$,
(ii) $A_{sp}(F_{jk}) = \hat{Q}_j(z)^{-1}\hat{Q}_k(z)$ whenever $\mathscr{S}_{jk} \neq \emptyset$.

Utilizing Definitions 6.9.4 and 6.9.5, we can rephrase Theorem 6.4.1 as follows.

THEOREM 6.9.7. *An asymptotic system $\{F_{jk}(z)\}$ is trivial at $z = \infty$ if $A_{sp}(F_{jk}) = I$ whenever $\mathscr{S}_{jk} \neq \emptyset$, where I is the $n \times n$ identity matrix.*

As a matter of fact, we can prove a more general result.

THEOREM 6.9.8. *An asymptotic system $\{F_{jk}(z)\}$ is trivial at $z = \infty$ if and only if it is formally trivial.*

PROOF. It is evident that the triviality at $z = \infty$ implies the formal triviality. Assume that $\{F_{jk}(z)\}$ is formally trivial. This means that there exist N matrices $\hat{Q}_j(z) \in GL(n, C[[z^{-1}]])$ $(j = 1, \cdots, N)$ such that

$$A_{sp}(F_{jk}) = \hat{Q}_j(z)^{-1}\hat{Q}_k(z) \qquad (\mathscr{S}_{jk} \neq \emptyset). \tag{6.9.6}$$

Since the map $\mathscr{A}_{sp} : GL(n, \mathscr{A}(\mathscr{S}_j)) \rightarrow GL(n, C[[z^{-1}]])$ is surjective, we can find N matrices $Q_j(z)$ $(j = 1, \cdots, N)$ such that

(i) $Q_j(z) \in GL(n, \mathscr{A}(\mathscr{S}_j))$,
(ii) $A_{sp}(Q_j) = \hat{Q}_j$.

Set

$$G_{jk}(z) = Q_j(z)F_{jk}(z)Q_k(z)^{-1} \qquad (\mathscr{S}_{jk} \neq \emptyset). \tag{6.9.7}$$

Then $\{G_{jk}(z)\}$ is an asymptotic system at $z = \infty$. In fact

$$\begin{aligned} G_{jh}G_{hk} &= Q_j F_{jh} Q_h^{-1} Q_h F_{hk} Q_k^{-1} \\ &= Q_j F_{jh} F_{hk} Q_k^{-1} = Q_j F_{jk} Q_k^{-1} = G_{jk}. \end{aligned}$$

Since

$$\begin{aligned} A_{sp}(G_{jk}) &= A_{sp}(Q_j) A_{sp}(F_{jk}) A_{sp}(Q_k^{-1}) \\ &= \hat{Q}_j A_{sp}(F_{jk}) \hat{Q}_k^{-1} = I, \end{aligned}$$

the system $\{G_{jk}(z)\}$ is trivial at $z = \infty$ by virtue of Theorem 6.9.7. Now, it is easy to show that $\{F_{jk}(z)\}$ is also trivial at $z = \infty$. □

In §6.2, we defined a Stokes phenomenon $\{\mathcal{S}_\ell, C_{k\ell}, \Gamma_0(z)\}$ in Definition 6.2.1. The main part of Definition 6.2.1 was that the matrix

$$F_{jk}(z) = \exp\{\,\Gamma_0(z)\}\; C_{jk}\; \exp\{\,-\Gamma_0(z)\,\} \tag{6.9.8}$$

defines an asymptotic system at $z = \infty$ and that $A_{sp}(F_{jk}) = I$. By virtue of Theorem 6.9.7 (or Theorem 6.4.1) there exists a cross section $\{P_j(z)\}$ of $\{F_{jk}(z)\}$ at $z = \infty$. In §6.3 we studied the relation between two cross sections of $\{F_{jk}(z)\}$. Generally speaking we can prove the result below.

THEOREM 6.9.9. *Assume that an asymptotic system $\{F_{jk}(z)\}$ is trivial at $z = \infty$. Let $\{P_j(z)\}$ and $\{\tilde{P}_j(z)\}$ be two cross sections of $\{F_{jk}(z)\}$ at $z = \infty$. Then there exists an $n \times n$ matrix $P(\zeta)$ such that*

(i) *$P(\zeta) \in GL(n, \mathscr{O}(\Delta(0, \frac{1}{R})))$ for some $R > 0$,*

(ii) $P_j(z) = P\left(\dfrac{1}{z}\right) \tilde{P}_j(z) \qquad (j = 1, \cdots, N).$

PROOF. Since $\{P_j(z)\}$ and $\{\tilde{P}_j(z)\}$ are cross sections of $\{F_{jk}(z)\}$, we have

$$F_{jk}(z) = P_j(z)^{-1} P_k(z) = \tilde{P}_j(z)^{-1} \tilde{P}_k(z) \qquad \text{whenever } \mathcal{S}_{jk} \neq \varnothing.$$

Hence $P_j(z)\tilde{P}_j(z)^{-1} = P_k(z)\tilde{P}_k(z)^{-1}$ whenever $\mathcal{S}_{jk} \neq \varnothing$. If we define

$$P(\zeta) = P_j\left(\frac{1}{\zeta}\right) \tilde{P}_j\left(\frac{1}{\zeta}\right)^{-1} \qquad \text{for } \frac{1}{\zeta} \in \mathcal{S}_j \qquad (j = 1, \cdots, N),$$

then P satisfies requirements (i) and (ii). □

§6.10. The Stokes structures

Now, we shall define Stokes phenomena for the general case. Let $\mathscr{E}$ be an equivalence class of differential equations with respect to the formal equivalence (cf. Definition 6.9.1). Select a differential equation

$$\frac{dW_0}{dz} = A_0(z) W_0 \tag{6.10.1}$$

from $\mathcal{E}$. Let

$$\frac{dW}{dz} = A(z)W \tag{6.10.2}$$

be another equation in $\mathcal{E}$. Then there exists a matrix $\hat{P}(z) \in GL(n, C[[z^{-1}]])$ such that the transformation

$$W = \hat{P}(z)W_0 \tag{6.10.3}$$

changes (6.10.2) to (6.10.1). This means that $\hat{P}(z)$ is a formal solution of the differential equation

$$\frac{dX}{dz} = A(z)X - XA_0(z). \tag{6.10.4}$$

We remark here that $\hat{P}(z)$ is not necessarily unique. Applying Remark 6.8.5 to differential equation (6.10.4), we derive the following result.

THEOREM 6.10.1. *There exist a finite number of sectorial domains $\mathcal{S}_1, \cdots, \mathcal{S}_N$ such that*

(i) *$\{\mathcal{S}_1, \cdots, \mathcal{S}_N\}$ is a covering at $z = \infty$, where the $\mathcal{S}_\ell$ are defined by* (6.1.6),

(ii) $0 < b_\ell - a_\ell < \pi \qquad (\ell = 1, \cdots, N)$,

(iii) *for every differential equation* (6.10.2) *in the equivalence class $\mathcal{E}$, there exist $n \times n$ matrices $P_1(z), \cdots, P_N(z)$ satisfying the conditions below:*

(1) $P_j(z) \in GL(n, \mathcal{A}(\mathcal{S}_j))$,

(2) *the transformation $W = P_j(z)W_0$ changes* (6.10.2) *to* (6.10.1) *on $\mathcal{S}_j$.*

We remark here that $A_{sp}(P_j)$ $(j = 1, \cdots, N)$ are not necessarily identical, i.e., $A_{sp}(P_j)$ may depend on j.

For each j, let $\Phi_j(z)$ be a fundamental solution of (6.10.1) on $\mathcal{S}_j$. This means that

(a) $\Phi_j(z) \in GL(n, \mathcal{O}(\mathcal{S}_j))$

(b) $\dfrac{d\Phi_j(z)}{dz} = A_0(z)\Phi_j(z)$ on $\mathcal{S}_j$.

Then $P_j(z)\Phi_j(z)$ is a fundamental solution of (6.10.2) on $\mathcal{S}_j$.

DEFINITION 6.10.2. We call $\{\mathcal{S}_j, P_j(z)\Phi_j(z)\ ;\ j = 1, \cdots, N\}$ a **system of fundamental solutions** of differential equation (6.10.2) in the neighborhood of $z = \infty$.

DEFINITION 6.10.3. We call $\{A_0(z), \{\mathcal{S}_\ell, \Phi_\ell(z)\}, C_{k\ell}\}$ a **Stokes phenomenon** at $z = \infty$ if the following conditions are satisfied:

(1) the differential equation (6.10.1) (i.e., $\dfrac{dW_0}{dz} = A_0(z)W_0$) belongs to the equivalence class $\mathcal{E}$,

(2) $\{\mathcal{S}_1, \cdots, \mathcal{S}_N\}$ is a covering at $z = \infty$ satisfying the conditions of Theorem 6.10.1,

(3) $\{\mathcal{S}_\ell, \Phi_\ell(z)\}$ is a system of fundamental solutions of (6.10.1) in the neighborhood of $z = \infty$,

(4) the $C_{k\ell}$ are matrices in $GL(n, C)$ which are defined whenever $\mathcal{S}_k \cap \mathcal{S}_\ell \neq \emptyset$,

(5) $C_{hk} C_{k\ell} = C_{h\ell}$ whenever $\mathcal{S}_h \cap \mathcal{S}_k \cap \mathcal{S}_\ell \neq \emptyset$,

(6) $\Phi_k(z)\ C_{k\ell}\ \Phi_\ell(z)^{-1} \in GL(n, \mathcal{A}(\mathcal{S}_k \cap \mathcal{S}_\ell))$ whenever $\mathcal{S}_k \cap \mathcal{S}_\ell \neq \emptyset$,

(7) the asymptotic system $\{\Phi_k(z)\ C_{k\ell}\ \Phi_\ell(z)^{-1}\}$ is trivial at $z = \infty$.

REMARK 6.10.4. Condition (7) can be replaced by

(7′) the asymptotic system $\{\Phi_k(z)\ C_{k\ell}\ \Phi_\ell(z)^{-1}\}$ is formally trivial.

In particular, if $A_{sp}(\Phi_k(z)\ C_{k\ell}\ \Phi_\ell(z)^{-1}) = I$ whenever $\mathcal{S}_k \cap \mathcal{S}_\ell \neq \emptyset$, condition (7) is satisfied.

Let $\{\mathcal{S}_\ell, P_\ell(z)\Phi_\ell(z)\}$ be a system of fundamental solutions of differential equation (6.10.2). Set

$$C_{k\ell} = (P_k(z)\Phi_k(z))^{-1}(P_\ell(z)\Phi_\ell(z)) \in GL(n, C) \tag{6.10.5}$$

whenever $\mathcal{S}_k \cap \mathcal{S}_\ell \neq \emptyset$. Then

$$\Phi_k(z)\ C_{k\ell}\ \Phi_\ell(z)^{-1} = P_k(z)^{-1}P_\ell(z) \tag{6.10.6}$$

on $\mathcal{S}_k \cap \mathcal{S}_\ell (\neq \emptyset)$. Hence $\{A_0(z), \{\mathcal{S}_\ell, \Phi_\ell(z)\}, C_{k\ell}\}$ is a Stokes phenomenon at $z = \infty$.

DEFINITION 6.10.5. The Stokes phenomenon $\{A_0(z), \{\mathcal{S}_\ell, \Phi_\ell(z)\}, C_{k\ell}\}$ constructed above is called the Stokes phenomenon defined at $z = \infty$ by the system $\{\mathcal{S}_\ell, P_\ell(z)\Phi_\ell(z)\}$ of fundamental solutions of (6.10.2).

Conversely, let $\{A_0(z), \{\mathcal{S}_\ell, \Phi_\ell(z)\}, C_{k\ell}\}$ be a Stokes phenomenon at $z = \infty$ and let $\{P_1(z), \cdots, P_N(z)\}$ be a cross section of the asymptotic system $\{\Phi_k(z)\ C_{k\ell}\ \Phi_\ell(z)^{-1}\}$ at $z = \infty$ (i.e., $P_\ell(z) \in GL(n, \mathcal{A}(\mathcal{S}_j))$ $(j = 1, \cdots, N)$ and (6.10.6) hold whenever $\mathcal{S}_k \cap \mathcal{S}_\ell \neq \emptyset$). This means that we have (6.10.5) and hence

$$P_\ell(z)\Phi_\ell(z) = P_k(z)\Phi_k(z)C_{k\ell} \qquad (\mathcal{S}_k \cap \mathcal{S}_\ell \neq \emptyset). \tag{6.10.7}$$

Since

$$\frac{d}{dz}(P_\ell\Phi_\ell) = \left(\frac{d}{dz}(P_k\Phi_k)\right) C_{k\ell},$$

it follows that

$$\left(\frac{d}{dz}(P_\ell\Phi_\ell)\right)(P_\ell\Phi_\ell)^{-1} = \left(\frac{d}{dz}(P_k\Phi_k)\right)(P_k\Phi_k)^{-1}$$

on $\mathscr{S}_k \cap \mathscr{S}_\ell$ $(\neq \emptyset)$. Therefore if we set

$$A(z) = \left(\frac{d}{dz}(P_k\Phi_k)\right)(P_k\Phi_k)^{-1} \qquad \text{on } \mathscr{S}_k \qquad (k = 1, \cdots, N), \quad (6.10.8)$$

then $A(z) \in gl(n, \mathscr{O}(\mathscr{D}(R)))$ for some positive number R (cf. (6.9.1)). Note that

$$A(z) = \frac{dP_k(z)}{dz}P_k(z)^{-1} + P_k(z)A_0(z)P_k(z)^{-1} \qquad \text{on } \mathscr{S}_k. \qquad (6.10.9)$$

Hence $A(z) \in gl(n, m(\mathscr{D}(R)))$ and $X = A_{sp}(P_k)$ is a formal solution of (6.10.4). This means that differential equation (6.10.2) with the matrix $A(z)$ defined by (6.10.8) belongs to the equivalence class $\mathscr{E}$ and $\{\mathscr{S}_\ell, P_\ell(z)\Phi_\ell(z)\}$ is a system of fundamental solutions of this differential equation. Moreover the system $\{\mathscr{S}_\ell, P_\ell(z)\Phi_\ell(z)\}$ defines the Stokes phenomenon $\{A_0(z), \{\mathscr{S}_\ell, \Phi_\ell(z)\}, C_{k\ell}\}$ at $z = \infty$ (cf. §6.4). Thus we proved the theorem below.

THEOREM 6.10.6. *For a given Stokes phenomenon* $\{A_0(z), \{\mathscr{S}_\ell, \Phi_\ell(z)\}, C_{k\ell}\}$ *there exists a differential equation* (6.10.2) *in the equivalence class* $\mathscr{E}$ *such that a system of fundamental solutions of* (6.10.2) *defines the given Stokes phenomenon at* $z = \infty$.

Let $\{\mathscr{S}_\ell, P_\ell(z)\Phi_\ell(z)\}$ and $\{\mathscr{S}_\ell, \tilde{P}_\ell(z)\Phi_\ell(z)\}$ be two systems of fundamental solutions of differential equation (6.10.2) which define two Stokes phenomena $\{A_0(z), \{\mathscr{S}_\ell, \Phi_\ell(z)\}, C_{k\ell}\}$ and $\{A_0(z), \{\mathscr{S}_\ell, \Phi_\ell(z)\}, \tilde{C}_{k\ell}\}$ respectively. This means that

(a) $P_\ell(z) \in GL(n, \mathscr{A}(\mathscr{S}_j))$, $\quad \tilde{P}_\ell(z) \in GL(n, \mathscr{A}(\mathscr{S}_j))$,

(b) $P_\ell(z)\Phi_\ell(z)$ and $\tilde{P}_\ell(z)\Phi_\ell(z)$ are fundamental solutions of (6.10.2) on $\mathscr{S}_\ell$,

(c) $\Phi_k(z)C_{k\ell}\ \Phi_\ell(z)^{-1} = P_k(z)^{-1}P_\ell(z)$, $\Phi_k(z)\ \tilde{C}_{k\ell}\ \Phi_\ell(z)^{-1} = \tilde{P}_k(z)^{-1}\tilde{P}_\ell(z)$ whenever $\mathscr{S}_k \cap \mathscr{S}_\ell \neq \emptyset$.

It follows from (b) that

$$\tilde{P}_\ell(z)\Phi_\ell(z) = P_\ell(z)\Phi_\ell(z)C_\ell \qquad \text{on } \mathscr{S}_\ell \qquad (\ell = 1, \cdots, N) \qquad (6.10.10)$$

for some $C_\ell \in GL(n, C)$. Hence from (c) we derive

$$\tilde{C}_{k\ell} = C_k^{-1}C_{k\ell}C_\ell \qquad (6.10.11)$$

whenever $\mathscr{S}_k \cap \mathscr{S}_\ell \neq \emptyset$. Furthermore (6.10.10) implies that

$$\Phi_\ell(z)C_\ell\Phi_\ell(z)^{-1} \in GL(n, \mathscr{A}(\mathscr{S}_\ell)) \qquad \text{on } \mathscr{S}_\ell. \qquad (6.10.12)$$

DEFINITION 6.10.7. Two Stokes phenomena $\{A_0(z), \{\mathscr{S}_\ell, \Phi_\ell(z)\}, C_{k\ell}\}$ and $\{A_0(z), \{\mathscr{S}_\ell, \Phi_\ell(z)\}, \tilde{C}_{k\ell}\}$ are said to be equivalent if there exist $n \times n$ matrices $C_1, \cdots, C_N$ such that

(i) $C_\ell \in GL(n, C)$,

(ii) $\Phi_\ell(z)C_\ell\Phi_\ell(z)^{-1} \in GL(n, \mathscr{A}(\mathscr{S}_\ell))$ on $\mathscr{S}_\ell$,

(iii) $\tilde{C}_{k\ell} = C_k^{-1}C_{k\ell}C_\ell$ whenever $\mathscr{S}_k \cap \mathscr{S}_\ell \neq \varnothing$.

This is an equivalence relation. Each equivalence class of Stokes phenomena is called **a Stokes structure** associated with $\{A_0(z), \{\mathscr{S}_\ell, \Phi_\ell(z)\}\}$.

THEOREM 6.10.8. *Given a Stokes structure $\mathscr{C}$ associated with $\{A_0(z), \{\mathscr{S}_\ell, \Phi_\ell(z)\}\}$, there exists a Stokes phenomenon $\{A_0(z), \{\mathscr{S}_\ell, \Phi_\ell(z)\}, C_{k\ell}\}$ in $\mathscr{C}$ such that*

$$A_{sp}(\Phi_k(z)C_{k\ell}\Phi_\ell(z)^{-1}) = I$$

whenever $\mathscr{S}_k \cap \mathscr{S}_\ell \neq \varnothing$, where I is the $n \times n$ identity matrix.

PROOF. By virtue of Theorem 6.10.6, there exists a differential equation (6.10.2) in $\mathscr{E}$ which defines the given Stokes structure $\mathscr{C}$. Since (6.10.2) belongs to $\mathscr{E}$, there exists a formal transformation (6.10.3) which changes (6.10.2) to (6.10.1). Now, applying Theorem 6.10.1, we construct a system $\{\mathscr{S}_\ell, P_\ell(z)\Phi_\ell(z)\}$ of fundamental solutions of (6.10.2) in the neighborhood of $z = \infty$. In doing this we can assume that $A_{sp}(P_\ell)$ is independent of ℓ. The system $\{\mathscr{S}_\ell, P_\ell(z)\Phi_\ell(z)\}$ defines a Stokes phenomenon $\{A_0(z), \{\mathscr{S}_\ell, \Phi_\ell(z)\}, C_{k\ell}\}$ which belongs to the structure $\mathscr{C}$, where

$$C_{k\ell} = (P_k(z)\Phi_k(z))^{-1}(P_\ell(z)\Phi_\ell(z))$$

and hence

$$A_{sp}(\Phi_k(z)C_{k\ell}\Phi_\ell(z)^{-1}) = A_{sp}(P_k(z)^{-1}P_\ell(z))$$
$$= A_{sp}(P_k(z))^{-1}A_{sp}(P_\ell(z)) = I. \quad \square$$

§6.11. Classification by means of the Stokes structures

Now, we can state our main result concerning the holomorphic classification of differential equations in $\mathscr{E}$.

THEOREM 6.11.1.

(I) *Every differential equation* (6.10.2) *in $\mathscr{E}$ defines a unique Stokes structure $\mathscr{C}$ associated with $\{A_0(z), \{\mathscr{S}_\ell, \Phi_\ell(z)\}\}$.*

(II) *For a given Stokes structure $\mathscr{C}$ associated with $\{A_0(z), \{\mathscr{S}_\ell, \Phi_\ell(z)\}\}$ there exists a differential equation* (6.10.2) *in $\mathscr{E}$ which defines the given Stokes structure $\mathscr{C}$.*

(III) *Two differential equations in $\mathscr{E}$ are holomorphically equivalent if and only if these two equations define the same Stokes structure associated with $\{A_0(z), \{\mathscr{S}_\ell, \Phi_\ell(z)\}\}$.*

PROOF OF (III). Consider two differential equations

$$\frac{dW}{dz} = A(z)W \tag{6.11.1}$$

and

$$\frac{dU}{dz} = B(z)U. \tag{6.11.2}$$

If these two equations are holomorphically equivalent, there exists a matrix $P(\zeta) \in GL(n, \mathcal{O}(\Delta(0, \frac{1}{R})))$ for some positive number R such that the transformation

$$W = P\left(\frac{1}{z}\right)U \tag{6.11.3}$$

changes (6.11.1) to (6.11.2) (cf. Definition 6.9.2). Let $\{\mathcal{S}_\ell, P_\ell(z)\Phi_\ell(z)\}$ be a system of fundamental solutions of (6.11.2) in the neighborhood of $z = \infty$. Then $\{\mathcal{S}_\ell, P(\frac{1}{z})P_\ell(z)\Phi_\ell(z)\}$ is a system of fundamental solutions of (6.11.1). Since

$$\begin{aligned}&(P_k(z)\Phi_k(z))^{-1}(P_\ell(z)\Phi_\ell(z))\\&= (P\left(\frac{1}{z}\right)P_k(z)\Phi_k(z))^{-1}(P\left(\frac{1}{z}\right)P_\ell(z)\Phi_\ell(z)) \in GL(n, C)\end{aligned}$$

whenever $\mathcal{S}_k \cap \mathcal{S}_\ell \neq \emptyset$, the two differential equations (6.11.1) and (6.11.2) define the same Stokes phenomenon at $z = \infty$. Hence the Stokes structures defined by (6.11.1) and (6.11.2) are identical.

Now suppose that (6.11.1) and (6.11.2) define the same Stokes structure $\mathcal{C}$ at $z = \infty$. Let $\{\mathcal{S}_\ell, P_\ell(z)\Phi_\ell(z)\}$ and $\{\mathcal{S}_\ell, \tilde{P}_\ell(z)\Phi_\ell(z)\}$ be systems of fundamental solutions of (6.11.1) and (6.11.2) respectively. Since (6.11.1) and (6.11.2) define the same Stokes structure $\mathcal{C}$, there exist $n \times n$ matrices $C_1, \cdots, C_N \in GL(n, C)$ such that

$$\begin{aligned}&(\tilde{P}_k(z)\Phi_k(z))^{-1}(\tilde{P}_\ell(z)\Phi_\ell(z))\\&= C_k^{-1}(P_k(z)\Phi_k(z))^{-1}(P_\ell(z)\Phi_\ell(z))C_\ell \qquad (\mathcal{S}_k \cap \mathcal{S}_\ell \neq \emptyset)\end{aligned} \tag{6.11.4}$$

and

$$\Phi_\ell(z)\ C_\ell\ \Phi_\ell(z)^{-1} \in GL(n, \mathcal{A}(\mathcal{S}_\ell)) \tag{6.11.5}$$

(cf. (6.10.5) and Definition 6.10.7). From (6.11.4) we derive

$$\begin{aligned}&\tilde{P}_k(z)^{-1}\tilde{P}_\ell(z)\\&= (P_k(z)\Phi_k(z)C_k\Phi_k(z)^{-1})^{-1}(P_\ell(z)\Phi_\ell(z)C_\ell\Phi_\ell(z)^{-1})\end{aligned} \tag{6.11.6}$$

whenever $\mathcal{S}_k \cap \mathcal{S}_\ell \neq \emptyset$. This means that $\{\tilde{P}_\ell(z)\}$ and $\{P_\ell(z)\Phi_\ell(z)C_\ell\Phi_\ell(z)^{-1}\}$ are cross sections of the same asymptotic system at $z = \infty$ (cf. Definition 6.9.5). Hence by virtue of Theorem 6.9.9, there exists an $n \times n$ matrix $P(\zeta) \in GL(n, \mathcal{O}(\Delta(0, \frac{1}{R})))$ such that

$$P_\ell(z)\Phi_\ell(z)C_\ell\Phi_\ell(z)^{-1} = P\left(\frac{1}{z}\right)\tilde{P}_\ell(z) \qquad (\ell = 1, \cdots, N), \tag{6.11.7}$$

where R is a suitable positive number. If we rewrite (6.11.7) in the form

$$P_\ell(z)\Phi_\ell(z)C_\ell = P\left(\frac{1}{z}\right)\tilde{P}_\ell(z)\Phi_\ell(z),$$

the left-hand side is a solution of (6.11.1), while $\tilde{P}_\ell(z)\Phi_\ell(z)$ is a solution of (6.11.2). Hence it follows immediately that transformation (6.11.3) changes (6.11.1) to (6.11.2). □

REMARK 6.11.2. To define a Stokes phenomenon $\{A_0(z), \{\mathscr{S}_\ell, \Phi_\ell(z)\}, C_{k\ell}\}$, we must choose $\{A_0(z), \{\mathscr{S}_\ell, \Phi_\ell(z)\}\}$. For each choice, we define Stokes phenomena and Stokes structures. Since holomorphic equivalence classes in $\mathscr{E}$ correspond to Stokes structures bijectively, Stokes structures defined in terms of two different choices of $\{A_0(z), \{\mathscr{S}_\ell, \Phi_\ell(z)\}\}$ correspond to each other bijectively. Certain particular choices were given by G. D. Birkhoff [14] and Balser–Jurkat–Lutz [48, 49]. [See also W. Balser [43].] We can hide $\{\mathscr{S}_\ell\}$ by constructing a sheaf over the unit circle S^1 (cf. B. Malgrange [99, 101]). In this regard, it is helpful to know the intrinsic meaning of $\{\Phi_k(z)C_{k\ell}\Phi_\ell(z)^{-1}\}$. In fact, $X = \Phi_k(z)C_{k\ell}\Phi_\ell(z)^{-1}$ is a general solution of the differential equation

$$\frac{dX}{dz} = A_0(z)X - XA_0(z). \tag{6.11.8}$$

An invertible matrix $X(z)$ satisfies (6.11.8) if and only if the transformation

$$W_0 = X(z)\tilde{W}_0 \tag{6.11.9}$$

changes (6.10.1) to itself, i.e. ,

$$\frac{d\tilde{W}_0}{dz} = A_0(z)\tilde{W}_0. \tag{6.10.1$'$}$$

Such transformations form a noncommutative group. If we construct a sheaf over S^1 by using this group, the set of all Stokes structures is given in terms of sheaf cohomologies. In this way, we can also hide $\{\Phi_\ell(z)\}$ (cf. Malgrange [99, 101]). We shall come back to this subject in Appendix 3, §A.3.2.

REMARK 6.11.3. Instead of holomorphic equivalence, we can study *meromorphic equivalence* in a similar way (cf. Y. Sibuya [134]).

APPENDIX 1

The Hukuhara–Turrittin Theorem in Terms of $D_{\mathscr{K}}$-Modules

§A.1.0. Preliminaries

As in §6.8, we denote by $\mathscr{K}$ the quotient field of $\mathbf{C}[[z^{-1}]]$, i.e.

$$\mathscr{K} = \mathbf{C}[[z^{-1}]][z].$$

Let V be an n-dimensional vector space over $\mathscr{K}$. The differential operator $\delta = x\frac{d}{dx}$ defines a map : $\mathscr{K} \rightarrow \mathscr{K}$ which is additive and satisfies the Leibniz condition

$$\delta(a\ b) = \delta(a)\ b\ +\ a\ \delta(b) \quad \text{for } a,\ b\ \in \mathscr{K}. \tag{A.1.0.1}$$

A map $T : V \rightarrow V$ is called a linear differential operator if

(i) $T(u+v) = T(u) + T(v)$ for $u, v \in V$,
(ii) $T(a\ u) = \delta(a)\ u + a\ T(u)$ for $a \in \mathscr{K}$, $u \in V$.

Given a basis $\{u_1, \cdots, u_n\}$ for V over $\mathscr{K}$, there exists a matrix $A(z) \in gl(n, \mathscr{K})$ such that

$$[\ T(u_1), \ \cdots,\ T(u_n)\ ] = [\ u_1,\ \cdots,\ u_n\ ]\ A(z). \tag{A.1.0.2}$$

Observe that a vector $u = \sum_{1\leq j\leq n} y_j\ u_j \in V$ $(y_j \in \mathscr{K})$ can be written in the form

$$u = [\ u_1,\ \cdots,\ u_n\ ]\ y, \qquad y = \begin{bmatrix} y_1 \\ y_2 \\ \cdot \\ \cdot \\ y_n \end{bmatrix}. \tag{A.1.0.3}$$

Hence

$$\begin{aligned} T(u) &= [u_1,\ \cdots,\ u_n\ ]\ \delta(y)\ +\ [\ T(u_1),\ \cdots,\ T(u_n)\ ]\ y \\ &= [\ u_1,\ \cdots,\ u_n]\ (\ \delta(y)\ +\ A(z)\ y\ ). \end{aligned} \tag{A.1.0.4}$$

It is easy to see that

$$\delta + A(z) : \mathscr{K}^n \to \mathscr{K}^n \tag{A.1.0.5}$$

is a linear differential operator : $\mathscr{K}^n \to \mathscr{K}^n$. (cf. (6.8.13)). We call $\delta + A(z)$ the *representation* of T with respect to the basis $\{ u_1, \cdots, u_n \}$.

For two bases $\{u_1, \cdots, u_n\}$ and $\{v_1, \cdots, v_n\}$ for V over $\mathscr{K}$, set

$$\begin{cases} [\, T(u_1), \cdots, T(u_n) \,] = [\, u_1, \cdots, u_n \,] \, A(z), \\ [\, T(v_1), \cdots, T(v_n) \,] = [\, v_1, \cdots, v_n \,] \, B(z), \\ [\, v_1, \cdots, v_n \,] = [\, u_1, \cdots, u_n \,] \, P(z). \end{cases} \tag{A.1.0.6}$$

Then $P(z)B(z) = (\, \delta + A(z) \,) \, P(z)$, or

$$\delta(P) + A\,P - P\,B = 0 \tag{A.1.0.7}$$

(cf. (6.8.15)).

A nonempty subset U of the vector space V is said to be *invariant* by T if $T(U) \subset U$. Suppose that V is the direct sum of two subspaces V_1 and V_2 which are invariant by T, i.e.,

$$V = V_1 + V_2, \qquad V_1 \cap V_2 = \{0\}, \qquad T(V_j) \subset V_j \qquad (j = 1, 2).$$

If $\dim_{\mathscr{K}} V_j = n_j$ $(j = 1, 2)$, we have $n_1 + n_2 = n$. A basis $\{u_{1,1}, \cdots, u_{1,n_1}\}$ for V_1 over $\mathscr{K}$ and a basis $\{u_{2,1}, \cdots, u_{2,n_2}\}$ for V_2 over $\mathscr{K}$ form a basis $\{u_{1,1}, \cdots, u_{1,n_1}, u_{2,1}, \cdots, u_{2,n_2}\}$ for V over $\mathscr{K}$. If we set

$$[T(u_{j,1}), \cdots, T(u_{j,n_j})] = [u_{j,1}, \cdots, u_{j,n_j}] \, A_j(z) \qquad (j = 1, 2),$$

then

$$[T(u_{1,1}), \cdots, T(u_{1,n_1}), T(u_{2,1}), \cdots, T(u_{2,n_2})]$$

$$= [u_{1,1}, \cdots, u_{1,n_1}, u_{2,1}, \cdots, u_{2,n_2}] \begin{bmatrix} A_1(z) & 0 \\ 0 & A_2(z) \end{bmatrix}.$$

From the observation above, we see that the conclusion of the Hukuhara–Turrittin theorem (cf. Theorem 6.8.1) means that V can be decomposed as a direct sum of suitable invariant subspaces of T. We can also explain this fact in terms of modules over a noncommutative ring.

§A.1.1. $D_{\mathscr{K}}$-modules

Let us consider a polynomial ring

$$\mathscr{K}[\sigma] = \left\{ \sum_{h=0}^{p} a_h \, \sigma^h \; ; \, p \in \mathbf{N}, \, a_h \in \mathscr{K} \right\}, \tag{A.1.1.1}$$

where N is the set of nonnegative integers. We define multiplication of σ and $a \in \mathscr{K}$ by

$$\sigma\, a - a\, \sigma = \delta(a). \tag{A.1.1.2}$$

Then $K[\sigma]$ becomes a noncommutative ring. We denote this ring by $D_{\mathscr{K}}$.

Let V be an n-dimensional vector space over $\mathscr{K}$ and a module over $D_{\mathscr{K}}$. In this case, we call V an n-dimensional $D_{\mathscr{K}}$-module. Given a basis $\{u_1, \cdots, u_n\}$ for V over $\mathscr{K}$, there exists a matrix $A(z) \in gl(n, \mathscr{K})$ such that

$$[\ \sigma u_1,\ \cdots,\ \sigma u_n\ ] = [\ u_1,\ \cdots,\ u_n\ ]\, A(z) \qquad \text{(A.1.1.3)}$$

and hence

$$\sigma\ ([\ u_1,\ \cdots,\ u_n\ ]\, y) = [\ u_1,\ \cdots,\ u_n\ ]\, (\ \delta + A(z)\ )\, y, \qquad \text{(A.1.1.4)}$$

where $y \in \mathscr{K}^n$.

Conversely, suppose that a differential operator $\delta + A(z) : \mathscr{K}^n \to \mathscr{K}^n$ and a vector space V over $\mathscr{K}$ are given. Then if, for a given basis $\{u_1, \cdots, u_n\}$ for V over $\mathscr{K}$, we define multiplication of σ and a vector in V by

$$\sigma\ ([\ u_1,\ \cdots,\ u_n\ ]\, y) = [\ u_1,\ \cdots,\ u_n\ ]\, (\ \delta + A(z)\ )\, y, \qquad \text{(A.1.1.5)}$$

where $y \in \mathscr{K}^n$, the space V becomes an n-dimensional $D_{\mathscr{K}}$-module. For example, let $V = \mathscr{K}$ and $\lambda(z) \in \mathscr{K}$. If we set

$$\sigma\, a = (\delta + \lambda(z))\, a \quad \text{for } a \in \mathscr{K}, \qquad \text{(A.1.1.6)}$$

then V becomes a one-dimensional $D_{\mathscr{K}}$-module. We denote this module by $E(\lambda)$.

Let V be an n-dimensional $D_{\mathscr{K}}$-module and $\lambda(z) \in \mathscr{K}$. Then the tensor product

$$E(\lambda) \bigotimes_{\mathscr{K}} V$$

becomes an n-dimensional $D_{\mathscr{K}}$-module if we set

$$\sigma(a \otimes u) = (\sigma a) \otimes u + a \otimes (\sigma u). \qquad \text{(A.1.1.7)}$$

For a given basis $\{u_1, \cdots, u_n\}$ for V over $\mathscr{K}$, there exists a matrix $A(z) \in gl(n, \mathscr{K})$ such that (A.1.1.3) holds. Also there exists a nonzero vector e in $E(\lambda)$ such that $\sigma\, e = \lambda(z)\, e$. Then $\{e \otimes u_1, \cdots, e \otimes u_n\}$ is a basis for $E(\lambda) \otimes V$ over $\mathscr{K}$ and we have

$$\sigma\ ([\ e \otimes u_1,\ \cdots,\ e \otimes u_n\ ]\, y) = [\, e\otimes u_1,\ \cdots,\ e\otimes u_n\ ]\, (\ \delta + \lambda(z) I + A(z)\ )\, y, \qquad \text{(A.1.1.8)}$$

where $y \in \mathscr{K}^n$ and I is the $n \times n$ identity matrix. In other words the differential operator

$$\delta + \lambda(z) I + A(z) : \mathscr{K}^n \to \mathscr{K}^n$$

represents the σ-multiplication with respect to the basis $\{e\otimes u_1, \cdots, e\otimes u_n\}$ for $E(\lambda) \otimes V$ over $\mathscr{K}$.

As above, let V be an n-dimensional $D_{\mathscr{K}}$-module. In order that a nonempty subset U of V be a $D_{\mathscr{K}}$-submodule of V, it is necessary and sufficient that U is a vector subspace of V over $\mathscr{K}$ and $\sigma\ U \subset U$. We can state the Hukuhara–Turrittin theorem as a decomposition of V into a direct sum of a finite number of irreducible $D_{\mathscr{K}}$-submodules of V. However, to do this, we need to extend the field $\mathscr{K}$ to

$$\mathscr{L} = \mathbf{C}[[\zeta^{-1}]][\zeta] \tag{A.1.1.9}$$

where $z = \zeta^s$ (cf. (6.8.2)). Since $s\ \delta\ =\ \zeta\frac{d}{d\zeta}$, there is a natural embedding $: D_{\mathscr{K}}\ \rightarrow\ D_{\mathscr{L}}$. Instead of the $D_{\mathscr{K}}$-module V, we shall look at the $D_{\mathscr{L}}$-module

$$\mathscr{L} \bigotimes_{\mathscr{K}} V.$$

§A.1.2. Fuchsian modules

So far we have given two definitions of regular singular points, i.e., one in §3.2 and the other in §5.2 (cf. Definition 5.2.1). The definition in §3.2 is algebraic, whereas the definition in §5.2 is analytic. The relation between these two definitions can be explained by Lemma 5.8.1 (case $\rho_0\ =\ 0$). In this section, we shall explain the algebraic definition in terms of $D_{\mathscr{K}}$-modules.

DEFINITION A.1.2.1. Let V be an n-dimensional $D_{\mathscr{K}}$-module. A $\mathbf{C}[[z^{-1}]]$-submodule $\mathscr{M}$ of V is called a **lattice** if there exists a basis $\{m_1, \cdots, m_n\}$ for V over $\mathscr{K}$ that generates $\mathscr{M}$ over $\mathbf{C}[[z^{-1}]]$.

DEFINITION A.1.2.2. An n-dimensional $D_{\mathscr{K}}$-module V is called a Fuchsian module if there exists a lattice $\mathscr{M}$ such that

$$\sigma\ \mathscr{M}\ \in \mathscr{M}. \tag{A.1.2.1}$$

Suppose that V is a Fuchsian module and let $\mathscr{M}$ be a lattice satisfying condition (A.1.2.1). Since $\mathscr{M}$ is a lattice, there exists a basis $\{m_1, \cdots, m_n\}$ for V over $\mathscr{K}$ that generates $\mathscr{M}$ over $\mathbf{C}[[z^{-1}]]$. Condition (A.1.2.1) implies that

$$\sigma\ [\ m_1,\ \cdots, m_n\ ]\ =\ [\ m_1, \cdots, m_n\ ]\ A(z)$$

for some $n \times n$ matrix $A(z) \in gl(n, \mathbf{C}[[z^{-1}]])$. Therefore by utilizing the proof of Theorem 3.9.4 (due to M. Hukuhara), we can prove the theorem below.

THEOREM A.1.2.3. *An n-dimensional $D_{\mathscr{K}}$-module V is a Fuchsian module if and only if there exists a basis $\{u_1, \cdots, u_n\}$ for V over $\mathscr{K}$ and an $n\times n$ constant matrix $A_0 \in gl(n, \mathbf{C})$ such that*

$$\sigma\ [\ u_1, \cdots, u_n\ ]\ =\ [\ u_1, \cdots, u_n\ ]\ A_0. \tag{A.1.2.2}$$

The matrix A_0 can be chosen so that any two distinct eigenvalues of A_0 do not differ by an integer.

If the matrix A_0 is reduced to its Jordan canonical form, we obtain the following theorem.

THEOREM A.1.2.4. *A Fuchsian module V can be decomposed into a direct sum of a finite number of irreducible Fuchsian submodules as follows:*

$$V = \bigoplus_{j=1}^{q} \bigoplus_{\ell=1}^{m_j} E(\lambda_j) \bigotimes_{\mathscr{K}} N_{j\ell}, \tag{A.1.2.3}$$

where

(i) $\lambda_j \in \mathbf{C}$ *and* $\lambda_j - \lambda_k \notin \mathbf{Z}$ *if* $j \neq k$,

(ii) $N_{j\ell}$ *is a* $D_{\mathscr{K}}$*-module which admits a basis* $\{v_{j\ell 1}, \cdots, v_{j\ell n_{j\ell}}\}$ *such that*

$$\begin{cases} \sigma [\, v_{j\ell 1}, \cdots, v_{j\ell n_{j\ell}} \,] = [\, v_{j\ell 1}, \cdots, v_{j\ell n_{j\ell}} \,] J_{j\ell}, \\ J_{j\ell} = \begin{bmatrix} 0 & 1 & 0 & \cdots & 0 & 0 \\ 0 & 0 & 1 & \cdots & 0 & 0 \\ \cdots & \cdots & \cdots & \cdots & \cdots & \cdots \\ 0 & 0 & 0 & \cdots & 0 & 1 \\ 0 & 0 & 0 & \cdots & 0 & 0 \end{bmatrix}. \end{cases} \tag{A.1.2.4}$$

REMARK A.1.2.5.

(1) The uniqueness of decomposition (A.1.2.3) can be proved by means of the Krull–Schmidt theorem (cf. N. Jacobson [72]).

(2) The ring $\mathbf{C}[[z^{-1}]]$ can be replaced by a complete discrete valuation ring whose residue field is algebraically closed of characteristic 0 (cf. I. J. Manin [104]).

(3) An n-dimensional $D_{\mathscr{K}}$-module V is a Fuchsian module if and only if every element $u \in V$ is contained in a finitely generated $\mathbf{C}[[z^{-1}]]$-submodule $\mathscr{M}(u)$ of V such that $\sigma \mathscr{M}(u) \in \mathscr{M}(u)$. This fact is closely related with Theorem 5.9.4 (due to D. Lutz). (See also Gérard–Levelt [16].)

§A.1.3. The general case

To state the Hukuhara–Turrittin theorem for the general case in terms of $D_{\mathscr{K}}$-modules, we must extend the field $\mathscr{K}$ to $\mathscr{L}$ by choosing a suitable positive integer s (cf. (6.8.2) and (A.1.1.9)).

THEOREM A.1.3.1. *Given an n-dimensional $D_{\mathscr{K}}$-module V, there exists a positive integer s such that the $D_{\mathscr{L}}$-module $\mathscr{L} \otimes_{\mathscr{K}} V$ can be decomposed into a direct sum of $D_{\mathscr{L}}$-submodules as follows:*

$$\mathscr{L} \bigotimes_{\mathscr{K}} V = \bigoplus_{j=1}^{p} \bigoplus_{\ell=1}^{m_j} E(\lambda_j) \bigotimes_{\mathscr{L}} N_{j\ell}, \tag{A.1.3.1}$$

where

(i) $\lambda_j \in \mathbf{C}[\zeta]$ *and* $\lambda_j(\zeta) - \lambda_k(\zeta) \notin \mathbf{Z}$ *if* $j \neq k$,

(ii) $N_{j\ell}$ *is a* $D_{\mathscr{L}}$*-module which admits a basis* $\{v_{j\ell 1}, \cdots, v_{j\ell n_{j\ell}}\}$ *such that*

$$\begin{cases} \sigma [\, v_{j\ell 1}, \cdots, v_{j\ell n_{j\ell}} \,] = \dfrac{1}{s} [\, v_{j\ell 1}, \cdots, v_{j\ell n_{j\ell}} \,] J_{j\ell}, \\ J_{j\ell} = \begin{bmatrix} 0 & 1 & 0 & \cdots & 0 & 0 \\ 0 & 0 & 1 & \cdots & 0 & 0 \\ \cdots & \cdots & \cdots & \cdots & \cdots & \cdots \\ 0 & 0 & 0 & \cdots & 0 & 1 \\ 0 & 0 & 0 & \cdots & 0 & 0 \end{bmatrix}. \end{cases} \tag{A.1.3.2}$$

REMARK A.1.3.2.

(1) The extension of $E(\lambda_j) \otimes_{\mathscr{L}} N_{j\ell}$ over any finite field extension $\mathscr{H}$ of $\mathscr{L}$ is an indecomposable $D_{\mathscr{H}}$-module.

(2) The uniqueness of decomposition (A.1.3.1) can be proved by means of the Krull–Schmidt theorem (cf. N. Jacobson [72]).

(3) The ring $\mathbf{C}[[z^{-1}]]$ can be replaced by a complete discrete valuation ring whose residue field is algebraically closed of characteristic 0 (cf. F. Baldassarri [42]).

§A.1.4. Cyclic vectors and Newton polygons

In the same way as we proved Lemma 5.7.1, we can prove the lemma below.

LEMMA A.1.4.1. *For a given n-dimensional $D_{\mathscr{K}}$-module V, there exists a vector $u_0 \in V$ such that*

$$u_0, \ \sigma u_0, \ \sigma^2 u_0, \ \cdots, \ \sigma^{n-1} u_0 \tag{A.1.4.1}$$

form a basis for V over $\mathscr{K}$.

As in §5.7, such a vector u_0 is called a *cyclic vector*.

Let u_0 be a cyclic vector. Then, for each vector $u \in V$, there exists a unique element $\sum_{0 \le h \le n-1} a_h \sigma^h \in D_{\mathscr{K}}$ such that

$$u = \left\{ \sum_{h=0}^{n-1} a_h \sigma^h \right\} u_0, \tag{A.1.4.2}$$

where $a_h \in \mathscr{K}$ $(h = 0, 1, \cdots, n-1)$. In particular there exist n elements $c_0, \cdots, c_{n-1} \in \mathscr{K}$ such that

$$\left\{ \sigma^n + \sum_{h=0}^{n-1} c_h \sigma^h \right\} u_0 = 0. \tag{A.1.4.3}$$

Set

$$\wp = \sigma^n + \sum_{h=0}^{n-1} c_h \sigma^h. \tag{A.1.4.4}$$

Utilizing a cyclic vector u_0, we can define a homomorphism ϕ of $D_{\mathscr{K}}$-modules : $D_{\mathscr{K}} \rightarrow V$ if we set

$$\phi(a) = a\, u_0 \quad \text{for } a \in D_{\mathscr{K}} \tag{A.1.4.5}$$

The homomorphism ϕ is surjective (cf. (A.1.4.2)) and

$$\operatorname{Kernel}(\phi) = D_{\mathscr{K}}\, \wp.$$

(cf. (A.1.4.3) and (A.1.4.4)). Hence we have an isomorphism of $D_{\mathscr{K}}$-modules

$$\tilde{\phi} : V \rightarrow D_{\mathscr{K}} / D_{\mathscr{K}}\, \wp. \tag{A.1.4.6}$$

For example, the irreducible Fuchsian modules $E(\lambda_j) \bigotimes_{\mathscr{K}} N_{j\ell}$ given in Theorem A.1.2.4 are isomorphic to $D_{\mathscr{K}} / D_{\mathscr{K}} (\sigma - \lambda_j)^{n_{j\ell}}$ respectively.

For $a = \sum \alpha_m z^{-m} \in \mathscr{K}$, we set

$$\nu(a) = \inf\{ m \,;\, \alpha_m \neq 0 \}.$$

For a given element

$$P = \sum_{h=0}^{N} a_h \sigma^h \in D_{\mathscr{K}} \qquad (a_N \neq 0), \tag{A.1.4.7}$$

consider $N+1$ points $(h, \nu(a_h))$ $(h = 0, 1, \cdots, N)$ on an (X, Y)-plane. Set

$$\begin{cases} \mathscr{P}_h = \{ (X, Y) \,;\, X \le h,\ Y \ge \nu(a_h) \}, \\[2ex] \mathscr{P} = \displaystyle\bigcup_{h=0}^{N} \mathscr{P}_h. \end{cases}$$

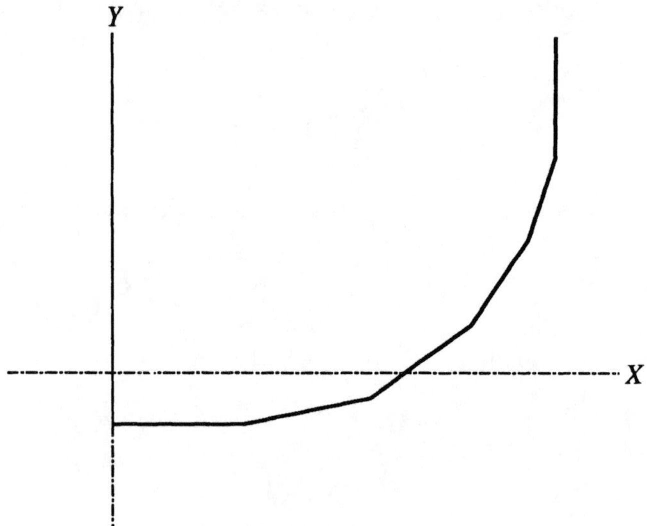

FIGURE 8

The boundary curve $\mathscr{C}$ of the smallest convex set containing $\mathscr{P}$ in the right half of the (X, Y)-plane is called the *Newton polygon* of the element P of $D_{\mathscr{K}}$. We denote by $\mathscr{N}(P)$ the Newton polygon of P. The polygon $\mathscr{N}(P)$ consists of a finite number of line segments with nonnegative slopes and two vertical rays (cf. Figure 8).

REMARK A.1.4.2. For a differential operator

$$\mathscr{L} = \sum_{h=0}^{N} a_h \delta^h,$$

where $a_h \in \mathscr{K}$ and $a_N \neq 0$, we can define the Newton polygon $\mathscr{N}(\mathscr{L})$ in exactly the same way as above. Let

$$0 \leq k_1 < k_2 < \cdots < k_p < +\infty \qquad \text{(A.1.4.8)}$$

be all of the distinct slopes of sides of the Newton polygon $\mathscr{N}(\mathscr{L})$ and let

$$\{ \ (X, Y_j + k_j(X - X_j)) \ ; \ X_j \leq X \leq X_{j+1} \ \}$$

be the side of $\mathscr{N}(\mathscr{L})$ whose slope is k_j, where $Y_{j+1} = Y_j + k_j(X_{j+1} - X_j)$ (cf. Figure 9). Note that the k_j are rational numbers.
Then

$$X_1 = 0, \ X_{p+1} = N, \quad \text{and} \quad X_{j+1} - X_j \in \mathbf{Z}^+,$$

where $\mathbf{Z}^+$ is the set of all positive integers. Set

$$n_j = X_{j+1} - X_j \qquad (j = 1, 2, \cdots, p).$$

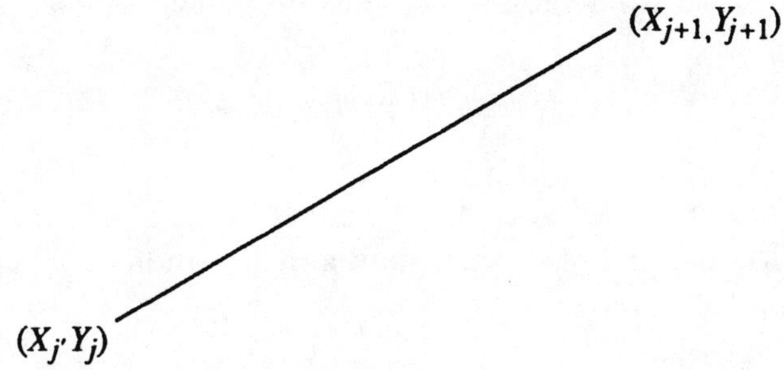

FIGURE 9

It is known that the differential equation

$$\mathcal{L}\, y = 0$$

admits n_j linearly independent formal solutions of the form

$$y_{j,h}(z) = z^{\mu_{j,h}} \exp[\lambda_{j,h} z^{k_j}\{1 + o(1)\}]\, f_{j,h}(z) \qquad (h = 1, \cdots, n_j), \quad \text{(A.1.4.9)}$$

where the $\mu_{j,h}$ are constants, the $\lambda_{j,h}$ are nonzero constants and $f_{j,h} \in \mathbf{C}[[z^{-\frac{1}{s}}]][z^{\frac{1}{s}}, \log z]$ for some positive integer s. Note that $\sum_{1 \le j \le p} n_j = N$. These formal solutions are essentially the same as those given in §6.8 (cf. (6.8.9), (6.8.10), and (6.8.11)). We can construct formal solutions (A.1.4.9) systematically by utilizing the Newton polygon $\mathcal{N}(\mathcal{L})$ (cf. for example, W. Sternberg [145]).

§A.1.5. Theorem A.1.3.1 and Newton polygons

We can prove the Baldassarri theorem (cf. Theorem A.1.3.1) also by utilizing the Newton polygon $\mathcal{N}(\wp)$ if a $D_{\mathcal{K}}$-module V is given by

$$V = D_{\mathcal{K}} / D_{\mathcal{K}}\, \wp \qquad \text{(A.1.5.1)}$$

(cf. B. Malgrange [100] , P. Robba [130] and Y. Sibuya [135]). The following observation will show a relation between a factorization of $\wp$ and a decomposition of module (A.1.5.1) into a direct sum of submodules.

OBSERVATION A.1.5.1. If $\wp = PQ$ for some P and Q in $D_{\mathcal{K}}$, we have an exact sequence

$$0 \to D_{\mathcal{K}} / D_{\mathcal{K}}\, P \to D_{\mathcal{K}} / D_{\mathcal{K}}\, \wp \to D_{\mathcal{K}} / D_{\mathcal{K}}\, Q \to 0 \qquad \text{(A.1.5.2)}$$

of homomorphisms of $D_{\mathcal{K}}$-modules. Exact sequence (A.1.5.2) splits if the Newton polygons $\mathcal{N}(P)$ and $\mathcal{N}(Q)$ do not have any common slopes. On the other hand, if all of the distinct slopes of the Newton polygon $\mathcal{N}(\wp)$ are given by (A.1.4.8), we have a factorization of $\wp$:

$$\wp = P_1 P_2 \cdots P_p$$

such that the Newton polygon $\mathcal{N}(P_j)$ has only one slope k_j. Hence

$$D_{\mathcal{K}}/D_{\mathcal{K}}\ \wp = \bigoplus_{j=1}^{p} D_{\mathcal{K}}/D_{\mathcal{K}}\ P_j. \tag{A.1.5.3}$$

Let us make more observations.

OBSERVATION A.1.5.2. We shall look at a $D_{\mathcal{K}}$-module V given by

$$V = D_{\mathcal{K}}/D_{\mathcal{K}}\ P, \tag{A.1.5.4}$$

where P is given by (A.1.4.7), assuming that the Newton polygon $\mathcal{N}(P)$ has only one slope k. Module (A.1.5.4) is a Fuchsian module if and only if $k = 0$. If $k > 0$, write

$$k = \frac{t}{s}, \tag{A.1.5.5}$$

where t and s are two relatively prime positive integers. Then s divides N and we can write P in the form

$$P = z^r \left\{ \sum_{\ell=0}^{\frac{N}{s}} \alpha_\ell\ (z^{-t}\sigma^s)^\ell + \sum_{\ell>kh} \alpha_{\ell h}\ z^{-\ell}\sigma^h \right\},$$

where $\alpha_\ell \in \mathbf{C}$, $\alpha_{\ell h} \in \mathbf{C}$, $r \in \mathbf{Z}$ and

$$\alpha_0 \neq 0, \qquad \alpha_{\frac{N}{s}} \neq 0.$$

The polynomial

$$\phi_P(\xi) = \sum_{\ell=0}^{\frac{N}{s}} \alpha_\ell\ \xi^\ell$$

is called the *determining polynomial* of P. Let

$$\phi_P(\xi) = \gamma \prod_{j=1}^{q} (\ \xi + \mu_j\ )^{m_j},$$

where the μ_j and γ are nonzero numbers in $\mathbf{C}$ and $\mu_j \neq \mu_h \quad (j \neq h)$.

The result below is a Hensel-type lemma (cf. P. Robba [130]).

OBSERVATION A.1.5.3. We have a factorization

$$P = P_1 \cdots P_q,$$

where, for each j, the Newton polygon $\mathcal{N}(P_j)$ has only one slope k and the determining polynomial of P_j is given by

$$\phi_{P_j}(\xi) = \gamma_j\ (\ \xi + \mu_j\ )^{m_j} \qquad (j = 1, \cdots, q),$$

where the γ_j are nonzero numbers in $\mathbf{C}$. Furthermore module (A.1.5.4) can be decomposed as

$$V = \bigoplus_{j=1}^{q} D_{\mathscr{K}}/D_{\mathscr{K}}\, P_j. \tag{A.1.5.6}$$

This decomposition of V is similar to the block-diagonalization which was explained in §5.5 (cf. Lemma 5.5.2). However in order to apply Lemma 5.5.2 to V, we must extend the field $\mathscr{K}$ to $\mathscr{L} = \mathbf{C}[[z^{-\frac{1}{s}}]][z^{\frac{1}{s}}]$. Decomposition (A.1.5.6) does not require such a field extension.

OBSERVATION A.1.5.4. Let us consider now the case when the Newton polygon $\mathscr{N}(P)$ has only one slope $k > 0$ and the determining polynomial of P is $\phi_P(\xi) = c(\xi + \mu)^m$, where c and μ are nonzero numbers in $\mathbf{C}$ and $m \in \mathbf{Z}^+$. Let us write $k = \frac{t}{s}$, where t and s are two relatively prime positive integers. We extend the field $\mathscr{K}$ to $\mathscr{L} = \mathbf{C}[[\zeta^{-1}]][\zeta]$, where $z = \zeta^s$. Then the Newton polygon $\mathscr{N}(P)$ of P as an element of $D_{\mathscr{L}}$ has only one slope t and the determining polynomial of P is $\tilde{\phi}_P(\eta) = c((\frac{\eta}{s})^s + \mu)^m$. Hence if $s > 1$, Observation A.1.5.3 applies. If $s = 1$ (i.e., $k = t$), we have

$$W = E(\mu z^t) \bigotimes_{\mathscr{K}} V \simeq D_{\mathscr{K}}/D_{\mathscr{K}} Q, \tag{A.1.5.7}$$

where

$$Q = \sum_{h=0}^{N} a_h\, (\sigma - \mu z^t)^h$$

and $\simeq$ means that the both sides of (A.1.5.7) are isomorphic to each other. Note that P is given by (A.1.4.7) and the one-dimensional $D_{\mathscr{K}}$-module $E(\lambda)$ was defined in §A.1.1. It can be proved that all the slopes of the Newton polygon $\mathscr{N}(Q)$ are strictly smaller than t. In this manner, we shall arrive at the same conclusion as Theorem A.1.3.1.

REMARK A.1.5.5. In some environments, a differential equation is not given *explicitly* as $\frac{dW}{dz} = A(z)W$, but a $D_{\mathscr{K}}$-module is given. In such a case computation of the matrix $A(z)$ is difficult. Even if $A(z)$ could be computed, the analytic expression of $A(z)$ might be hideously complicated. In any case, the method based on the theory of $D_{\mathscr{K}}$-modules is more intrinsic. Through such an approach, we find many helpful ideas in the theory of commutative algebra (cf. M. F. Atiyah and I. G. MacDonald [36]). To study formal structure of a completely integrable Pfaffian system at singularities, the best approach is to look at the problem as a study of D_R-modules, where $R = \mathbf{C}[[z_1, \cdots, z_m]]$ for example (cf. H. Charrière [52], H. Charrière and R. Gérard [53], and A. R. P. van den Essen and A. H. M. Levelt [56]). [The translator finds that book [120] of F. Pham is helpful to understand the theory of differential equations at singularities in terms of D_R-modules, where R is a differential ring.]

APPENDIX 2

Gevrey Asymptotics

§A.2.0. Preliminaries

In §5.1 we introduced a concept of asymptotic expansions. We defined there a map

$$A_{sp} : \mathscr{A}(R_0, a, b) \rightarrow \mathbf{C}[[z^{-1}]] \qquad \text{(A.2.0.1)}$$

as follows:

$$A_{sp}(f) = p \qquad \text{(A.2.0.2)}$$

means that

(i) $f \in \mathscr{O}(R_0, a, b)$,
(ii) $p = \sum_{m\geq 0} a_m z^{-m} \in \mathbf{C}[[z^{-1}]]$,
(iii) an inequality

$$\left| f(z) - \sum_{m=0}^{N-1} a_m z^{-m} \right| \leq K_{N,R,\alpha,\beta} \, |z|^{-N} \qquad \text{(A.2.0.3)}$$

holds on $\mathscr{D}(R, \alpha, \beta)$ for every positive integer N and every (R, α, β) satisfying the inequalities $0 < R_0 < R$ and $a < \alpha < \beta < b$, where $K_{N,R,\alpha,\beta}$ is a nonnegative number determined by (N, R, α, β). Also,

$$\begin{cases} \mathscr{D}(R_0, a, b) = \{ z \, ; \, |z| > R_0, \ a < \arg z < b \}, \\ \\ \mathscr{O}(R_0, a, b) = \mathscr{O}(\mathscr{D}(R_0, a, b)). \end{cases}$$

The domain of definition of A_{sp} is denoted by $\mathscr{A}(R_0, a, b)$. This definition of asymptotic expansions was originally given by H. Poincaré [122].

As we indicated in (4) of §5.1 , A_{sp} is surjective but not injective. Let us denote the kernel of A_{sp} by $\mathscr{A}_{+\infty,0}(R_0, a, b)$, i.e.,

$$\mathscr{A}_{+\infty,0}(R_0, a, b) = \{ f \in \mathscr{A}(R_0, a, b) \, ; \, A_{sp}(f) = 0 \}. \qquad \text{(A.2.0.4)}$$

If we examine various cases of asymptotic solutions of differential equations, we notice that $\mathbf{C}[[z^{-1}]]$ and $\mathscr{A}_{+\infty,0}(R_0, a, b)$ are too large. The following theorem due to E.Maillet suggests a suitable restriction of $\mathbf{C}[[z^{-1}]]$.

THEOREM A.2.0.1 (E. Maillet [94]). *Let $F(x, y_0, y_1, \cdots, y_n)$ be a nonzero polynomial in $y_0, y_1, \cdots, y_n$ with coefficients belonging to $\mathbf{C}\{x\}$, and let $f = \sum_{m\geq 0} a_m x^m \in \mathbf{C}[[x]]$ be a formal solution of the differential equation:*

$$F\left(x, y, \frac{dy}{dx}, \cdots, \frac{d^n y}{dx^n}\right) = 0. \tag{A.2.0.5}$$

Then there exist nonnegative numbers K, p, and A such that

$$|a_m| \leq K(m!)^p A^m \qquad (m \geq 0). \tag{A.2.0.6}$$

N. Watson [148] and F. Nevanlinna [111] developed a theory of asymptotic series under restriction (A.2.0.6). J.-P. Ramis reorganized and developed further such a theory systematically (cf. J.-P.Ramis [123, 124]). Condition (A.2.0.6) is also related with *ultradistributions* with support at the origin (cf. H. Komatsu [81, 82, 83]). The initial motivation for the Ramis theory was the duality between the space of such ultradistributions and the space of formal power series satisfying condition (A.2.0.6) (cf. J.-P. Ramis [123, 125]). Thus he called his theory of asympotic series the **Gevrey asymptotics.**

The Gevrey asymptotics are based on the two definitions below.

DEFINITION A.2.0.2. Let s be a nonnegative number. A formal power series $f = \sum_{m\geq 0} a_m z^{-m} \in \mathbf{C}[[z^{-1}]]$ is said to be of **Gevrey order** s if there exist two nonnegative numbers C and A such that

$$|a_m| \leq C(m!)^s A^m \quad \text{for } m \in \mathbf{N}. \tag{A.2.0.7}$$

We denote by $\mathbf{C}[[z^{-1}]]_s$ the set of all power series of Gevrey order s.

DEFINITION A.2.0.3. A function f of z is said to admit an asymptotic expansion of **Gevrey order** s $(s \geq 0)$ as $z \to \infty$ on a sectorial domain $\mathscr{D}(R_0, a, b)$ if

(i) $f \in \mathscr{A}(R_0, a, b)$,
(ii) $A_{sp}(f) = \sum_{m\geq 0} a_m z^{-m} \in \mathbf{C}[[z^{-1}]]_s$,
(iii) an inequality

$$\left| f(z) - \sum_{m=0}^{N-1} a_m z^{-m} \right| \leq K_{R,\alpha,\beta}\, (N!)^s\, (B_{R,\alpha,\beta})^N\, |z|^{-N} \tag{A.2.0.8}$$

holds on $\mathscr{D}(R, \alpha, \beta)$ for every positive integer N and every (R, α, β) satisfying the inequalities $0 < R_0 < R$ and $a < \alpha < \beta < b$, where $K_{R,\alpha,\beta}$ and $B_{R,\alpha,\beta}$ are nonnegative numbers determined by (R, α, β). We denote by $\mathscr{A}_s(R_0, a, b)$ the set of all functions admitting asymptotic expansions of Gevrey order s as $z \to \infty$ on the sectorial domain $\mathscr{D}(R_0, a, b)$.

As we shall explain in §A.2.2, the kernel of the map $A_{sp} : \mathscr{A}_s(R_0, a, b) \to \mathbf{C}[[z^{-1}]]_s$ is characterized by the theorem below.

THEOREM A.2.0.4. *For $s > 0$, set*

$$\mathscr{A}_{s,0}(R_0, a, b) = \mathscr{A}_s(R_0, a, b) \cap \mathscr{A}_{+\infty,0}(R_0, a, b). \tag{A.2.0.9}$$

Then $f \in \mathscr{A}_{s,0}(R_0, a, b)$ if and only if

(i) $f \in \mathscr{O}(R_0, a, b)$,

(ii) *an inequality*

$$|f(z)| \leq \mu(R, \alpha, \beta) \exp\left\{ -\lambda(R, \alpha, \beta)\, |z|^{\frac{1}{s}} \right\} \tag{A.2.0.10}$$

holds on $\mathscr{D}(R, \alpha, \beta)$ for every (R, α, β) satisfying the inequalities $0 < R_0 < R$ and $a < \alpha < \beta < b$, where $\mu(R, \alpha, \beta)$ is a non-negative number and $\lambda(R, \alpha, \beta)$ is a positive number, and they are determined by (R, α, β).

As in §6.1, if N sectorial domains

$$\mathscr{S}_\ell = \mathscr{D}(R, a_\ell, b_\ell) \qquad (\ell = 1, \cdots, N) \tag{A.2.0.11}$$

satisfy the condition

$$\bigcup_{\ell=1}^{N} \mathscr{S}_\ell = \{\, z \; ; \; |z| > R \,\}, \tag{A.2.0.12}$$

we call $\{ \mathscr{S}_1, \cdots, \mathscr{S}_N \}$ **a covering at** $z = \infty$. Furthermore we call $\{ \mathscr{S}_1, \cdots, \mathscr{S}_N \}$ a **good covering** at $z = \infty$ if

(i) $a_\ell < a_{\ell+1}$ $(\ell = 1, \cdots, N)$ where $a_{N+1} = a_1 + 2\pi$,

(ii) $b_\ell - a_\ell < \pi$ $(\ell = 1, \cdots, N)$,

(iii) $\mathscr{S}_\ell \cap \mathscr{S}_{\ell+1} \neq \emptyset$ $(\ell = 1, \cdots, N)$ and $\mathscr{S}_\ell \cap \mathscr{S}_k = \emptyset$ otherwise if $\ell \neq k$, where $\mathscr{S}_{N+1} = \mathscr{S}_1$.

In order to improve various results of the Poincaré asymptotics by means of the Gevrey asymptotics, the following lemma is fundamental.

LEMMA A.2.0.5. *Assume that a covering $\{ \mathscr{S}_1, \cdots, \mathscr{S}_N \}$ at $z = \infty$ is good and that N functions $\phi_1(z), \cdots, \phi_N(z)$ satisfy the conditions*

(1) $\phi_\ell(z) \in \mathscr{O}(R, a_\ell, b_\ell)$,

(2) *$\phi_\ell(z)$ is bounded on $\mathscr{D}(R, a_\ell, b_\ell)$,*

(3) *we have*

$$|\phi_\ell(z) - \phi_{\ell+1}(z)| \leq \gamma \exp\left\{ -\lambda\, |z|^k \right\} \quad \text{on } \mathscr{S}_\ell \cap \mathscr{S}_{\ell+1}, \tag{A.2.0.13}$$

where $\gamma \geq 0$, $\lambda > 0$, and $k > 0$ are suitable numbers.

Then

$$\phi_\ell \in \mathcal{A}_{\frac{1}{k}}(R, a_\ell, b_\ell) \qquad (\ell = 1, \cdots, N). \tag{A.2.0.14}$$

[The author of this book also found a similar result (cf. Theorem A.2.5.2) through his study of the Matkowsky condition for the Ackerberg–O'Malley resonance (cf. Ackerberg–O'Malley [35], Y. Sibuya [136], B. J. Matkowsky [107], N. Kopell [84, 85] and C.-H. Lin [87]).]

In this appendix, we shall explain some basic properties of $\mathbf{C}[[z^{-1}]]_s$ and $\mathcal{A}_s(R_0, a, b)$, the existence and uniqueness of Gevrey asymptotic solutions, applications of Lemma A.2.0.5, and some results similar to Lemma A.2.0.5.

§A.2.1. Properties of $\mathbf{C}[[z^{-1}]]_s$

In this section, we shall explain some basic properties of $\mathbf{C}[[z^{-1}]]_s$

PROPERTY A.2.1.1. For $s \geq 0$, $\mathbf{C}[[z^{-1}]]_s$ is a subalgebra of $\mathbf{C}[[z^{-1}]]$ as a commutative differential algebra over $\mathbf{C}$. In particular, $\mathbf{C}[[z^{-1}]]_0 = \mathbf{C}\{z^{-1}\}$.

Actually we can prove that

(i) $f + g \in \mathbf{C}[[z^{-1}]]_s$ if f and $g \in \mathbf{C}[[z^{-1}]]_s$,

(ii) $fg \in \mathbf{C}[[z^{-1}]]_s$ if f and $g \in \mathbf{C}[[z^{-1}]]_s$,

(iii) $cf \in \mathbf{C}[[z^{-1}]]_s$ if $c \in \mathbf{C}$ and $f \in \mathbf{C}[[z^{-1}]]_s$,

(iv) $\frac{df}{dz} \in \mathbf{C}[[z^{-1}]]_s$ if $f \in \mathbf{C}[[z^{-1}]]_s$,

(v) $\int_\infty^z f dz \in \mathbf{C}[[z^{-1}]]_s$ if $f \in z^{-2}\mathbf{C}[[z^{-1}]]_s$.

It is easy to prove (i), (iii), (iv), and (v). Property (ii) follows from the lemma below (cf. G. N. Watson [148]).

LEMMA A.2.1.2. *For $s > 0$ and any integer $h \geq 2$, we have*

$$\sum_{p_1+\cdots+p_h=n,\ p_j\in\mathbf{N}} (p_1!)^s(p_2!)^s\cdots(p_h!)^s \leq (\gamma(s))^{h-1}(n!)^s \qquad (n \in \mathbf{N}), \tag{A.2.1.1}$$

where $\mathbf{N}$ is the set of all nonnegative integers and $\gamma(s)$ is a positive number depending only on s.

PROPERTY A.2.1.3. Let $s_1, \cdots, s_n$ and $\sigma_1, \cdots, \sigma_n$ be nonnegative real numbers. Assume also that a formal power series in n variables $y_1, \cdots, y_n$,

$$F(y_1, \cdots, y_n) = \sum_{(p_1,\cdots,p_n)\in\mathbf{N}^n} \alpha_{p_1\cdots p_n} y_1^{p_1}\cdots y_n^{p_n},$$

with coefficients $\alpha_{p_1\cdots p_n}$ in $\mathbf{C}$ satisfy the conditions

$$|\alpha_{p_1\cdots p_n}| \leq C(p_1!)^{s_1}\cdots(p_n!)^{s_n}A_1^{p_1}\cdots A_n^{p_n} \qquad ((p_1, \cdots, p_n) \in \mathbf{N}^n)$$

for some nonnegative numbers $C, A_1, \cdots, A_n$ and that

$$f_j = \sum_{m\geq 1} z^{-m} \in \mathbf{C}[[z^{-1}]]_{\sigma_j} \qquad (j = 1, \cdots, n).$$

Set

$$s = \max\{s_1, \cdots, s_n, \sigma_1, \cdots, \sigma_n\}. \tag{A.2.1.2}$$

Then

$$F(f_1, \cdots, f_n) \in \mathbf{C}[[z^{-1}]]_s. \tag{A.2.1.3}$$

In particular, if $f = \sum_{m\geq 0} b_m z^{-m} \in \mathbf{C}[[z^{-1}]]_s$ $(s \geq 0)$ and if $b_0 \neq 0$, then

$$\frac{1}{f} \in \mathbf{C}[[z^{-1}]]_s \tag{A.2.1.4}$$

and

$$f^{\frac{1}{p}} \in \mathbf{C}[[z^{-1}]]_s \tag{A.2.1.5}$$

for every positive integer p.

This is a good exercise (cf. M. Gevrey [59]).

EXAMPLE A.2.1.4. The power series $f = \sum_{m\geq 0} m! z^{-m}$ belongs to $\mathbf{C}[[z^{-1}]]_1$. Hence $\frac{1}{f} \in \mathbf{C}[[z^{-1}]]_1$ and $f^{\frac{1}{p}} \in \mathbf{C}[[z^{-1}]]_1$ for every positive integer p. The power series f is a formal solution of the linear differential equation

$$\frac{d}{dz}\left[z \left(\frac{d}{dz} + 1 \right) \right] z^{-1} y = 0.$$

However, $\frac{1}{f}$ and $f^{\frac{1}{p}}$ $(p \geq 2)$ do not satisfy any linear ordinary differential equations of the form

$$\sum_{h=0}^{N} a_h(z) \frac{d^h y}{dz^h} = 0 \qquad (a_h \in \mathbf{C}\{z^{-1}\})$$

(cf. W. A. Harris–Y. Sibuya [64, 65] , S. Sperber [144] and M. Singer [143]).

For two real numbers $s \geq 0$ and $A > 0$ and for a formal power series $f = \sum_{m\geq 0} a_m z^{-m} \in \mathbf{C}[[z^{-1}]]$ we set

$$||f||_{s,A} = \sum_{m\geq 0} \frac{|a_m|}{(m!)^s A^m}. \tag{A.2.1.6}$$

Note that

$$||f||_{s_1,A} \leq ||f||_{s_2,A} \qquad \text{if} \qquad s_2 \leq s_1. \tag{A.2.1.7}$$

In order that $f \in \mathbf{C}[[z^{-1}]]_s$, it is necessary and sufficient that $||f||_{s,A} < +\infty$ for some $A > 0$. Let us denote by $E(s,A)$ the set of all $f \in \mathbf{C}[[z^{-1}]]$ such that $||f||_{s,A} < +\infty$, i.e.,

$$E(s,A) = \{f \in \mathbf{C}[[z^{-1}]];\ ||f||_{s,A} < +\infty\}. \tag{A.2.1.8}$$

It is easy to show that $E(s,A)$ is a Banach space with the norm $||f||_{s,A}$.

PROPERTY A.2.1.5. If f and $g \in E(s,A)$, then $fg \in E(s,A)$ and

$$||fg||_{s,A} \leq ||f||_{s,A}||g||_{s,A}. \tag{A.2.1.9}$$

Also if $f \in E(s,A)$, then $f \in E(s,B)$ for all $B \geq A$ and

$$\begin{cases} ||f||_{s,B} \leq ||f||_{s,A}, \\ \\ ||f||_{s,B} - |f(0)| \leq \dfrac{A}{B}\ \{||f||_{s,A} - |f(0)|\}, \end{cases} \tag{A.2.1.10}$$

for $B \geq A$. Furthermore, if $f \in E(s,A)$ and if q is an integer not less than $\frac{1}{s}$, then

$$\left|\left| z^{1-q}\frac{df}{dz}\right|\right|_{s,A} \leq \frac{1}{A^q}\ ||f||_{s,A}. \tag{A.2.1.11}$$

PROPERTY A.2.1.6. Consider a formal power series

$$\vec{F}(z,\vec{y},\vec{u}) = \sum_{|\vec{p}|+|\vec{q}|\geq 0} \vec{f}_{\vec{p},\vec{q}}\ \vec{y}^{\vec{p}}\ \vec{u}^{\vec{q}}, \tag{A.2.1.12}$$

where $\vec{f}_{\vec{p},\vec{q}} \in \mathbf{C}[[z^{-1}]]^n$, $\vec{p} = (p_1,\cdots,p_n) \in \mathbf{N}^n$, $\vec{q} = (q_1,\cdots,q_m) \in \mathbf{N}^m$, $|\vec{p}| = \sum_{1\leq h\leq n} p_h$, $|\vec{q}| = \sum_{1\leq h\leq m} q_h$, $\vec{y}^{\vec{p}} = y_1^{p_1}\cdots y_n^{p_n}$, $\vec{u}^{\vec{q}} = u_1^{q_1}\cdots u_m^{q_m}$ and

$$\vec{y} = \begin{bmatrix} y_1 \\ \cdot \\ \cdot \\ \cdot \\ y_n \end{bmatrix}, \qquad \vec{u} = \begin{bmatrix} u_1 \\ \cdot \\ \cdot \\ \cdot \\ u_m \end{bmatrix}.$$

Let us denote by $\vec{F}_{\vec{y}}(z,\vec{y},\vec{u})$ the Jacobian matrix of $\vec{F}$ with respect to $\vec{y}$. Then we can prove an **implicit function theorem.**

THEOREM A.2.1.7. *Assume that power series* (A.2.1.12) *satisfies the following conditions*:

(1) $\vec{f}_{\vec{0},\vec{0}} \in z^{-1}\mathbf{C}[[z^{-1}]]^n$,

(2) $\vec{F}_{\vec{y}}(z,\vec{0},\vec{0}) - \Phi \in z^{-1}gl(n,\mathbf{C}[[z^{-1}]])$ *for some invertible matrix* $\Phi \in GL(n,\mathbf{C})$,

(3) $\vec{f}_{\vec{p},\vec{q}} \in E(s,A)^n$ *for some* $s \geq 0$ *and* $A > 0$,

(4) $\sum_{|\vec{p}|\geq 0, |\vec{q}|\geq 0} ||\vec{f}_{\vec{p},\vec{q}}||_{s,A}\ \vec{y}^{\vec{p}}\ \vec{u}^{\vec{q}}$ *is a convergent power series in* $\vec{y}$ *and* $\vec{u}$, *where*

$$||\vec{f}||_{s,A} = \max_{1\leq j\leq n} ||f_j||_{s,A}$$

for

$$\vec{f} = \begin{bmatrix} f_1 \\ f_2 \\ \cdot \\ \cdot \\ f_n \end{bmatrix} \in E(s,A)^n.$$

Then, there exists a unique power series in z^{-1} *and* $\vec{u}$

$$\vec{\phi}(z,\vec{u}) = \sum_{|\vec{q}|\geq 0} \vec{\phi}_{\vec{q}}\ \vec{u}^{\vec{q}} \tag{A.2.1.13}$$

such that

(i)
$$\begin{cases} \vec{\phi}_{\vec{q}} \in \mathbf{C}[[z^{-1}]]^n, \qquad \vec{\phi}_{\vec{0}} \in z^{-1}\mathbf{C}[[z^{-1}]]^n, \\ \vec{F}(z,\vec{\phi}(z,\vec{u}),\vec{u}) = 0, \end{cases}$$

(ii) $\vec{\phi}_{\vec{q}} \in E(s,B)^n$ *for some* $B > 0$, *and*

(iii) *the power series* $\sum_{|\vec{q}|\geq 0} ||\vec{\phi}_{\vec{q}}||_{s,B}\ \vec{u}^{\vec{q}}$ *is a convergent power series in* $\vec{u}$.

This is also a good exercise. To verify this result, it suffices to construct a suitable majorizing series.

PROPERTY A.2.1.8. Let us look at a differential equation

$$\vec{y} = \vec{g}_0(z) + z^{1-q}\ \Phi(z)\ \frac{d\vec{y}}{dz} + \sum_{|\vec{p}|\geq 2} \vec{g}_{\vec{p}}(z)\ \vec{y}^{\vec{p}}, \tag{A.2.1.14}$$

where

$$\vec{g}_0 \in z^{-1}\ E(s,A)^n, \qquad \Phi \in gl(n, E(s,A)), \qquad \vec{g}_{\vec{p}} \in E(s,A)^n$$

for some $s \geq 0$ and $A > 0$ and $\sum_{|\vec{p}|\geq 2} ||g_{\vec{p}}||_{s,A}\ \vec{y}^{\vec{p}}$ is a convergent power series in $\vec{y}$.

Utilizing (A.2.1.7), (A.2.1.9), (A.2.1.10), and (A.2.1.11), we can prove Theorem A.2.1.9.

THEOREM A.2.1.9. *Differential equation* (A.2.1.14) *admits a unique solution* $\vec{\phi}(z) \in z^{-1} E(\sigma, B)^n$, *where*

$$\sigma = \max\left\{ s, \frac{1}{q} \right\} \tag{A.2.1.15}$$

and B *is a sufficiently large positive number.*

Note that, if B is sufficiently large, the right-hand side of differential equation (A.2.1.14) defines a contraction map : $E(\sigma, B)^n \to E(\sigma, B)^n$.

Theorem A.2.1.9 is a result better than the Maillet theorem (cf. Theorem A.2.0.1), although it applies only to differential equation (A.2.1.14). In §A.2.4, we shall explain by means of Lemma A.2.0.5 a generalization of Theorem A.2.1.9 that applies to algebraic differential equation (A.2.0.5) .

§A.2.2. Properties of $\mathscr{A}_s(R_0, a, b)$

In this section, we shall explain some basic properties of $\mathscr{A}_s(R_0, a, b)$.

PROPERTY A.2.2.1. In §5.1, we indicated in (1) and (2) that $\mathscr{A}(R_0, a, b)$ is a commutative differential algebra over $\mathbf{C}$ and that the map

$$A_{sp} : \mathscr{A}(R_0, a, b) \to \mathbf{C}[[z^{-1}]] \tag{A.2.2.1}$$

is a homomorphism of commutative differential algebras over $\mathbf{C}$. To verify a similar result for $\mathscr{A}_s(R_0, a, b)$ $(s \geq 0)$ the following characterization of $\mathscr{A}_s(R_0, a, b)$ is helpful.

LEMMA A.2.2.2. *Let* s *be a nonnegative number. Then a function* $f(z)$ *belongs to* $\mathscr{A}_s(R_0, a, b)$ *if and only if*

(i) $f \in \mathscr{O}(R_0, a, b)$ *and*

(ii) *for every* (R, α, β) *satisfying the condition*

$$0 < R_0 < R, \qquad a < \alpha < \beta < b, \tag{A.2.2.2}$$

there exist two nonnegative numbers $C_{R,\alpha,\beta}$ *and* $A_{R,\alpha,\beta}$ *such that*

$$\left| \frac{d^p f\left(\frac{1}{x}\right)}{dx^p} \right| \leq C_{R,\alpha,\beta} \, (p!)^{s+1} \, \left(A_{R,\alpha,\beta}\right)^p \tag{A.2.2.3}$$

for every $p \in \mathbf{N}$ *and* $z = \frac{1}{x} \in \mathscr{D}(R, \alpha, \beta)$.

By virtue of Lemma A.2.2.2, we can show for $s \geq 0$ that $\mathscr{A}_s(R_0, a, b)$ is a subalgebra of $\mathscr{A}(R_0, a, b)$ as a commutative differential algebra over $\mathbf{C}$ and that $\mathscr{A}_0(R_0, a, b)$ is the set of all functions in $\mathscr{O}(R_0, a, b)$ that are holomorphic at $z = \infty$. The map

$$A_{sp} : \mathscr{A}_s(R_0, a, b) \to \mathbf{C}[[z^{-1}]]_s \tag{A.2.2.4}$$

is a restriction of map (A.2.2.1) and hence is a homomorphism of commutative differential algebras over **C**. We can also show that

$$\int_{\infty}^{z} f(\zeta)d\zeta \in \mathscr{A}_s(R_0, a, b) \qquad \text{if } f \in z^{-2}\mathscr{A}_s(R_0, a, b).$$

PROPERTY A.2.2.3. Consider a map

$$\frac{1}{f} : \mathscr{D}(R_1, a_1, b_1) \rightarrow \mathscr{D}(R_2, a_2, b_2)$$

such that

$$\begin{cases} f \in \mathscr{A}_{s_1}(R_1, a_1, b_1), \\ f \notin \mathscr{A}_{s_1,0}(R_1, a_1, b_1), \\ A_{sp}(f) \in z^{-1}\mathbf{C}[[z^{-1}]]_{s_1}. \end{cases}$$

Then, if $g \in \mathscr{A}_{s_2}(R_2, a_2, b_2)$, we have

$$\begin{cases} g \circ \left(\frac{1}{f}\right) \in \mathscr{A}_{\max(s_1,s_2)}(R_1, a_1, b_1), \\ A_{sp}\left(g \circ \left(\frac{1}{f}\right)\right) = (A_{sp}(g)) \circ \left(\frac{1}{A_{sp}(f)}\right). \end{cases}$$

This can be proved basically by using Lemma A.2.2.2.

PROPERTY A.2.2.4. We define a norm for functions in $\mathscr{O}(R_0, a, b)$ by

$$\|f\|_{s,A,V} = \sup_{\frac{1}{x} \in V} \sum_{p=0}^{+\infty} \frac{1}{(p!)^{s+1}A^p} \left| \frac{d^p f\left(\frac{1}{x}\right)}{dx^p} \right| \qquad (f \in \mathscr{O}(R_0, a, b)),$$

(A.2.2.5)

where $s \geq 0$, $A > 0$ and $V = \mathscr{D}(R_0, a, b)$. Then Lemma A.2.2.2 implies that $f \in \mathscr{O}(R_0, a, b)$ belongs to $\mathscr{A}_s(R_0, a, b)$ if and only if

$$\|f\|_{s,A_W,W} < +\infty \tag{A.2.2.6}$$

for every $W = \mathscr{D}(R, \alpha, \beta)$ satisfying condition (A.2.2.2) and some positive constant A_W.

PROPERTY A.2.2.5. Set

$$E(s, A, V) = \{ f \in \mathscr{O}(R_0, a, b) \, ; \, \|f\|_{s,A,V} < +\infty \}. \tag{A.2.2.7}$$

Then $E(s, A, V)$ is a Banach space with the norm $\|f\|_{s,A,V}$. The following theorem is a good exercise.

THEOREM A.2.2.6. *Let*

$$F(z, \vec{y}) = \sum_{|\vec{p}| \geq 0} f_{\vec{p}}(z)\, \vec{y}^{\vec{p}} \tag{A.2.2.8}$$

be a power series in $\vec{y} = (y_1, \cdots, y_n)$ *with coefficients* $f_{\vec{p}} \in \mathscr{A}_s(R_0, a, b)$. *Assume that, for every* $W = \mathscr{D}(R, \alpha, \beta)$ *satisfying condition* (A.2.2.2), *there exist two positive numbers* A_W *and* r_W *such that the power series*

$$\sum_{|\vec{p}| \geq 0} ||f_{\vec{p}}(z)||_{s,A_W,W}\ \vec{y}^{\vec{p}}$$

is uniformly and absolutely convergent for $|y_j| \leq r_W$ $(j = 1, \cdots, n)$. *Let* $\vec{g} = (g_1, \cdots, g_n) \in \mathscr{A}_s(R_0, a, b)^n$ *be such that*

$$\begin{cases} A_{sp}(g_j) \in z^{-1}\mathbf{C}[[z^{-1}]]_s, \\ ||g_j||_{s,A_W,W} \leq r_W \end{cases} \tag{A.2.2.9}$$

for $j = 1, \cdots, n$. *Then*

$$\begin{cases} F(z, \vec{g}) \in \mathscr{A}_s(R_0, a, b), \\ A_{sp}(F(x, \vec{g})) = \displaystyle\sum_{|\vec{p}| \geq 0} (A_{sp}(f_{\vec{p}}))(A_{sp}(\vec{g}))^{\vec{p}}. \end{cases} \tag{A.2.2.10}$$

To prove Theorem A.2.2.6, we need the lemma below.

LEMMA A.2.2.7.

(i) *If* $f \in E(s, A, V)$, *then* $f \in E(s, B, V)$ *for all* $B \geq A$ *and*

$$\begin{cases} \sup_{z \in V} |f(z)| \leq ||f||_{s,B,V} \leq ||f||_{s,A,V}, \\ ||f||_{s,B,V} \leq \sup_{z \in V} |f(z)| + \dfrac{A}{B}\, ||f||_{s,A,V} \end{cases} \tag{A.2.2.11}$$

for $B \geq A$.

(ii) *If* f *and* g *belong to* $E(s, A, V)$, *then* fg *also belongs to* $E(s, A, V)$ *and*

$$||fg||_{s,A,V} \leq ||f||_{s,A,V}\ ||g||_{s,A,V}. \tag{A.2.2.12}$$

REMARK A.2.2.8. In general, the condition: $||g_j||_{s,A_W,W} \leq r_W$ might not be satisfied by a given $\vec{g} \in \mathscr{A}_s(R_0, a, b)^n$. In such a case, we can find a positive-valued continuous function $\rho(\theta)$ on the interval $a < \theta < b$ such that $F(z, \vec{g}) \in \mathscr{A}_s(\rho(\theta), a, b)$, where

$$\mathscr{D}(\rho(\theta), a, b) = \{z\ ;\ a < \arg z < b,\ |z| > \rho(\arg z)\ \}. \tag{A.2.2.13}$$

For example, if $f \in \mathscr{A}_s(R_0, a, b)$ satisfies the condition $A_{sp}(f) \notin z^{-1}\mathbf{C}[[z^{-1}]]_s$, then there exists a positive-valued continuous function $\rho(\theta) > R_0$ on the interval $a < \theta < b$ such that $\frac{1}{f} \in \mathscr{A}_s(\rho(\theta), a, b)$. We can also prove an implicit function theorem similar to Theorem A.2.1.7.

THEOREM A.2.2.9. *Consider a power series*

$$\vec{F}(z,\vec{y},\vec{u}) = \vec{f}_0(z) + \Phi(z)\vec{y} + \Psi(z)\vec{u} + \sum_{|\vec{p}|+|\vec{q}|\geq 2} \vec{f}_{\vec{p},\vec{q}}\ \vec{y}^{\vec{p}}\ \vec{u}^{\vec{q}},$$

where $\vec{f}_0 \in \mathscr{A}_s(R_0,a,b)^n$, $\vec{f}_{\vec{p},\vec{q}} \in \mathscr{A}_s(R_0,a,b)^n$, $\Phi \in gl(n,\mathscr{A}_s(R_0,a,b))$, Ψ *is an* $n \times m$ *matrix with entries in* $\mathscr{A}_s(R_0,a,b)$, *and*

$$\vec{y} = \begin{bmatrix} y_1 \\ \cdot \\ \cdot \\ \cdot \\ y_n \end{bmatrix}, \qquad \vec{u} = \begin{bmatrix} u_1 \\ \cdot \\ \cdot \\ \cdot \\ u_m \end{bmatrix}.$$

Assume that the following conditions are satisfied:

(1) $A_{sp}(\vec{f}_0) \in z^{-1}\mathbf{C}[[z^{-1}]]_s{}^n$,

(2) $A_{sp}(\Phi) - \Phi_0 \in z^{-1}gl(n,\mathscr{A}_s(R_0,a,b))$ *for some invertible matrix* $\Phi_0 \in GL(n,\mathbf{C})$,

(3) *for every* $W = \mathscr{D}(R,\alpha,\beta)$ *satisfying the condition*: $R_0 < R$, $a < \alpha < \beta < b$ *there exists a positive number* A_W *such that the power series*

$$\sum_{|\vec{p}|+|\vec{q}|\geq 2} ||\vec{f}_{\vec{p},\vec{q}}||_{s,A_W,W}\ \vec{y}^{\vec{p}}\ \vec{u}^{\vec{q}},$$

is convergent in $\vec{y}$ *and* $\vec{u}$.

Then, there exists a unique power series in $\vec{u}$

$$\vec{\phi}(z,\vec{u}) = \sum_{|\vec{q}|\geq 0} \vec{\phi}_{\vec{q}}\ \vec{u}^{\vec{q}}$$

such that

(i)

$$\begin{cases} \vec{\phi}_{\vec{q}} \in \mathscr{A}_s(\rho(\theta),a,b)^n, \\ A_{sp}(\vec{\phi}_{\vec{q}}) \in z^{-1}\mathbf{C}[[z^{-1}]]^n \qquad if \qquad |\vec{q}| = 0, \end{cases}$$

for a suitable positive-valued continuous function $\rho(\theta) > R_0$ *on the interval* $a < \theta < b$,

(ii) *if* $W = \mathscr{D}(R,\alpha,\beta)$ *satisfies the condition*

$$a < \alpha < \beta < b, \qquad \max_{\alpha\leq\theta\leq\beta} \rho(\theta) < R$$

there exist two positive numbers B_W *and* r_W *such that the series*

$$\sum_{|\vec{q}|\geq 0} ||\vec{\phi}_{\vec{q}}||_{s,B_W,W}\ (r_W)^{|\vec{q}|}$$

is convergent,

(iii) $\vec{F}(z, \vec{\phi}(z, \vec{u}), \vec{u}) = 0$ *for* $z \in W$ *and* $||u|| \leq r_W$ *if* W *satisfies the condition given in* (ii).

PROPERTY A.2.2.10. Let us investigate the surjectivity of homomorphism (A.2.2.4). The homomorphism A_{sp} is not always surjective. For example, since $\mathbf{C}[[z^{-1}]]_0 = \mathbf{C}\{z^{-1}\}$, but $\mathcal{A}_0(R_0, a, b)$ is the set of all functions in $\mathcal{O}(R_0, a, b)$ that are holomorphic at $z = \infty$, homomorphism (A.2.2.4) is not surjective if $s = 0$. As a matter of fact, we can prove the following theorem.

THEOREM A.2.2.11. *Homomorphism* (A.2.2.4) *is surjective if and only if*

$$b - a \leq s\,\pi. \tag{A.2.2.14}$$

To prove the surjectivity of A_{sp}, we utilize the *incomplete Leroy transforms* $\ell_{r,k}$ *of order* k of a convergent power series $\phi \in \mathbf{C}\{z^{-1}\}$ which are defined by

$$\ell_{r,k}(\phi) = k\ z^k \int_0^r \phi(\frac{1}{t}) \exp\left\{-(tz)^k\right\} t^{k-1}\, dt, \tag{A.2.2.15}$$

where r and k are two positive numbers. Note that $\phi\left(\frac{1}{t}\right) \in \mathbf{C}\{t\}$. Hence, if r is smaller than the radius of convergence of $\phi(\frac{1}{t})$ and if $|\arg z| \leq \frac{\pi}{2k}$, integral (A.2.2.15) is well defined. The "if" part of Theorem A.2.2.11 is a corollary of the lemma below.

LEMMA A.2.2.12. *Given a convergent power series* $\phi = \sum_{m\geq 0} c_m\, z^{-m} \in \mathbf{C}\{z^{-1}\}$, *we have*

(i) $\ell_{r,k}(\phi) \in \mathcal{A}_{\frac{1}{k}}(R, -\frac{\pi}{2k}, \frac{\pi}{2k})$ *for any positive numbers* k *and* R,

(ii) $A_{sp}(\ell_{r,k}(\phi)) = \sum_{m\geq 0} \Gamma(1 + \frac{m}{k})\, c_m\, z^{-m}$.

Here we assume that r *is smaller than the radius of convergence of* $\phi(\frac{1}{t})$.

We shall come back to the "only if" part of Theorem A.2.2.11 in Appendix 4.

PROPERTY A.2.2.13. Let us investigate the injectivity of homomorphism (A.2.2.4). The kernel $\mathcal{A}_{s,0}(R_0, a, b)$ of A_{sp} consists of all functions f in $\mathcal{O}(R_0, a, b)$ such that

$$|f(z)| \leq K_{R,\alpha,\beta}(N!)^s (B_{R,\alpha,\beta})^N |z|^{-N} \tag{A.2.2.16}$$

holds on $\mathcal{D}(R, \alpha, \beta)$ for every positive integer N and every (R, α, β) satisfying condition (A.2.2.2), where $K_{R,\alpha,\beta}$ and $B_{R,\alpha,\beta}$ are nonnegative numbers determined by (R, α, β). (cf. Definition A.2.0.3). The characterization of $\mathscr{A}_{s,0}(R_0, a, b)$ given by (A.2.0.10) of Lemma A.2.0.4 follows from the characterization given by (A.2.2.16.) (cf. N. Watson [148]).

We can derive the following theorem from Lemma A.2.0.4.

THEOREM A.2.2.14. *Homomorphism* (A.2.2.4) *is injective if and only if*

$$b - a > s\pi. \tag{A.2.2.17}$$

The "if" part of this theorem is a corollary of the lemma below.

LEMMA A.2.2.15 (N. Watson [148]). *Suppose that a function f is holomorphic on $\mathscr{D}(R_0, a, b)$ and satisfies the condition*

$$|f(z)| \leq \mu \exp\{-\lambda |z|^k\} \tag{A.2.2.18}$$

for $z \in \mathscr{D}(R_0, a, b)$, where μ is a nonnegative number and λ is a positive number. In this case, if a, b, and $s = \frac{1}{k}$ satisfy condition (A.2.2.17), *then $f(z) = 0$ identically on $\mathscr{D}(R_0, a, b)$.*

The "only if" part of Theorem A.2.2.14 is evident.

§A.2.3. Gevrey asymptotic solutions

For a given linear differential operator

$$P = \sum_{h=0}^{n} a_h(z)\, \delta^h, \tag{A.2.3.1}$$

where $a_h \in \mathscr{A}_s(\rho(\theta), a, b)$ and $\delta = z\frac{d}{dz}$, we write

$$A_{sp}(P) = \sum_{h=0}^{n} A_{sp}(a_h)\, \delta^h. \tag{A.2.3.2}$$

Here $\rho(\theta)$ is a positive-valued continuous function on the interval $a < \theta < b$. The operator P is a **C**-linear map : $\mathscr{A}_s(\rho(\theta), a, b) \to \mathscr{A}_s(\rho(\theta), a, b)$ and the operator $A_{sp}(P)$ is a **C**-linear map : $\mathbf{C}[[z^{-1}]]_s \to \mathbf{C}[[z^{-1}]]_s$. We define the Newton polygon $\mathscr{N}(P)$ of the operator P by

$$\mathscr{N}(P) = \mathscr{N}(A_{sp}(P)), \tag{A.2.3.3}$$

where $\mathscr{N}(A_{sp}(P))$ is the Newton polygon of $A_{sp}(P)$ (cf. Remark A.1.4.2 of §A.1.4). If a given operator P satisfies the condition $A_{sp}(a_n) \neq 0$, there exists a nonnegative integer $M(P)$ such that $\mathscr{N}(P + z^{-M}R) = \mathscr{N}(P)$ for any integer M not less than $M(P)$ and any linear differential operator $R = \sum_{0 \leq h \leq n} b_h(z)\, \delta^h \quad (b_h \in \mathscr{A}_s(\rho(\theta), a, b))$.

In this section we shall explain the existence and uniqueness of Gevrey asymptotic solutions of a differential equation

$$P[y] = z^{-M}\, G(z, y, \delta y, \cdots, \delta^{n-1} y), \tag{A.2.3.4}$$

where

$$\begin{cases} P = \sum_{h=0}^{n} a_h(z)\,\delta^h, \\ G(z, y_0, y_1, \cdots, y_{n-1}) = G_0(z) + \sum_{|\vec{p}|\geq 1} G_{\vec{p}}(z)\,\vec{y}^{\vec{p}} \end{cases} \tag{A.2.3.5}$$

and $\vec{p} = (p_0, \cdots, p_{n-1}) \in \mathbf{N}^n$, $|\vec{p}| = \sum_{0\leq h\leq n-1} p_h$, $\vec{y} = (y_0, y_1, \cdots, y_{n-1})$.

ASSUMPTION A.2.3.1.

(i) $a_h \in \mathscr{A}_s(\rho(\theta), a, b)$, $G_0 \in \mathscr{A}_s(\rho(\theta), a, b)$, *and* $G_{\vec{p}} \in \mathscr{A}_s(\rho(\theta), a, b)$,
(ii) $A_{sp}(a_n) \neq 0$,
(iii) *for any* $W = \mathscr{D}(R, \alpha, \beta)$ *satisfying the condition*

$$a < \alpha < \beta < b, \qquad \max_{\alpha\leq\theta\leq\beta} \rho(\theta) < R, \tag{A.2.3.6}$$

there exist two positive numbers A_W *and* r_W *such that the power series*

$$\sum_{|\vec{p}|\geq 1} ||G_{\vec{p}}||_{s,A_W,W}\,\vec{y}^{\vec{p}}$$

is convergent uniformly and absolutely for $|y_j| < r_W$ $(j = 0, 1, \cdots, n)$,
(iv) *the integer* M *is not less than* $M(P)$ *(i.e.,* $M \geq M(P)$*),*
(v) *the differential equation*

$$A_{sp}(P)[y] = z^{-M}\left\{A_{sp}(G_0) + \sum_{|\vec{p}|\geq 1} A_{sp}(G_{\vec{p}})\,y^{p_0}\,(\delta y)^{p_1}\cdots(\delta^{n-1}y)^{p_{n-1}}\right\} \tag{A.2.3.7}$$

admits a solution

$$f(z) = \sum_{1\leq m<+\infty} c_m\,z^{-m} \in z^{-1}\mathbf{C}[[z^{-1}]]_s. \tag{A.2.3.8}$$

Under this assumption we can prove the following theorem.

THEOREM A.2.3.2. *For any given direction* $\arg z = \theta$ *such that* $a < \theta < b$ *there exist a sectorial domain* $W = \mathscr{D}(R, \alpha, \beta)$ *and a function* $\phi(z) \in \mathscr{A}_s(R, \alpha, \beta)$ *such that*

(i) W *satisfies condition* (A.2.3.6),
(ii) $\alpha < \theta < \beta$,
(iii) $A_{sp}(\phi) = f$,
(iv) *the function* ϕ *satisfies differential equation* (A.2.3.4) *on* W.

SKETCH OF PROOF. Since the case $s = 0$ is a well-known case, we assume that $s > 0$. Set

$$\psi = \sum_{m\geq 1} \frac{c_m}{\Gamma(1+sm)} (e^{i\theta} z)^{-m} \quad \text{and} \quad \phi_0(e^{i\theta} z) = \ell_{r,\frac{1}{s}}(\psi).$$

Then

$$\psi \in \mathbf{C}\{z^{-1}\}, \qquad \phi_0 \in \mathscr{A}_s(R, \theta - \frac{s\pi}{2}, \theta + \frac{s\pi}{2}), \qquad A_{sp}(\phi_0) = f \quad \text{(A.2.3.9)}$$

for any positive number R (cf. Property A.2.2.10 of §A.2.2). Here we choose r smaller than the radius of convergence of ψ.

Change the unknown y by

$$y = u + \phi_0(z). \qquad \text{(A.2.3.10)}$$

Then differential equation (A.2.3.4) becomes

$$P[u] = z^{-M} G(z, u + \phi_0, \delta[u + \phi_0], \cdots, \delta^{n-1}[u + \phi_0]) - P[\phi_0].$$

Set

$$\begin{cases} F_0(z) &= z^{-M} G(z, \phi_0, \delta[\phi_0], \cdots, \delta^{n-1}[\phi_0]) - P[\phi_0], \\ F(z, \vec{u}) &= G(z, u_0 + \phi_0, u_1 + \delta[\phi_0], \cdots, u_n + \delta^{n-1}[\phi_0]) \\ &\quad - G(z, \phi_0, \delta[\phi_0], \cdots, \delta^{n-1}[\phi_0]) \end{cases} \qquad \text{(A.2.3.11)}$$

Then differential equation (A.2.3.4) can be written in the form

$$P[u] = F_0(z) + z^{-M} F(z, u, \delta u, \cdots, \delta^{n-1} u). \qquad \text{(A.2.3.12)}$$

Note that

$$\begin{aligned} &A_{sp}(F_0) \\ &= z^{-M} \left\{ A_{sp}(G_0) + \sum_{|\vec{p}|\geq 1} A_{sp}(G_{\vec{p}}) \, (A_{sp}(\phi_0))^{p_0} \cdots (\delta^{n-1}[A_{sp}(\phi_0)])^{p_{n-1}} \right\} \\ &\quad - A_{sp}(P)[A_{sp}(\phi_0)] \\ &= z^{-M} \left\{ A_{sp}(G_0) + \sum_{|\vec{p}|\geq 1} A_{sp}(G_{\vec{p}}) \, f^{p_0} \cdots (\delta^{n-1}[f])^{p_{n-1}} \right\} \\ &\quad - A_{sp}(P)[f] \\ &= 0. \end{aligned}$$

Hence $F_0(z)$ belongs to $\mathscr{A}_{s,0}(R, a, b)$ for every $W = \mathscr{D}(R, \alpha, \beta)$ satisfying conditions (A.2.3.6) and

$$\theta - \frac{s\pi}{2} < \alpha < \beta < \theta + \frac{s\pi}{2}. \qquad \text{(A.2.3.13)}$$

This is an application of Theorem A.2.2.6. Therefore, there exist a nonnegative number $\mu(R, \alpha, \beta)$ and a positive number $\lambda(R, \alpha, \beta)$ such that

$$|F_0(z)| \leq \mu(R, \alpha, \beta)\exp\{-\lambda(R, \alpha, \beta)\ |z|^{\frac{1}{s}}\} \tag{A.2.3.14}$$

on a sectorial domain $W = \mathscr{D}(R, \alpha, \beta)$ if W satisfies conditions (A.2.3.6) and (A.2.3.13) (cf. Theorem A.2.0.4).

Note also that $F(z, \vec{u})$ is holomorphic and bounded in $(z, \vec{u})$ on any domain

$$z \in W = \mathscr{D}(R, \alpha, \beta), \qquad |u_j| < \hat{r}_W \qquad (j = 0, 1, \cdots, n-1) \tag{A.2.3.15}$$

if W satisfies conditions (A.2.3.6) and (A.2.3.13) and $\hat{r}_W$ is a sufficiently small positive number. Furthermore

$$F(z, \vec{0}) = 0. \tag{A.2.3.16}$$

Now we can complete the proof of Theorem A.2.3.2 by means of the lemma below.

LEMMA A.2.3.3 (J.-P. Ramis–Y. Sibuya [129]). *There exist a sectorial domain* $W = \mathscr{D}(R, \alpha, \beta)$ *and a function* $\phi_1(z) \in \mathscr{A}_{s,0}(R, \alpha, \beta)$ *such that*

(i) W *satisfies conditions* (A.2.3.6) *and* (A.2.3.13),

(ii) $\alpha < \theta < \beta$,

(iii) *the function* ϕ_1 *satisfies differential equation* (A.2.3.12) *on* W.

In fact, if we set

$$\phi(z) = \phi_0(z) + \phi_1(z),$$

on W of Lemma A.2.3.3, the sectorial domain W and the function ϕ satisfy all of the requirements given in Theorem A.2.3.2. □

REMARK A.2.3.4.

(1) The hardest part of the proof of Lemma A.2.3.3 is the choice of the sectorial domain W. The reader will find a profound analysis concerning such a sectorial domain W in M. Hukuhara [67] (cf. also M. Iwano [70] and Ramis–Sibuya [129]). In Ramis–Sibuya [129], Lemma A.2.3.3 was reduced to the problem of finding a bounded solution by means of the Phragmén-Lindelöf theorem.

(2) If $G(z, \vec{y})$ of differential equation (A.2.3.4) is holomorphic and bounded in $(z, \vec{y})$ on a domain

$$|z| > R_0, \qquad |y_j| < r_0 \qquad (j = 0, 1, \cdots, n-1),$$

then G satisfies condition (iii) of Assumption A.2.3.1 for any nonnegative number s.

(3) If $H(z, y_0, \cdots, y_n)$ is a polynomial in $(y_0, \cdots, y_n)$ with coefficients in $\mathscr{A}_s(R_0, a, b)$, the differential equation

$$P[y] = z^{-\hat{M}}\, H(z, y, \delta y, \cdots, \delta^{n-1} y, \delta^n y) \qquad \text{(A.2.3.17)}$$

can be changed to differential equation (A.2.3.4) by means of the implicit function Theorem A.2.2.9 if $A_{sp}(a_n) \neq 0$ and the integer $\hat{M}$ is sufficiently large. Furthermore G satisfies condition (iii) of Assumption A.2.3.4. Hence Theorem A.2.3.2 applies to (A.2.3.17) if $A_{sp}(a_n) \neq 0$ and $\hat{M}$ is sufficiently large.

(4) Let

$$F(z, y_0, y_1, \ldots, y_n) = \sum_{0 \le |\vec{p}| \le N} F_{\vec{p}}(z)\, \vec{y}^{\vec{p}}$$

be a polynomial in $y_0, \cdots, y_n$ with coefficients $F_{\vec{p}}(z)$ in $\mathscr{A}_s(R_0, a, b)$. Set

$$\hat{F}(z, \vec{y}) = \sum_{0 \le |\vec{p}| \le N} A_{sp}(F_{\vec{p}})\, \vec{y}^{\vec{p}}.$$

Suppose that the differential equation

$$\hat{F}(z, v, \delta v, \cdots, \delta^n v) = 0$$

has a solution $f = \sum_{m \ge 0} c_m z^{-m} \in \mathbf{C}[[z^{-1}]]_s$ such that

$$\frac{\partial \hat{F}}{\partial y_n}(z, f, \delta f, \cdots, \delta^n f) \neq 0.$$

Then by means of a transformation $v = \sum_{0 \le m \le q} c_m\, z^{-m} + z^{-r}\, y$, we can change the differential equation

$$F(z, v, \delta v, \cdots, \delta^n v) = 0 \qquad \text{(A.2.3.18)}$$

to differential equation (A.2.3.17). Furthermore we can make $\hat{M}$ of (A.2.3.17) as large as we want by choosing q and r in a suitable way. The operator P of (A.2.3.17) depends on the choice of q and r. However the shape of the Newton polygon $\mathscr{N}(P)$ becomes independent of q and r if q and r are sufficiently large (cf. Y. Sibuya–S. Sperber [141, 142] and K. Mahler [92, 93]).

(5) In condition (v) of Assumption A.2.3.1, we required that f should be in $z^{-1}\mathbf{C}[[z^{-1}]]_s$. In the Poincaré asymptotics, f is supposed to be in $z^{-1}\mathbf{C}[[z^{-1}]]$. To understand the relation between asymptotic solutions in the Poincaré sense and Gevrey asymptotic solutions, we must study the Maillet Theorem A.2.0.1 more carefully (cf. Theorem A.2.4.2).

Let $\phi_1(z)$ and $\phi_2(z)$ be two solutions of differential equation (A.2.3.4) satisfying all of the requirements of Theorem A.2.3.2. Set

$$\psi(z) = \phi_1(z) - \phi_2(z)$$

on $W = \mathscr{D}(R, \alpha, \beta)$. Then

$$\begin{aligned} &P[\psi] \\ &= z^{-M}\{G(z, \phi_2 + \psi, \delta[\phi_2 + \psi], \cdots, \delta^{n-1}[\phi_2 + \psi]) \\ &\quad - G(z, \phi_2, \delta[\phi_2], \cdots, \delta^{n-1}[\phi_2])\} \\ &= z^{-M} \sum_{0 \leq h \leq n-1} \left[\int_0^1 \frac{\partial G}{\partial y_h}(z, \cdots, \delta_h[\phi_2 + t\psi], \cdots) dt\right] \delta^h \psi \\ &= z^{-M} \sum_{0 \leq h \leq n-1} \left[\int_0^1 \frac{\partial G}{\partial y_h}(z, \cdots, \delta^h[t\phi_1 + (1-t)\phi_2], \cdots) dt\right] \delta^h \psi. \end{aligned}$$

If we put

$$R = \sum_{0 \leq h \leq n-1} \left[\int_0^1 \frac{\partial G}{\partial y_h}(z, \cdots, \delta^h[t\phi_1 + (1-t)\phi_2], \cdots) dt\right] \delta^h,$$

then

$$\begin{cases} (P - z^{-M}R)[\psi] = 0, \\ \mathscr{N}(P - z^{-M}R) = \mathscr{N}(P), \\ \psi \in \mathscr{A}_{s,0}(R, \alpha, \beta). \end{cases} \tag{A.2.3.19}$$

Let

$$0 \leq k_1 < k_2 < \cdots < k_p < +\infty \tag{A.2.3.20}$$

be all of the distinct slopes of the Newton polygon $\mathscr{N}(P)$ and let

$$\{ (X, Y_j + k_j(X - X_j)) \ ; \ X_j \leq X \leq X_{j+1} \}$$

be the side of $\mathscr{N}(P)$ whose slope is k_j, where $Y_{j+1} = Y_j + k_j(X_{j+1} - X_j)$. Note that $X_1 = 0$, $X_{p+1} = N$, and $X_{j+1} - X_j \in \mathbf{Z}^+$. Set $n_j = X_{j+1} - X_j$ ($j = 1, 2, \cdots, p$). Then, as it was mentioned in Remark A.1.4.2, the differential equation $A_{sp}(P - z^{-M}R)[y] = 0$ admits n_j linearly independent solutions of the form

$$y_{j,h}(z) = z^{\mu_{j,h}} \exp[\lambda_{j,h} z^{k_j}\{1 + o(1)\}] f_{j,h}(z) \qquad (h = 1, \cdots, n_j), \tag{A.2.3.21}$$

where the $\mu_{j,h}$ are constants, the $\lambda_{j,h}$ are nonzero constants and $f_{j,h} \in \mathbf{C}[[z^{-\frac{1}{\sigma}}]][z^{\frac{1}{\sigma}}, \log z]$ for some positive integer σ. On the other hand, there exist a nonnegative number $\mu(R, \alpha, \beta)$ and a positive number $\lambda(R, \alpha, \beta)$ such that

$$|\psi(z)| \leq \mu(R, \alpha, \beta) \exp\{-\lambda(R, \alpha, \beta) \, |z|^{\frac{1}{s}}\}$$

for $z \in W = \mathscr{D}(R, \alpha, \beta)$ (cf. Theorem A.2.0.4).

Let ℓ be the total number of pairs (j, h) such that

$$k_j \geq \frac{1}{s}, \qquad \Re(\lambda_{j,h}\, z^{k_j}) < 0 \quad \text{on} \quad W. \tag{A.2.3.22}$$

Now concerning the uniqueness of Gevrey asymptotic solutions, we can state the theorem below.

THEOREM A.2.3.5 (Ramis–Sibuya [129]). *The general solution ϕ of differential equation* (A.2.3.4) *that satisfies all of the requirements of Theorem* A.2.3.2 *on a sectorial domain $W = \mathscr{D}(R, \alpha, \beta)$ contains ℓ arbitrary constants.*

§A.2.4. Applications of Lemma A.2.0.5

In this section, we shall explain a Gevrey-asymptotics version of Theorem 6.4.1 (or Theorem 6.9.7) and a result concerning the Gevrey properties of formal solutions which is more precise than the Maillet Theorem A.2.0.1. In both cases, we shall utilize Lemma A.2.0.5.

THEOREM A.2.4.1. *Let s be a positive number. Assume that N sectorial domains*

$$\mathscr{S}_\ell = \mathscr{D}(R, a_\ell, b_\ell) \qquad (\ell = 1, \cdots, N) \tag{A.2.4.1}$$

form a covering at $z = \infty$ and that $n \times n$ matrices $F_{k\ell}(z) \in gl(n, \mathscr{A}_s(\mathscr{S}_k \cap \mathscr{S}_\ell))$ are defined for all (k, ℓ) satisfying the condition

$$\mathscr{S}_k \cap \mathscr{S}_\ell \neq \emptyset. \tag{A.2.4.2}$$

Also assume that, if

$$\mathscr{S}_h \cap \mathscr{S}_k \cap \mathscr{S}_\ell \neq \emptyset, \tag{A.2.4.3}$$

the relation

$$F_{hk}(z) F_{k\ell}(z) = F_{h\ell}(z) \qquad (z \in \mathscr{S}_h \cap \mathscr{S}_k \cap \mathscr{S}_\ell) \tag{A.2.4.4}$$

holds and that

$$A_{sp}(F_{k\ell}) = I, \tag{A.2.4.5}$$

where I is the $n \times n$ identity matrix. Then there exist a positive number R' $(\geq R)$ and $n \times n$ matrices $Q_\ell(z)$ $(\ell = 1, \cdots, N)$ such that

(i) *$Q_\ell(z)$ and $Q_\ell(z)^{-1}$ belong to $gl(n, \mathscr{A}_s(R', a_\ell, b_\ell))$,*
(ii) *$\lim_{z \to \infty} Q_\ell(z) = I$,*
(iii) *the relation*

$$F_{k\ell}(z) = Q_k(z)^{-1} Q_\ell(z) \qquad (z \in \mathscr{S}'_k \cap \mathscr{S}'_\ell) \tag{A.2.4.6}$$

holds whenever condition (A.2.4.2) *is satisfied, where $\mathscr{S}'_\ell = \mathscr{D}(R', a_\ell, b_\ell)$.*

SKETCH OF PROOF. We may assume that $\{\mathcal{S}_1, \mathcal{S}_2, \cdots, \mathcal{S}_N\}$ is a good covering at $z = \infty$. Condition (A.2.4.5) implies that there exist a nonnegative number γ and a positive number λ such that the inequalities

$$\begin{cases} |F_{\ell\ \ell+1}(z) - I| \leq \gamma \exp\left\{-\lambda\, |z|^{\frac{1}{s}}\right\}, \\ |F_{\ell+1\ \ell}(z) - I| \leq \gamma \exp\left\{-\lambda\, |z|^{\frac{1}{s}}\right\} \end{cases} \tag{A.2.4.7}$$

hold on $\mathcal{S}_\ell \cap \mathcal{S}_{\ell+1}$ respectively.

By virtue of Theorem 6.4.1, there exist a positive number R' ($\geq R$) and $n \times n$ matrices $Q_\ell(z) \in gl(n, \mathscr{A}(R', a_\ell, b_\ell))$ ($\ell = 1, \cdots, N$) such that

(1) $Q_\ell(z)^{-1} \in gl(n, \mathscr{A}(R', a_\ell, b_\ell))$,
(2) $\lim_{z\to\infty} Q_\ell(z) = I$,
(3) the relations

$$\begin{cases} F_{\ell+1\ \ell}(z) = Q_{\ell+1}(z)^{-1} Q_\ell(z), \\ F_{\ell\ \ell+1}(z) = Q_\ell(z)^{-1} Q_{\ell+1}(z) \end{cases}$$

hold on $\mathcal{S}'_\ell \cap \mathcal{S}'_{\ell+1}$, where $\mathcal{S}'_\ell = \mathscr{D}(R', a_\ell, b_\ell)$.

Observe that

$$\begin{cases} Q_\ell(z) - Q_{\ell+1}(z) = Q_\ell(z)\ \{\ I - F_{\ell\ \ell+1}(z)\ \}, \\ Q_\ell(z)^{-1} - Q_{\ell+1}(z)^{-1} = \{\ I - F_{\ell+1\ \ell}(z)\ \}\ Q_\ell(z)^{-1}. \end{cases} \tag{A.2.4.8}$$

Hence

$$\begin{cases} |Q_\ell(z) - Q_{\ell+1}(z)| \leq |Q_\ell(z)|\ \gamma \exp\left\{-\lambda\, |z|^{\frac{1}{s}}\right\}, \\ |Q_\ell(z)^{-1} - Q_{\ell+1}(z)^{-1}| \leq |Q_\ell(z)^{-1}|\ \gamma \exp\left\{-\lambda\, |z|^{\frac{1}{s}}\right\} \end{cases} \tag{A.2.4.9}$$

on $\mathcal{S}'_\ell \cap \mathcal{S}'_{\ell+1}$. Therefore by utilizing Lemma A.2.0.5, we conclude that $Q_\ell(z)$ and $Q_\ell(z)^{-1}$ belong to $gl(n, \mathscr{A}_s(R', a_\ell, b_\ell))$. □
[This proof is due to B. Malgrange.]

Consider a differential equation of form (A.2.3.17)

$$P[y] = z^{-M}\ H(z, y, \delta y, \cdots, \delta^{n-1} y, \delta^n y), \tag{A.2.4.10}$$

where

$$\begin{cases} P = \displaystyle\sum_{h=0}^{n} a_h(z)\ \delta^h \qquad (a_h \in \mathbf{C}\{z^{-1}\}), \\ a_h \neq 0, \\ H(z, y_0, y_1, \cdots, y_{n-1}, y_n) \in \mathbf{C}\{z^{-1}\}[y_0, y_1, \cdots, y_{n-1}, y_n], \\ M \in \mathbf{N}. \end{cases} \tag{A.2.4.11}$$

Let

$$0 < k_1 < k_2 < \cdots < k_p < +\infty \tag{A.2.4.12}$$

be all of the positive slopes of the Newton polygon $\mathcal{N}(P)$. As in §A.2.3, we denote by $M(P)$ the nonnegative integer such that $\mathcal{N}(P+z^{-M}R) = \mathcal{N}(P)$ for any integer M not less than $M(P)$ and any linear differential operator $R = \sum_{0\leq h\leq n} b_h(z)\,\delta^h \quad (b_h \in \mathbf{C}[[z^{-1}]])$.

THEOREM A.2.4.2. *If* $f \in \mathbf{C}[[z^{-1}]]$ *satisfies differential equation* (A.2.4.10) *and if* $M \geq M(P)$, *then*

$$f \in \mathbf{C}[[z^{-1}]]_s, \tag{A.2.4.13}$$

where

$$s \in \left\{0, \frac{1}{k_1}, \frac{1}{k_2}, \cdots, \frac{1}{k_p}\right\}. \tag{A.2.3.14}$$

REMARK A.2.4.3.

(1) Suppose that a formal power series $f = \sum_{m\geq 0} c_m\, z^{-m} \in \mathbf{C}[[z^{-1}]]$ satisfies the differential equation

$$F(z, v, \delta v, \cdots, \delta^n v) = 0, \tag{A.2.4.15}$$

where

$$F(z, y_0, y_1, \ldots, y_n) \in \mathbf{C}\{z^{-1}\}[y_0, y_1, \ldots, y_n].$$

In such a case, if

$$\frac{\partial F}{\partial y_n}(z, f, \delta f, \cdots, \delta^n f) \neq 0, \tag{A.2.4.16}$$

we can change differential equation (A.2.4.15) to differential equation (A.2.4.10) by means of a transformation $v = \sum_{0\leq m\leq q} c_m\, z^{-m} + z^{-r}\, y$. Furthermore we can make M of (A.2.4.10) as large as we want by choosing q and r in a suitable way. Also if we set

$$P_0 = \sum_{h=0}^{n} \left(\frac{\partial F}{\partial y_h}(z, f, \delta f, \cdots, \delta^n f)\right) \delta^h, \tag{A.2.4.17}$$

then the Newton polygon $\mathcal{N}(P)$ of the operator P of differential equation (A.2.4.10) and $\mathcal{N}(P_0)$ have the same shape if q and r are sufficiently large (cf. (4) of Remark A.2.3.4). If condition (A.2.4.16) is not satisfied, we can replace F by another F satisfying (A.2.4.16) (cf. Y. Sibuya–S. Sperber [141, 142], K. Mahler [92, 93], and B. Malgrange [103]). Therefore, Theorem A.2.4.2 is a refinement of the Maillet Theorem A.2.0.1.

(2) J.-P. Ramis [123] proved Theorem A.2.4.2 in the case when differential equation (A.2.4.10) is linear. See, also, H. Gingold [60].
(3) B. Malgrange [103] showed that $s \leq \frac{1}{k_1}$. [For the general case, see Y. Sibuya [140].]
(4) Conclusion (A.2.4.14) of Theorem A.2.4.2 means that if $s < \frac{1}{k_j}$, then $s \leq \frac{1}{k_{j+1}}$.
(5) In the case when $a_h \in \mathbf{C}[[z^{-1}]]_s$ and $H \in \mathbf{C}[[z^{-1}]]_s[y_0, \cdots, y_n]$, set

$$\sigma_j = \max\left\{s, \frac{1}{k_j}\right\} \qquad (j = 1, \cdots, p)$$

then we can conclude that

$$f \in \mathbf{C}[[z^{-1}]]_\sigma,$$

where

$$0 \leq \sigma \leq s \quad \text{or} \quad \sigma \in \{0, \sigma_1, \cdots, \sigma_p\}$$

(cf. Y. Sibuya [140]; see also Theorem A.2.1.9).

SKETCH OF PROOF OF THEOREM A.2.4.2. Utilizing the Poincaré asymptotics, we can find

(a) a good covering $\{\mathscr{S}_1, \cdots, \mathscr{S}_p\}$ at $z = \infty$,
(b) p functions $\phi_1(z), \cdots, \phi_p(z)$ such that

$$\begin{cases} \phi_j \in \mathscr{A}(\mathscr{S}_j), \\ A_{sp}(\phi_j) = f, \\ P[\phi_j] = z^{-M} \; H(z, \phi_j, \delta\phi_j, \cdots, \delta^n\phi_j) \quad \text{on } \mathscr{S}_j, \end{cases} \qquad (j = 1, \cdots, p).$$

Set

$$u_j = \phi_j - \phi_{j+1} \qquad \text{on} \qquad \mathscr{S}_j \cap \mathscr{S}_{j+1}.$$

Then

$$\begin{cases} u_j \in \mathscr{A}_{+\infty,0}(\mathscr{S}_j \cap \mathscr{S}_{j+1}) \quad (\text{i.e., } A_{sp}(u_j) = 0), \\ (P - z^{-M}R)[u_j] = 0 \qquad \text{on} \qquad \mathscr{S}_j \cap \mathscr{S}_{j+1}, \end{cases} \qquad (j = 1, \cdots, p),$$

where

$$R = \sum_{0 \leq h \leq n} \left[\int_0^1 \frac{\partial H}{\partial y_h}(z, \cdots, \delta^h[t\phi_j + (1-t)\phi_{j+1}], \cdots) dt\right] \delta^h.$$

(cf. §A.2.3). Since $\mathscr{N}(P - z^{-M}R) = \mathscr{N}(P)$, we must have

$$|u_j(z)| \leq \gamma \exp\left\{-\lambda \, |z|^k\right\} \qquad \text{on} \qquad \mathscr{S}_j \cap \mathscr{S}_{j+1}$$

for a nonnegative number γ and a positive number λ, where

$$k \in \{k_1, \; \cdots, \; k_p\}.$$

Now, by utilizing Lemma A.2.0.5 we conclude that

$$\phi_j \in \mathscr{A}_s(\mathscr{S}_j) \qquad (j = 1, \cdots, p),$$

where

$$s \in \left\{0, \frac{1}{k_1}, \cdots, \frac{1}{k_p}\right\}.$$

Hence

$$f = A_{sp}(\phi) \in \mathbf{C}[[z^{-1}]]_s. \quad \square$$

REMARK A.2.4.4. The proof of Theorem A.2.4.2 also showed that every asymptotic solution of differential equation (A.2.4.10) in the Poincaré sense is a Gevrey asymptotic solution.
[The sketch of the proof given above involves some ambiguities. Such ambiguities will be removed by choosing the covering $\{\mathscr{S}_1, \cdots, \mathscr{S}_N\}$ prudently.]

§A.2.5. Phragmén–Lindelöf theorem in a cohomological form

Lemma A.2.0.5 is a corollary of the theorem below.

THEOREM A.2.5.1. *Let* $\{\mathscr{S}_1, \mathscr{S}_2, \cdots, \mathscr{S}_p\}$ *be a good covering at* $z = \infty$. *Assume that p functions* $\delta_1(z), \delta_2(z), \cdots, \delta_p(z)$ *satisfy the following conditions*:

(i) $\delta_j(z) \in \mathscr{O}(\mathscr{S}_j \cap \mathscr{S}_{j+1})$,
(ii) $|\delta_j(z)| \leq \gamma \exp\{-\lambda\, |z|^k\}$ *on* $\mathscr{S}_j \cap \mathscr{S}_{j+1}$, *where* $\gamma \geq 0$, $\lambda > 0$ *and* $k > 0$ *are suitable numbers.*

Then, there exist p functions $\psi_1(z), \psi_2(z), \cdots, \psi_p(z)$ *such that*

(a) $\psi_j(z) \in \mathscr{A}_{\frac{1}{k}}(\mathscr{S}_j)$.
(b) $\delta_j(z) = \psi_j(z) - \psi_{j+1}(z)$ *on* $\mathscr{S}_j \cap \mathscr{S}_{j+1}$.

The following is a sketch of the proof of Lemma A.2.0.5 by means of Theorem A.2.5.1.
Applying Theorem A.2.5.1 to N functions $\delta_\ell(z) = \phi_\ell(z) - \phi_{\ell+1}(z)$, we derive

$$\phi_\ell(z) - \phi_{\ell+1}(z) = \psi_\ell(z) - \psi_{\ell+1}(z) \qquad \text{on } \mathscr{S}_\ell \cap \mathscr{S}_{\ell+1}.$$

Hence

$$\phi_\ell(z) - \psi_\ell(z) = \phi_{\ell+1}(z) - \psi_{\ell+1}(z) \qquad \text{on } \mathscr{S}_\ell \cap \mathscr{S}_{\ell+1}.$$

Set

$$\phi(z) = \phi_\ell(z) - \psi_\ell(z) \qquad \text{on } \mathscr{S}_\ell \qquad (\ell = 1, \cdots, N).$$

Then $\phi(z)$ is holomorphic and bounded on $\cup_{1\leq\ell\leq N}\mathcal{S}_\ell$. Therefore

$$\phi_\ell = \psi_\ell + \phi \in \mathcal{A}_{\frac{1}{k}}(\mathcal{S}_\ell) \qquad (\ell = 1, \cdots, N). \quad \square$$

In Appendix 3, we shall restate Theorem A.2.5.1 in terms of cohomologies (cf. Theorem A.3.0.4).

If we replace conditions (1) and (2) of Lemma A.2.0.5 by the condition

$$\phi_\ell \in \mathcal{A}_{+\infty,0}(\mathcal{S}_\ell) \qquad (\ell = 1,\ 2,\ \cdots,\ N),$$

then we conclude that

$$\phi_\ell \in \mathcal{A}_{\frac{1}{k},0}(\mathcal{S}_\ell) \qquad (\ell = 1,\ 2,\ \cdots,\ N).$$

Actually we can derive a more precise conclusion.

THEOREM A.2.5.2 (Y. Sibuya [136]). *Assume that a covering* $\{\ \mathcal{S}_\ell\ ;\ \ell = 1,\ \cdots,\ N\ \}$ *at* $z=\infty$ *is good and that* N *functions* $\phi_1(z),\ \cdots,\ \phi_N(z)$ *satisfy the conditions*

(1) $\phi_\ell(z) \in \mathcal{A}_{+\infty,0}(\mathcal{S}_\ell)$ *and bounded on* $\mathcal{S}_\ell$,

(2)

$$|\phi_\ell(z) - \phi_{\ell+1}(z)| \leq \gamma \exp\{-\lambda\,|z|^k\} \qquad on \quad \mathcal{S}_\ell \cap \mathcal{S}_{\ell+1},$$

where $\gamma \geq 0$, $\lambda > 0$ *and* $k > 0$ *are suitable numbers.*

Then there exists a nonnegative number μ *such that*

$$|\phi_\ell(z)| \leq \mu \exp\{-\lambda\,|z|^k\} \qquad on \quad \mathcal{S}_\ell \qquad (\ell = 1, \cdots, N). \tag{A.2.5.1}$$

REMARK A.2.5.3.

(1) To understand the Matkowsky condition for the Ackerberg–O'Malley resonance (cf. Ackerberg-O'Malley [35], Y. Sibuya [136], B. J. Matkowsky [107], N. Kopell [84, 85], and C.-H. Lin [87]), we must estimate flat functions precisely. In particular, we must show an exponential decay of such functions. In such a study, it is important that λ of (A.2.5.1) and λ of condition (2) of Theorem A.2.5.2 are the same.

(2) The constant μ of (A.2.5.1) depends on λ. However for a sufficiently small λ we can choose μ independent of λ so that μ remains constant as $\lambda \to 0$ (cf. C.-H. Lin [86]).

So far $\{\mathcal{S}_1,\ \cdots,\ \mathcal{S}_N\}$ has been a covering at $z=\infty$. This means that $\cup_{1\leq\ell\leq N}\mathcal{S}_\ell = \{\ z;\ |z| > R_0\ \}$. We can prove a theorem similar to Theorem A.2.5.2 also in the case when $\cup_{1\leq\ell\leq N}\mathcal{S}_\ell$ is a sectorial domain.

ASSUMPTION A.2.5.4. *A set of sectorial domains* $\{\mathcal{S}_\ell = \mathcal{D}(R_0, a_\ell, b_\ell)\ ;\ \ell = 1, 2, \cdots, N\}$ *satisfies the following conditions*:

(i) $a_\ell < a_{\ell+1} \quad (\ell = 1, \cdots, N-1)$,

(ii) $b_\ell - a_\ell < \pi \quad (\ell = 1, \cdots, N)$,

(iii) $\mathcal{S}_\ell \cap \mathcal{S}_{\ell+1} \neq \varnothing \quad (\ell = 1, \cdots, N-1)$ *and* $\mathcal{S}_\ell \cap \mathcal{S}_k = \varnothing$ *otherwise if* $\ell \neq k$,

(iv) $-\pi < a_1 < b_N < \pi$.

Note that

$$\bigcup_{1 \leq \ell \leq N} \mathcal{S}_\ell = \mathcal{D}(R_0, a_1, b_N) \tag{A.2.5.2}$$

THEOREM A.2.5.5 (C.-H. Lin [86]). *Let* $\{\mathcal{S}_1, \cdots, \mathcal{S}_N\}$ *be a set of sectorial domains satisfying assumption* A.2.5.4. *Assume that* N *functions* $\phi_1(z), \cdots, \phi_N(z)$ *satisfy the following conditions*:

(i) $\phi_\ell(z) \in \mathcal{A}_{+\infty,0}(\mathcal{S}_\ell)$ *and bounded on* $\mathcal{S}_\ell$,

(ii) $\phi_1(z)$ *is continuous and*

$$|\phi_1(z)| \leq K_m\, |z|^{-m} \qquad (m = 0, 1, 2, \cdots)$$

on a sectorial domain

$$|z| > R_0, \quad a_1 \leq \arg z \leq a_1 + \varepsilon, \tag{A.2.5.3}$$

where the K_m *are suitable nonnegative numbers and* ε *is a small positive number*,

(iii) $\phi_N(z)$ *is continuous and*

$$|\phi_N(z)| \leq K_m\, |z|^{-m} \qquad (m = 0, 1, 2, \cdots)$$

on a sectorial domain

$$|z| > R_0, \quad b_N - \varepsilon \leq \arg z \leq b_N, \tag{A.2.5.4}$$

(iv)

$$|\phi_\ell(z) - \phi_{\ell+1}(z)| \leq \gamma \exp\{-\lambda\, |z|^k\} \qquad on \quad \mathcal{S}_\ell \cap \mathcal{S}_{\ell+1},$$

where $\mu \geq 0$, $\lambda > 0$, *and* $k > 0$ *are suitable numbers*,

(v)

$$\begin{cases} |\phi_1(t\, e^{i\, a_1})| \leq \gamma \exp\{-\lambda\, t^k\}, \\ |\phi_N(t\, e^{i\, b_N})| \leq \gamma \exp\{-\lambda\, t^k\}, \end{cases} \quad for \quad t \geq R_0. \tag{A.2.5.5}$$

Then there exists a nonnegative number μ *such that*

$$|\phi_\ell(z)| \leq \ \mu \exp\{-\lambda\, |z|^k\} \qquad on \quad \mathcal{S}_\ell \qquad (\ell = 1, \cdots, N). \tag{A.2.5.6}$$

Furthermore, for a sufficiently small λ *we can choose* μ *independent of* λ *so that* μ *remains constant as* $\lambda \to 0$.

C.-H. Lin [86] proved the following theorem by using Theorem A.2.5.5.

PHRAGMÉN–LINDELÖF THEOREM IN A COHOMOLOGICAL FORM (C.-H. Lin). *Let* $\{\mathcal{S}_1, \cdots, \mathcal{S}_N\}$ *be a set of sectorial domains satisfying Assumption A.2.5.4. Assume that*

$$-\frac{\pi}{2} < a_1 < b_N < \frac{\pi}{2}.$$

Assume also that N *functions* $\phi_1(z), \cdots, \phi_N(z)$ *satisfy the following conditions*:

(i) $\phi_\ell(z) \in \mathcal{O}(\mathcal{S}_\ell)$ *and continuous on the closure of* $\mathcal{S}_\ell$,

(ii)
$$|\phi_\ell(z)| \leq K \exp\{c|z|\} \qquad on \quad \mathcal{S}_\ell$$
for positive numbers K *and* c,

(iii)
$$|\phi_\ell(z) - \phi_{\ell+1}(z)| \leq L_0 \qquad on \quad \mathcal{S}_\ell \cap \mathcal{S}_{\ell+1},$$
where L_0 *is a positive number,*

(iv)
$$\begin{cases} |\phi_1(t\, e^{i\, a_1})| \leq L_0, \\ |\phi_N(t\, e^{i\, b_N})| \leq L_0, \end{cases} \qquad for \quad t \geq R_0.$$

Then there exists a positive number L *such that*

$$|\phi_\ell(z)| \leq L \qquad on \quad \mathcal{S}_\ell \qquad (\ell = 1, \cdots, N).$$

APPENDIX 3

Cohomological Methods

§A.3.0. Preliminaries

In §6.11, we gave our main result concerning the holomorphic classification of differential equations within an equivalence class with respect to the formal equivalence (cf. Theorem 6.11.1). B. Malgrange [99, 101] explained such a classification in terms of sheaf-cohomologies, utilizing various sheaves over the unit-circle S^1 in the complex plane.

OBSERVATION A.3.0.1. For an open connected arc V on S^1, let

$$\mathscr{S}(\rho(\theta), V) = \{z\ ; \arg z \in V,\ |z| > \rho(\arg z)\},$$

where $\rho(\theta)$ is a positive-valued continuous function on the arc V. We set

$$\mathscr{A}_s(V) = \{f \in \mathscr{A}_s(\mathscr{S}(\rho(\theta), V)) \quad \text{for some } \rho(\theta) > 0\},$$

where $0 \leq s \leq +\infty$ and $\mathscr{A}_{+\infty}(\mathscr{S}) = \mathscr{A}(\mathscr{S})$. Similarly we set

$$\mathscr{A}_{s,0}(V) = \{f \in \mathscr{A}_{s,0}(\mathscr{S}(\rho(\theta), V)) \quad \text{for some } \rho(\theta) > 0\}.$$

Thus we define two kinds of sheaves $\mathscr{A}_s$ and $\mathscr{A}_{s,0}$ over S^1. We can regard these sheaves either as sheaves of commutative differential algebras over $\mathbf{C}$ or as sheaves of $\mathbf{C}$-linear spaces. For a point θ of S^1, we denote by $\mathscr{A}_s(\theta)$ and $\mathscr{A}_{s,0}(\theta)$ the stalks of the sheaves $\mathscr{A}_s$ and $\mathscr{A}_{s,0}$ at θ respectively.

OBSERVATION A.3.0.2. Utilizing the following homomorphisms of commutative differential algebras over $\mathbf{C}$:

$$\begin{cases} A_{sp} : \mathscr{A}_s(V) \to \mathbf{C}[[z^{-1}]]_s, \\ \mathrm{id}_{\mathscr{A}_{s,0}(V)} : \mathscr{A}_{s,0}(V) \to \mathscr{A}_s(V), \end{cases}$$

we can define exact sequences

$$0 \to \mathscr{A}_{s,0}(V) \to \mathscr{A}_s(V) \to \mathbf{C}[[z^{-1}]]_s \to 0 \tag{A.3.0.1}$$

if the size of the arc V is properly restricted, where $0 \leq s \leq +\infty$ and $\mathbf{C}[[z^{-1}]]_{+\infty} = \mathbf{C}[[z^{-1}]]$. In particular, we have exact sequences

$$0 \to \mathscr{A}_{s,0}(\theta) \to \mathscr{A}_s(\theta) \to \mathbf{C}[[z^{-1}]]_s \to 0 \tag{A.3.0.2}$$

for every point θ of S^1. Hence we have exact sequences of homomorphisms of sheaves

$$0 \to \mathscr{A}_{s,0} \to \mathscr{A}_s \to \mathbf{C}[[z^{-1}]]_s \to 0, \tag{A.3.0.3}$$

where $\mathbf{C}[[z^{-1}]]_s$ are constant sheaves over S^1 with $\mathbf{C}[[z^{-1}]]_s(V) = \mathbf{C}[[z^{-1}]]_s$ for every open connected arc V of S^1 respectively.

OBSERVATION A.3.0.3. Since S^1 is a real one-dimensional manifold,

$$H^k(S^1, \mathscr{A}_{s,0}) = 0, \quad H^k(S^1, \mathscr{A}_s) = 0, \quad H^k(S^1, \mathbf{C}[[z^{-1}]]_s) = 0 \quad \text{for } k \geq 2.$$

Also we have

$$H^0(S^1, \mathscr{A}_{s,0}) = 0, \quad H^0(S^1, \mathscr{A}_s) = \mathbf{C}\{z^{-1}\}, \quad H^0(S^1, \mathbf{C}[[z^{-1}]]_s) = \mathbf{C}[[z^{-1}]]_s.$$

Thus we derive exact sequences

$$\begin{aligned} 0 \to \mathbf{C}\{z^{-1}\} \to \mathbf{C}[[z^{-1}]]_s \to H^1(S^1, \mathscr{A}_{s,0}) &\to H^1(S^1, \mathscr{A}_s) \\ &\to H^1(S^1, \mathbf{C}[[z^{-1}]]_s) \to 0 \end{aligned} \tag{A.3.0.4}$$

of $\mathbf{C}$-linear maps.

Now let us rephrase theorem A.2.5.1 as follows.

THEOREM A.3.0.4. *For* $0 \leq s \leq +\infty$, *we have*

$$\operatorname{Image}(H^1(S^1, \mathscr{A}_{s,0}) \to H^1(S^1, \mathscr{A}_s)) = 0. \tag{A.3.0.5}$$

[The case when $s = +\infty$ was explained in §6.5.] By virtue of Theorem A.3.0.4, we can derive from (A.3.0.4) exact sequences

$$0 \to \mathbf{C}\{z^{-1}\} \to \mathbf{C}[[z^{-1}]]_s \to H^1(S^1, \mathscr{A}_{s,0}) \to 0. \tag{A.3.0.6}$$

Thus we arrive at the result below.

THEOREM A.3.0.5 (B. Malgrange [99] and J.-P. Ramis [123]). *We have*

$$H^1(S^1, \mathscr{A}_{s,0}) \simeq \mathbf{C}[[z^{-1}]]_s / \mathbf{C}\{z^{-1}\} \tag{A.3.0.7}$$

for $0 \leq s \leq +\infty$, *where* $\simeq$ *means that the both sides of* (A.3.0.7) *are isomorphic to each other as* $\mathbf{C}$*-linear spaces.*

§A.3.1. Irregularity

Let us consider a differential operator

$$P = \sum_{h=0}^{n} a_h(z)\, \delta^h \qquad (a_h \in \mathbf{C}\{z^{-1}\};\ a_n \neq 0).$$

We can regard P as a $\mathbf{C}$-linear map in various ways:

$$P \begin{cases} : \mathbf{C}[[z^{-1}]]_s \rightarrow \mathbf{C}[[z^{-1}]]_s, \\ : \mathscr{A}_s(V) \rightarrow \mathscr{A}_s(V), \\ : \mathscr{A}_{s,0}(V) \rightarrow \mathscr{A}_{s,0}(V). \end{cases} \qquad (0 \le s \le +\infty).$$

Setting

$$\mathscr{K}_s(V) = \mathrm{Kernel}(P : \mathscr{A}_{s,0}(V) \rightarrow \mathscr{A}_{s,0}(V)),$$

we define sheaves $\mathscr{K}_s$ over S^1 for $0 \le s \le +\infty$. Then, if the size of V is properly restricted, we have exact sequences

$$0 \rightarrow \mathscr{K}_s(V) \rightarrow \mathscr{A}_{s,0}(V) \rightarrow \mathscr{A}_{s,0}(V) \rightarrow 0 \qquad (0 \le s \le +\infty).$$

In particular we have exact sequences

$$0 \rightarrow \mathscr{K}_s(\theta) \rightarrow \mathscr{A}_{s,0}(\theta) \rightarrow \mathscr{A}_{s,0}(\theta) \rightarrow 0 \qquad (0 \le s \le +\infty)$$

at every point θ of S^1. Thus we arrive at exact sequences of homomorphisms of sheaves

$$0 \rightarrow \mathscr{K}_s \rightarrow \mathscr{A}_{s,0} \rightarrow \mathscr{A}_{s,0} \rightarrow 0 \qquad (0 \le s \le +\infty). \tag{A.3.1.1}$$

Now (A.3.1.1) implies that

$$0 \rightarrow H^1(S^1, \mathscr{K}_s) \rightarrow H^1(S^1, \mathscr{A}_{s,0}) \rightarrow H^1(S^1, \mathscr{A}_{s,0}) \rightarrow 0 \tag{A.3.1.2}$$

are exact. Utilizing Theorem A.3.0.5 we rewrite exact sequences (A.3.1.2) as

$$0 \rightarrow H^1(S^1, \mathscr{K}_s) \rightarrow \mathbf{C}[[z^{-1}]]_s/\mathbf{C}\{z^{-1}\} \rightarrow \mathbf{C}[[z^{-1}]]_s/\mathbf{C}\{z^{-1}\} \rightarrow 0. \tag{A.3.1.3}$$

We can summarize what we have explained above in the theorem below.

THEOREM A.3.1.1. *For* $0 \le s \le +\infty$, *we have*

$$H^1(S^1, \mathscr{K}_s) \simeq \mathrm{Kernel}\left(P : \mathbf{C}[[z^{-1}]]_s/\mathbf{C}\{z^{-1}\} \rightarrow \mathbf{C}[[z^{-1}]]_s/\mathbf{C}\{z^{-1}\}\right) \tag{A.3.1.4}$$

as $\mathbf{C}$-*linear spaces.*

Set

$$\ell_s = \dim_{\mathbf{C}} H^1(S^1, \mathscr{K}_s) \qquad (0 \le s \le +\infty). \tag{A.3.1.5}$$

THEOREM A.3.1.2 (J.-P. Ramis [123]). *Let*

$$0 < k_1 < k_2 < \cdots < k_p \tag{A.3.1.6}$$

be all of the positive slopes of the Newton polygon $\mathscr{N}(P)$. *Then* ℓ_s *satisfies the conditions below*:

(i) $0 \le \ell_s \le \ell_{+\infty} < +\infty \quad (0 \le s \le +\infty)$,

(ii) $\ell_{s_1} \leq \ell_{s_2}$ *if* $s_1 < s_2$,
(iii)

$$\ell_s = \begin{cases} 0 & if \ s < \frac{1}{k_p}, \\ \ell_{\frac{1}{k_{j+1}}} & if \ \frac{1}{k_{j+1}} \leq s < \frac{1}{k_j}, \\ \ell_{\frac{1}{k_1}} & if \ \frac{1}{k_1} \leq s \leq +\infty. \end{cases} \tag{A.3.1.7}$$

OBSERVATION A.3.1.3. Let $J : \mathbf{C}[[z^{-1}]]_s \to \mathbf{C}[[z^{-1}]]_s/\mathbf{C}\{z^{-1}\}$ be the natural $\mathbf{C}$-linear map such that $f \in J(f)$ for every $f \in \mathbf{C}[[Z^{-1}]]_s$. Then $f \in \mathbf{C}[[z^{-1}]]_s$ satisfies the condition:

$$J(f) \in \text{Kernel}(P : \mathbf{C}[[z^{-1}]]_s/\mathbf{C}\{z^{-1}\} \to \mathbf{C}[[z^{-1}]]_s/\mathbf{C}\{z^{-1}\}),$$

if and only if

$$P[f] \in \mathbf{C}\{z^{-1}\}. \tag{A.3.1.8}$$

Condition (A.3.1.8) means that f satisfies a linear differential equation $P[y] = \phi$ for some ϕ in $\mathbf{C}\{z^{-1}\}$. Note that the dimension ℓ_s of

$$\text{Kernel}(P : \mathbf{C}[[z^{-1}]]_s/\mathbf{C}\{z^{-1}\} \to \mathbf{C}[[z^{-1}]]_s/\mathbf{C}\{z^{-1}\})$$

is continuous in s except possibly for $s \in \{\frac{1}{k_1}, \cdots, \frac{1}{k_p}\}$. By utilizing this fact, J.-P. Ramis proved Theorem A.2.4.2 in the case when differential equation (A.2.4.10) is linear (cf. (2) of remark A.2.4.3).

OBSERVATION A.3.1.4. There are two ways to show that $\ell_{+\infty}$ is finite.

METHOD 1. Let us look at

$$\ell_{+\infty} = \dim_{\mathbf{C}} \text{Kernel}\left(P : \mathbf{C}[[z^{-1}]]/\mathbf{C}\{z^{-1}\} \to \mathbf{C}[[z^{-1}]]/\mathbf{C}\{z^{-1}\}\right).$$

To evaluate $\ell_{+\infty}$, we start with the commutative diagram below:

$$\begin{array}{ccccccccc} 0 \to & \mathbf{C}\{z^{-1}\} \to & \mathbf{C}[[z^{-1}]] \to & \mathbf{C}[[z^{-1}]]/\mathbf{C}\{z^{-1}\} \to & 0 \\ & \downarrow P & \downarrow P & \downarrow P & \\ 0 \to & \mathbf{C}\{z^{-1}\} \to & \mathbf{C}[[z^{-1}]] \to & \mathbf{C}[[z^{-1}]]/\mathbf{C}\{z^{-1}\} \to & 0 \end{array}.$$

Set

$$\begin{cases} K_1 = \text{Kernel}(P : \mathbf{C}\{z^{-1}\} \to \mathbf{C}\{z^{-1}\}), \\ K_2 = \text{Kernel}(P : \mathbf{C}[[z^{-1}]] \to \mathbf{C}[[z^{-1}]]), \\ K_3 = \text{Kernel}(P : \mathbf{C}[[z^{-1}]]/\mathbf{C}\{z^{-1}\} \to \mathbf{C}[[z^{-1}]]/\mathbf{C}\{z^{-1}\}), \\ CK_1 = \text{Cokernel}(P : \mathbf{C}\{z^{-1}\} \to \mathbf{C}\{z^{-1}\}), \\ CK_2 = \text{Cokernel}(P : \mathbf{C}[[z^{-1}]] \to \mathbf{C}[[z^{-1}]]), \\ CK_3 = \text{Cokernel}(P : \mathbf{C}[[z^{-1}]]/\mathbf{C}\{z^{-1}\} \to \mathbf{C}[[z^{-1}]]/\mathbf{C}\{z^{-1}\}), \end{cases}$$

Then

$$0 \to K_1 \to K_2 \to K_3 \to CK_1 \to CK_2 \to CK_3 \to 0$$

is exact. Note that K_1 and K_2 are finite-dimensional over $\mathbf{C}$ and that $CK_3 = 0$ (cf. (A.3.1.3)). Also it is not difficult to show that CK_2 is finite-dimensional over $\mathbf{C}$. Hence if CK_1 is finite-dimensional over $\mathbf{C}$, we can conclude that $\ell_{+\infty}$ is finite. To show that CK_1 is finite-dimensional over $\mathbf{C}$, we look at the diagram below:

$$\begin{array}{ccccccc} 0 \to & z^{-N}\mathbf{C}\{z^{-1}\} \to & \mathbf{C}\{z^{-1}\} \to & \mathbf{C}\{z^{-1}\}/z^{-N}\mathbf{C}\{z^{-1}\} \to & 0 \\ & \downarrow P & \downarrow P & \downarrow P & \\ 0 \to & z^{-M}\mathbf{C}\{z^{-1}\} \to & \mathbf{C}\{z^{-1}\} \to & \mathbf{C}\{z^{-1}\}/z^{-M}\mathbf{C}\{z^{-1}\} \to & 0 \end{array},$$

where M and N are suitable positive integers such that $M \geq N$. It can be shown that there exists a nonnegative integer N_0 such that

$$\text{Kernel}(P : z^{-N}\mathbf{C}\{z^{-1}\} \to z^{-N-N_0}\mathbf{C}\{z^{-1}\}) = 0$$

for all sufficiently large integers N. By utilizing a method due to W. A. Harris–Y. Sibuya–L. Weinberg [63] we can show that, if a positive integer N is sufficiently large,

$$\dim_{\mathbf{C}} \text{Cokernel}(P : z^{-N}\mathbf{C}\{z^{-1}\} \to z^{-N-N_0}\mathbf{C}\{z^{-1}\})$$

is finite and can be evaluated explicitly. Hence CK_1 is finite-dimensional over $\mathbf{C}$.

METHOD 2. We look at

$$\ell_{+\infty} = \dim_{\mathbf{C}} H^1(S^1, \mathscr{K}_{+\infty}).$$

Set

$$\ell_{+\infty}(\theta) = \dim_{\mathbf{C}} \mathscr{A}_{+\infty,0}(\theta) \tag{A.3.1.9}$$

Then the function $\ell_{+\infty}(\theta)$ is piecewise continuous on S^1 with a finite number of jumps. The discontinuity of $\ell_{+\infty}(\theta)$ takes place only at those values of θ which define the so-called *Stokes lines*. By virtue of an idea due to P. Deligne [55] we can show that

$$\ell_{+\infty} = \frac{1}{2}\,\text{Var}(\theta; \ell_{+\infty}(\theta)), \tag{A.3.1.10}$$

where $\text{Var}(\theta; \ell_{+\infty}(\theta))$ denotes the total sum of the absolute values of jumps of $\ell_{+\infty}(\theta)$ over S^1 (cf. J.-P. Ramis [123], D. Bertrand [51], and D. G. Babbitt–V. S. Varadarajan [40]).

J.-P. Ramis [123] also gave the formulas

$$\ell_s = \frac{1}{2}\,\text{Var}(\theta; \ell_s(\theta)) \tag{A.3.1.11}$$

for $0 \leq s < +\infty$, where $\ell_s(\theta) = \dim_{\mathbf{C}} \mathscr{A}_{s,0}(\theta)$.

REMARK A.3.1.5.

(1) Method 2 utilizes the theory of asymptotic solutions, but Method 1 does not. Note that our proof of the refined theorem of Maillet (cf. Theorem A.2.4.2) of §A.2.4 utilizes the theory of asymptotic solutions. However the proofs given by E. Maillet, K. Mahler, H.Gingold and B. Malgrange (cf. Remark A.2.4.3) do not use such asymptotic analysis.
(2) The evaluation of ℓ_s amounts to the evaluation of indices of the operator P in various situations such as shown at the beginning of this section (cf. B. Malgrange [98]). Generally speaking such a study can be done by two methods, one of which is based on asymptotic analysis but the other is not. J.-P. Ramis treated the indices of P by means of asymptotic analysis in his paper [123] and without utilizing aysmptotic analysis in his paper [125]. Those indices are explicitly given by means of the Newton polygon $\mathcal{N}(P)$ of P.
(3) The main purpose of the work of W. A. Harris–Y. Sibuya–L. Weinberg [63] was to estimate the number of linearly independent convergent power-series solutions of a system of linear differential equations at $z = 0$ (cf. O. Perron [121], F. Lettenmeyer [90]; see also P. Hartman [66]). Their method does not involve asymptotic analysis. M. Hukuhara–M. Iwano [68] and M. Iwano [69] studied similar problems by utilizing asymptotic analysis. The idea of P. Deligne is quite similar to the idea of Hukuhara–Iwano.

DEFINITION A.3.1.6 (B. Malgrange [98]). The integer $\ell_{+\infty}$ is called the **irregularity** of the operator P at $z = \infty$.

REMARK A.3.1.7. The point at $z = \infty$ is a regular singular point of P if and only if $\ell_{+\infty} = 0$.

§A.3.2. Classification by means of cohomologies

To explain Malgrange's idea of classifying analytic linear differential equations by means of cohomologies, we shall make some observations.

OBSERVATION A.3.2.1. For a given open connected arc V of S^1, let us denote by $\mathcal{G}_s(V)$ the set of all $n \times n$ matrices F such that $F \in gl(n, \mathcal{A}_s(V))$ and $F^{-1} \in gl(n, \mathcal{A}_s(V))$. Also set $\mathcal{G}_{s,0}(V) = \{F \in \mathcal{G}_s(V);\ A_{sp}(F) = I_n\}$, where I_n is the $n \times n$ identity matrix. Let $\mathcal{G}_s$ and $\mathcal{G}_{s,0}$ be the sheaves over S^1 defined by $\mathcal{G}_s(V)$ and $\mathcal{G}_{s,0}(V)$, respectively. We denote by $\mathcal{G}_s(\theta)$ and $\mathcal{G}_{s,0}(\theta)$ the stalks of $\mathcal{G}_s$ and $\mathcal{G}_{s,0}$ at θ, respectively. If we use the following homomorphisms of noncommutative groups:

$$\begin{cases} A_{sp} : \mathcal{G}_s(V) & \rightarrow & GL(n, \mathbf{C}[[z^{-1}]]_s), \\ \mathrm{id}_{\mathcal{G}_{s,0}(V)} : \mathcal{G}_{s,0}(V) & \rightarrow & \mathcal{G}_s(V), \end{cases}$$

we can derive exact sequences

$$0 \rightarrow \mathcal{G}_{s,0}(V) \rightarrow \mathcal{G}_s(V) \rightarrow GL(n, \mathbf{C}[[z^{-1}]]_s) \rightarrow 0$$

for a sufficiently small arc V and

$$0 \rightarrow \mathcal{G}_{s,0}(\theta) \rightarrow \mathcal{G}_s(\theta) \rightarrow GL(n, \mathbf{C}[[z^{-1}]]_s) \rightarrow 0$$

for every $\theta \in S^1$. Hence we have exact sequences of homomorphisms of sheaves

$$0 \rightarrow \mathcal{G}_{s,0} \rightarrow \mathcal{G}_s \rightarrow GL(n, \mathbf{C}[[z^{-1}]]_s) \rightarrow 0$$

for $0 \leq s \leq +\infty$, where $GL(n, \mathbf{C}[[z^{-1}]]_s)$ are constant sheaves.

OBSERVATION A.3.2.2. Since S^1 is a real one-dimensional manifold,

$$\begin{cases} H^k(S^1, \mathcal{G}_{s,0}) = 0, \quad H^k(S^1, \mathcal{G}_s) = 0, \\ H^k(S^1, GL(n, \mathbf{C}[[z^{-1}]]_s)) = 0, \end{cases} \qquad (k \geq 2),$$

and also

$$\begin{cases} H^0(S^1, \mathcal{G}_{s,0}) = 0, \quad H^0(S^1, \mathcal{G}_s) = GL(n, \mathbf{C}\{z^{-1}\}), \\ H^0(S^1, GL(n, \mathbf{C}[[z^{-1}]]_s)) = GL(n, \mathbf{C}[[z^{-1}]]_s). \end{cases}$$

For each s ($0 \leq s \leq +\infty$), utilizing the results above, we can define a map

$$J_s : GL(n, \mathbf{C}[[z^{-1}]]_s)/GL(n, \mathbf{C}\{z^{-1}\}) \rightarrow H^1(S^1, \mathcal{G}_{s,0}) \qquad \text{(A.3.2.1)}$$

as follows: Let $\{V_1, \cdots, V_N\}$ be a covering of S^1 by open connected arcs of reasonable sizes. Given $F \in GL(n, \mathbf{C}[[z^{-1}]]_s)$, there exist $F_\ell \in \mathcal{G}_s(V_\ell)$ ($\ell = 1, \cdots, N$) such that $A_{sp}(F_\ell) = F$. Set $F_{\ell k} = F_\ell F_k^{-1}$ whenever $V_\ell \cap V_k \neq \varnothing$. Then $F_{\ell k} \in \mathcal{G}_{s,0}(V_\ell \cap V_k)$ and the assignment

$$F \longmapsto \{F_{\ell k};\ V_\ell \cap V_k \neq \varnothing\} \qquad \text{(A.3.2.2)}$$

induces map (A.3.2.1).

THEOREM A.3.2.3 (B. Malgrange [99] and J.-P. Ramis [123]). *The maps J_s are bijective.*

In particular, the surjectivity of the J_s follows from Theorem 6.4.1 (or Theorem 6.9.7) and Theorem A.2.4.1.

OBSERVATION A.3.2.4. Let us consider two differential equations

$$\frac{dW}{dz} = A_0(z)W \qquad \text{(A.3.2.3)}$$

and

$$\frac{dV}{dz} = A(z)V, \tag{A.3.2.4}$$

where W and V are $n \times n$ unknown matrices and $A_0(z)$ and $A(z) \in gl(n, \mathscr{L})$. Here $\mathscr{L}$ denotes $\mathbf{C}\{z^{-1}\}[z]$.

DEFINITION A.3.2.5. Two differential equations (A.3.2.3) and (A.3.2.4) are said to be **formally equivalent of Gevrey order** s if there exists an $n \times n$ matrix $\hat{P}(z) \in GL(n, \mathbf{C}[[z^{-1}]]_s)$ such that the transformation

$$W = \hat{P}(z)V \tag{A.3.2.5}$$

changes (A.3.2.3) to (A.3.2.4).

Since $A = \hat{P}^{-1}A_0\hat{P} - \hat{P}^{-1}\frac{d\hat{P}}{dz}$, let us look at

$$G_s[A_0] = \{\hat{P} \in GL(n, \mathbf{C}[[z^{-1}]]_s);\ \hat{P}[A_0] \in gl(n, \mathscr{L})\}, \tag{A.3.2.6}$$

where

$$\hat{P}[A_0] = \hat{P}^{-1}A_0\hat{P} - \hat{P}^{-1}\frac{d\hat{P}}{dz}. \tag{A.3.2.7}$$

Note that $G_s[A_0]$ is a subset of $GL(n, \mathbf{C}[[z^{-1}]]_s)$ such that

$$G_s[A_0]GL(n, \mathbf{C}\{z^{-1}\}) \subset G_s[A_0].$$

Hence the map J_s (cf. (A.3.2.1)) can be restricted to $G_s[A_0]/GL(n, \mathbf{C}\{z^{-1}\})$. We must investigate

$$\text{Image}(J_s : G_s[A_0]/GL(n, \mathbf{C}\{z^{-1}\}) \ \rightarrow\ H^1(S^1, \mathscr{G}_{s,0})).$$

Let $\{V_1, \cdots, V_N\}$ be a covering of S^1 by open connected arcs of reasonable sizes. Given $F \in G_s[A_0]$, we can choose $F_\ell \in \mathscr{G}_s(V_\ell)$ $(\ell = 1, \cdots, N)$ such that

$$\begin{cases} A_{sp}(F_\ell) = F, \\ F[A_0] = F_\ell^{-1}A_0F_\ell - F_\ell^{-1}\dfrac{dF_\ell}{dz}. \end{cases} \tag{A.3.2.8}$$

Set $F_{\ell k} = F_\ell F_k^{-1}$ whenever $V_\ell \cap V_k \neq \emptyset$. Then $F_{\ell k}$ belongs to $\mathscr{G}_{s,0}(V_\ell \cap V_k)$ and $X = F_{\ell k}$ satisfy the condition

$$A_0 = X^{-1}A_0X - X^{-1}\frac{dX}{dz}. \tag{A.3.2.9}$$

Note that condition (A.3.2.9) and (6.11.8) are the same. Note also that if X satisfies condition (A.3.2.9) and if $n \times n$ matrices P and Q satisfy the condition $XP = Q$, then

$$P^{-1}A_0P - P^{-1}\frac{dP}{dz} = Q^{-1}A_0Q - Q^{-1}\frac{dQ}{dz}.$$

For an open connected arc V of S^1 we set

$$\mathcal{G}_{s,0}[A_0](V) = \left\{ F \in \mathcal{G}_{s,0}(V) \ ; \ A_0 = F^{-1}A_0F - F^{-1}\frac{dF}{dz} \right\} \qquad \text{(A.3.2.10)}$$

and we define sheaves $\mathcal{G}_{s,0}[A_0]$ over S^1. Then B. Malgrange [99] and J.-P. Ramis [124] gave the theorem below.

THEOREM A.3.2.6. *For $0 \le s \le +\infty$, there are bijective maps*

$$J_s[A_0] : G_s[A_0]/GL(n, \mathbf{C}\{z^{-1}\}) \ \to \ H^1(S^1, \mathcal{G}_{s,0}[A_0]) . \qquad \text{(A3.2.11)}$$

This is the classification theorem due to the idea of Malgrange.

OBSERVATION A.3.2.7. Let $\{V_1, \cdots, V_N\}$ be a covering of S^1 by open connected arcs of reasonable sizes and $\{F_{\ell k} \in \mathcal{G}_{s,0}[A_0](V_\ell \cap V_k); V_\ell \cap V_k \ne \varnothing\}$ be a one-cocycle defined by the sheaf $\mathcal{G}_{s,0}[A_0]$. Also let $\{\mathcal{S}_\ell, \Phi_\ell(z)\}$ be a system of fundamental solutions of (A.3.2.3) in the neighborhood of $z = \infty$, where

$$\mathcal{S}_\ell = \mathcal{S}(\rho(\theta), V_\ell) = \{z \ ; \arg z \in V_\ell, \ |z| > \rho(\arg z)\}.$$

Here $\rho(\theta)$ is a positive-valued continuous function on the arc V. Condition (A.3.2.9) implies that there exist $n \times n$ matrices $C_{\ell k} \in GL(n, \mathbf{C})$ such that

$$F_{\ell k}(z) = \Phi_\ell(z) C_{\ell k} \Phi_k(z)^{-1} \qquad \text{on} \quad \mathcal{S}_\ell \cap \mathcal{S}_k.$$

Since $\{F_{\ell k}\}$ is a one-cocycle, the matrices $C_{\ell k}$ satisfy the condition below:

$$C_{hk} C_{k\ell} = C_{h\ell} \qquad \text{whenever} \quad \mathcal{S}_h \cap \mathcal{S}_k \cap \mathcal{S}_\ell \ne \varnothing.$$

Furthermore, $F_{\ell k} \in \mathcal{G}_{s,0}[A_0](V_\ell \cap V_k)$ means that

$$A_{sp}(F_{\ell k}) \ = \ I_n, \qquad \text{(A.3.2.12)}$$

where I_n is the $n \times n$ identity matrix. This implies that $\{F_{\ell k}\}$ defines a trivial asymptotic system at $z = \infty$ (cf. Definition 6.9.4, Definition 6.9.5 and Theorem 6.9.7). Let us compare Definition 6.10.3 of Stokes phenomena and what we have explained above.

DEFINITION 6.10.3. We call $\{A_0(z), \{\mathcal{S}_\ell, \Phi_\ell(z)\}, C_{k\ell}\}$ a **Stokes phenomenon** at $z = \infty$ if the following conditions are satisfied:

(1) the differential equation (6.10.1) (i.e., $\frac{dW_0}{dz} = A_0(z)W_0$) belongs to the equivalence class $\mathcal{E}$,

(2) $\{\mathcal{S}_1, \cdots, \mathcal{S}_N\}$ is a covering at $z = \infty$ satisfying the conditions of Theorem 6.10.1,

(3) $\{\mathcal{S}_\ell, \Phi_\ell(z)\}$ is a system of fundamental solutions of (6.10.1) in the neighborhood of $z = \infty$,

(4) the $C_{k\ell}$ are matrices in $GL(n, \mathbf{C})$ which are defined whenever $\mathcal{S}_k \cap \mathcal{S}_\ell \ne \varnothing$,

(5) $C_{hk}C_{k\ell} = C_{h\ell}$ whenever $\mathscr{S}_h \cap \mathscr{S}_k \cap \mathscr{S}_\ell \neq \emptyset$,
(6) $\Phi_k(z)\ C_{k\ell}\ \Phi_\ell(z)^{-1} \in GL(n, \mathscr{A}(\mathscr{S}_k \cap \mathscr{S}_\ell))$ whenever $\mathscr{S}_k \cap \mathscr{S}_\ell \neq \emptyset$,
(7) the asymptotic system $\{\Phi_k(z)\ C_{k\ell}\ \Phi_\ell(z)^{-1}\}$ is trivial at $z = \infty$.

Right after this definition, we made the following remark.

REMARK 6.10.4. Condition (7) can be replaced by

(7′) the asymptotic system $\{\Phi_k(z)\ C_{k\ell}\ \Phi_\ell(z)^{-1}\}$ is formally trivial.

In particular, if $A_{sp}(\Phi_k(z)\ C_{k\ell}\ \Phi_\ell(z)^{-1}) = I$ whenever $\mathscr{S}_k \cap \mathscr{S}_\ell \neq \emptyset$, condition (7) is satisfied.

Note that we did not assume condition (A.3.2.12). Instead, we assumed condition (7). To go from $G_s[A_0]/GL(n, \mathbf{C}\{z^{-1}\})$ to the moduli space of analytic linear differential equations with respect to the holomorphic classification, we must classify $G_s[A_0]/GL(n, \mathbf{C}\{z^{-1}\})$ further. For details, see D. G. Babbitt–V. S. Varadarajan [39, 40].

REMARK A.3.2.8. The idea of B. Malgrange applies also to nonlinear differential equations (cf. J. Martinet–J.-P. Ramis [105, 106]).

§A.3.3. Analytization

The main problem of Chapter 3 was to explain the following theorem of G. D. Birkhoff.

THEOREM 3.3.1 (G. D. Birkhoff). *Given an integer k and a matrix $A(\zeta) \in gl(n, \mathscr{O}(\Delta(0, \frac{1}{R_0})))$, we consider the differential equation*

$$\frac{dW}{dz} = z^k A\left(\frac{1}{z}\right) W. \tag{3.3.3}$$

Then, there exists a matrix $P(\zeta) \in GL(n, \mathscr{O}(\Delta(0, \frac{1}{R_0})))$ such that the transformation

$$W = P\left(\frac{1}{z}\right) V \tag{3.3.4}$$

changes (3.3.3) *to a differential equation*

$$\frac{dV}{dz} = z^k B\left(\frac{1}{z}\right) V \tag{3.3.5}$$

with a matrix $B(\frac{1}{z})$ whose entries are polynomials in $\frac{1}{z}$ with complex constant coefficients. Moreover we can choose such a $P(\zeta)$ so that $z = 0$ is at worst a regular singular point of (3.3.5).

In this section we shall explain how to relax the requirement that A should belong to $\in gl(n, \mathscr{O}(\Delta(0, \frac{1}{R_0})))$. The main result is the following theorem.

THEOREM A.3.3.1 (J.-P. Ramis [124]). *Given an s $(0 \le s \le +\infty)$ and an operator $L = \frac{d}{dz} - z^k \hat{A}(z)$, where k is an integer and $\hat{A}$ is an $n \times n$ matrix $\in gl(n, \mathbf{C}[[z^{-1}]]_s)$, there exists an $n \times n$ matrix $\hat{P} \in GL(n, \mathbf{C}[[z^{-1}]]_s)$ such that*

$$\hat{P}^{-1} \circ L \circ \hat{P} = \frac{d}{dz} - z^k A(z) \qquad \textit{with} \quad A \in gl(n, \mathbf{C}\{z^{-1}\}). \qquad \text{(A.3.3.1)}$$

REMARK A.3.3.2. $\hat{P}^{-1} \circ L \circ \hat{P}$ denotes the composition of three operators $\hat{P}^{-1}$, L and $\hat{P}$.

SKETCH OF PROOF OF THEOREM A.3.3.1. Let $\{V_1, \cdots, V_N\}$ be a *good* covering of S^1 by open connected arcs of reasonable sizes. Then there exist $A_\ell \in gl(n, \mathcal{A}_s(V_\ell))$ and $P_{\ell h} \in \mathcal{G}_{s,0}(V_\ell \cap V_h)$, $(V_\ell \cap V_h \neq \emptyset)$ such that

$$\begin{cases} A_{sp}(A_\ell) = \hat{A} & (\ell = 1, \cdots, N), \\ \dfrac{dP_{\ell h}(z)}{dz} = z^k\{A_\ell(z)P_{\ell h}(z) - P_{\ell h}(z)A_h(z)\} & (z \in V_\ell \cap V_h \neq \emptyset), \\ P_{\ell j}P_{jh} = P_{\ell h} & \text{if} \quad V_\ell \cap V_j \cap V_h \neq \emptyset. \end{cases}$$

Set $\mathcal{L}_\ell = \frac{d}{dz} - z^k A_\ell(z)$. Then

$$\mathcal{L}_h = P_{\ell h}^{-1} \circ \mathcal{L}_\ell \circ P_{\ell h} \qquad \text{if} \quad V_\ell \cap V_h \neq \emptyset.$$

By virtue of Theorem 6.9.7 and Theorem A.2.4.1, there exist $P_\ell \in \mathcal{G}_s(V_\ell)$, $(\ell = 1, \cdots, N)$ such that

$$\begin{cases} A_{sp}(P_\ell - I_n) \in z^{-1} gl(n, \mathbf{C}[[z^{-1}]]_s), \\ P_{\ell h}(z) = P_\ell(z) P_h(z)^{-1} \qquad \text{if} \quad V_\ell \cap V_h \neq \emptyset. \end{cases}$$

Hence

$$P_\ell^{-1} \circ \mathcal{L}_\ell \circ P_\ell = P_h^{-1} \circ \mathcal{L}_h \circ P_k \qquad \text{on} \quad V_\ell \cap V_h \neq \emptyset.$$

Now if we set $A_{sp}(P_\ell) = \hat{P}$ and $\hat{P}^{-1} \circ \mathcal{L} \circ \hat{P} = \frac{d}{dz} - z^k A(z)$, we have

$$\begin{cases} \hat{P} \in GL(n, \mathbf{C}[[z^{-1}]]_s), \\ A \in gl(n, \mathbf{C}\{z^{-1}\}). \quad \square \end{cases}$$

[The idea of this proof is due to B. Malgrange (cf. J.-P. Ramis [124]). The author of this book also wrote a paper [138] on analytization which is not quite complete but not totally useless.]

§A.3.4. Injectivity of D-modules $\mathcal{A}_{s,0}(\theta)$

Set

$$D = \left\{ \sum_{h=0}^{m} a_h(z)\, \delta^h \; ; \; a_h \in \mathbf{C}\{z^{-1}\}, \; m \in \mathbf{N} \right\}. \qquad \text{(A.3.4.1)}$$

We can regard D as a noncommutative ring. Then, for every $\theta \in S^1$, the stalks $\mathcal{A}_{s,0}(\theta)$, $\mathcal{A}_s(\theta)$, and $\mathbf{C}[[z^{-1}]]_s$ are D-modules and we have exact sequences of homomorphisms of D-modules

$$0 \to \mathcal{A}_{s,0}(\theta) \to \mathcal{A}_s(\theta) \to \mathbf{C}[[z^{-1}]]_s \to 0, \tag{A.3.4.2}$$

where

$$\begin{cases} A_{sp} : \mathcal{A}_s(\theta) & \to \mathbf{C}[[z^{-1}]]_s, \\ \mathrm{id}_{\mathcal{A}_{s,0}(\theta)} : \mathcal{A}_{s,0}(\theta) & \to \mathcal{A}_s(\theta). \end{cases}$$

The main result of this section is the theorem below.

THEOREM A.3.4.1 (Y. Sibuya [137]). *Exact sequences* (A.3.4.2) *of homomorphisms of D-modules split.*

The proof of this theorem is based on the observations below:

OBSERVATION A.3.4.2.

I: A D-module M is said to be *divisible* if the following conditions are satisfied:

(i) if $z^{-1}a = 0$ for an $a \in M$, then $a = 0$,

(ii) for every $a \in M$ and every $P \in D$, there exists $b \in M$ such that $Pb = a$.

II: A D-module M is said to be *injective* if for every diagram

$$\begin{array}{ccccc} 0 & \longrightarrow & M_1 & \xrightarrow{\ i\ } & M_2 \\ & & \big\downarrow \lambda & & \\ & & M & & \end{array}$$

of homomorphisms of D-modules, where the horizontal sequence is exact, there exists $\mu \in \mathrm{Hom}_D(M_2, M)$ such that $\lambda = \mu \circ i$.

III: If a D-module M is divisible , then M is injective (cf. R. Godement [61] and Y. Sibuya [137]).

OBSERVATION A.3.4.3. Lemma A.2.3.3 implies that the D-modules $\mathcal{A}_{s,0}(\theta)$ are divisible and hence injective. In particular, if we consider the diagram below

$$\begin{array}{ccccc} 0 & \longrightarrow & \mathcal{A}_{s,0}(\theta) & \xrightarrow{\ \mathrm{id}\ } & \mathcal{A}_s(\theta) \\ & & \big\downarrow \mathrm{id} & & \\ & & \mathcal{A}_{s,0}(\theta) & & \end{array}$$

there exist homomorphisms $\mu_{s,\theta} : \mathcal{A}_s(\theta) \to \mathcal{A}_{s,0}(\theta)$ satisfying the requirement

$$\mu_{s,\theta}\Big|_{\mathcal{A}_{s,0}(\theta)} = \mathrm{id}_{\mathcal{A}_{s,0}(\theta)}.$$

Note that $\mu_{s,\theta}$ are not unique (cf. Y. Sibuya [137]).

OBSERVATION A.3.4.4. For a given $\hat{f} \in \mathbf{C}[[z^{-1}]]_s$, choose $f \in \mathcal{A}_s(\theta)$ such that $A_{sp}(f) = \hat{f}$ and set

$$\phi_{s,\theta}(\hat{f}) = f - \mu_{s,\theta}(f). \tag{A.3.4.3}$$

Then $\phi_{s,\theta}(\hat{f})$ are independent of the choice of f. Furthermore the maps

$$\phi_{s,\theta} : \mathbf{C}[[z^{-1}]]_s \to \mathcal{A}_s(\theta)$$

are homomorphisms of D-modules such that

$$A_{sp} \circ \phi_{s,\theta} = \mathrm{id}_{\mathbf{C}[[z^{-1}]]_s}. \tag{A.3.4.4}$$

This proves Theorem A.3.4.1. Note that

$$\begin{cases} \mu_{s,\theta}\Big|_{\mathbf{C}\{z^{-1}\}} = 0, & \phi_{s,\theta}\Big|_{\mathbf{C}\{z^{-1}\}} = \mathrm{id}_{\mathbf{C}\{z^{-1}\}}, \\ \mu_{s,\theta} \circ \phi_{s,\theta} = 0, & \phi_{s,\theta} \circ A_{sp} = \mathrm{id}_{\mathcal{A}_s(\theta)} - \mu_{s,\theta}. \end{cases} \tag{A.3.4.5}$$

APPLICATION A.3.4.5:. Let us consider the following system of linear differential equations of higher order (cf. Y. Sibuya [137]):

$$\sum_{h \in N(j)} p_{jh}\, u_h = f_j \qquad (j \in \mathcal{J}), \tag{A.3.4.6}$$

where $\mathcal{H}$ and $\mathcal{J}$ are nonempty sets, $N(j)$ are nonempty subsets of $\mathcal{H}$, $p_{jh} \in D$, $u_h \in \mathcal{A}_s(\theta)$ are unknowns, and $f_j \in \mathcal{A}_s(\theta)$. We shall look at system (A.3.4.6) under the assumption below.

ASSUMPTION A.3.4.6. *There exist* $\hat{g}_h \in \mathbf{C}[[z^{-1}]]_s$ $(h \in \mathcal{H})$ *such that*

$$\sum_{h \in N(j)} p_{jh}\, \hat{g}_h = A_{sp}(f_j) \qquad (j \in \mathcal{J}), \tag{A.3.4.7}$$

i.e., $u_h = \hat{g}_h$ $(h \in \mathcal{H})$ *is a formal solution of system* (A.3.4.6).

First of all, the following theorem is a basic result.

THEOREM A.3.4.7. *System* (A.3.4.6) *admits a solution*

$$u_h = \eta_h \in \mathcal{A}_s(\theta) \qquad (h \in \mathcal{H})$$

satisfying the condition

$$A_{sp}(\eta_h) = \hat{g}_h \qquad (h \in \mathcal{H}) \tag{A.3.4.8}$$

if and only if the system

$$\sum_{h \in N(j)} p_{jh}\, v_h = \mu_{s,\theta}(f_j) \qquad (j \in \mathcal{J}) \tag{A.3.4.9}$$

admits a solution

$$v_h = \sigma_h \in \mathscr{A}_{s,0}(\theta) \qquad (h \in \mathscr{H}).$$

SKETCH OF PROOF. Assumption (A.3.4.7) implies that

$$\sum_{h \in N(j)} p_{jh}\, \phi_{s,\theta}(\hat{g}_h) = f_j - \mu_{s,\theta}(f_j) \qquad (j \in \mathscr{J}).$$

Hence

$$\eta_h = \phi_{s,\theta}(\hat{g}_h) + \sigma_h \qquad (h \in \mathscr{H}). \tag{A.3.4.10}$$

□

Since $\mu_{s,\theta}(f) = 0$ if $f \in \mathbf{C}\{z^{-1}\}$, we derive the following corollary of Theorem A.3.4.7.

COROLLARY A.3.4.8. *If* $f_j \in \mathbf{C}\{z^{-1}\}$ $(j \in \mathscr{J})$, *then*

$$\eta_h = \phi_{s,\theta}(\hat{g}_h) \in \mathscr{A}_s(\theta) \qquad (h \in \mathscr{H})$$

satisfy system (A.3.4.6) *and condition* (A.3.4.8).

To investigate system (A.3.4.9), the following lemma is helpful.

LEMMA A.3.4.9. *If* $f_j \in \mathscr{A}_{s,0}(\theta)$ $(j \in \mathscr{J})$, *then system* (A.3.4.6) *admits a solution* $u_h \in \mathscr{A}_{s,0}(\theta)$ $(h \in \mathscr{H})$ *if and only if every finite subsystem of* (A.3.4.6) *admits a solution in* $\mathscr{A}_{s,0}(\theta)$.

This lemma is a particular case of the theorem below.

THEOREM A.3.4.10 (S. Gacsalyi [58]). *Let R be an associative ring* (*with unit*) *and M be an injective R-module. A system of equations*

$$\sum r_{jk} X_k = m_j, \qquad (r_{jk} \in R,\ m_j \in M),$$

where, for each j, $r_{jk} = 0$ *for almost all* k, *has a solution in* M *if and only if any finite subsystem has a solution in* M.

A proof of this theorem is found in L. Fuchs [57] (cf. also Y. Sibuya [137]).

APPENDIX 4

k-Summability

§A.4.0. Preliminaries

In Appendix 2 we studied injectivity and surjectivity of the maps $A_{sp} : \mathscr{A}_s(V) \to \mathbf{C}[[z^{-1}]]_s$ where V is an open connected arc of the unit circle S^1 in the complex plane (cf. Theorems A.2.2.11 and A.2.2.14). In this appendix we shall study the case when V is large so that A_{sp} is injective but not surjective. One of the basic results is a theorem (cf. Theorem A.4.0.3) due to F. Nevanlinna [111] (see also G. N. Watson [148]).

Observation A.4.0.1. For a given $f \in \mathscr{A}_s\left(R_0, -\frac{\beta_0}{2}, \frac{\beta_0}{2}\right)$, set $A_{sp}(f) = \sum_{m\geq 0} a_m z^{-m} \in \mathbf{C}[[z^{-1}]]_s$. This means that, for every positive number R greater than R_0 and every positive number β less than β_0, there exist two nonnegative numbers $K_{R,\beta}$ and $B_{R,\beta}$ such that

$$\left| f(z) - \sum_{m=0}^{N} a_m z^{-m} \right| \leq K_{R,\beta} \, ((N+1)!)^s \, (B_{R,\beta})^{N+1} \, |z|^{-(N+1)}$$

on the sectorial domain $\mathscr{D}(R, -\frac{\beta}{2}, \frac{\beta}{2})$ for all nonnegative integers N. Throughout this appendix, we assume that

$$+\infty > \beta_0 > s\pi > 0. \tag{A.4.0.1}$$

Also we denote $\frac{1}{s}$ by k, i.e.

$$k = \frac{1}{s}. \tag{A.4.0.2}$$

Observation A.4.0.2. For a given $g = \sum_{m\geq 0} a_m z^{-m} \in \mathbf{C}[[z^{-1}]]$ let us write

$$\phi_g(z) = \sum_{m\geq 0} \frac{a_m}{\Gamma\left(1 + \frac{m}{k}\right)} z^{-m}. \tag{A.4.0.3}$$

Then $g \in \mathbf{C}[[z^{-1}]]_s$ if and only if $\phi_g \in \mathbf{C}\{z^{-1}\}$. F. Nevanlinna [111] characterized

$$\text{Image}\left(A_{sp} : \mathscr{A}_s\left(R_0, -\frac{\beta_0}{2}, \frac{\beta_0}{2}\right) \to \mathbf{C}[[z^{-1}]]_s \right)$$

by means of the following result.

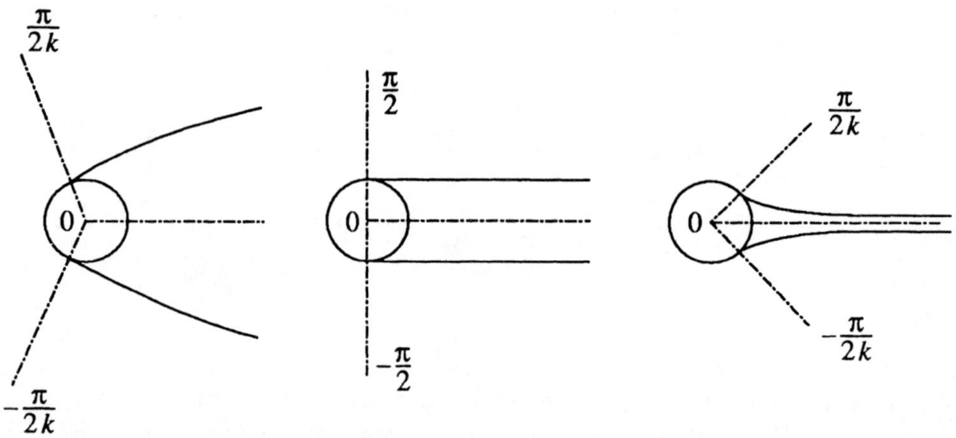

FIGURE 10

THEOREM A.4.0.3 (F. Nevanlinna). *Assume that* β_0 *satisfies condition* (A.4.0.1). *Suppose also that* $f \in \mathscr{A}_s\left(R_0, -\frac{\beta_0}{2}, \frac{\beta_0}{2}\right)$ *and* $g \in \mathbf{C}[[z^{-1}]]_s$ *satisfy the condition*

$$A_{sp}(f) = g.$$

Then $\phi_g(\frac{1}{t})$ *is holomorphic and satisfies the estimate*

$$\left|\phi_g\left(\frac{1}{t}\right)\right| \leq C\exp\left\{\lambda^k \Re(t^k)\right\} \tag{A.4.0.4}$$

on a domain

$$\{t\ ;\ |t| < r_0\} \cup \{t\ ;\ |\Im(t^k)| < r_0^k,\ \Re(t^k) > 0\}, \tag{A.4.0.5}$$

where $C \geq 0$, $\lambda > 0$ *and* $r_0 > 0$ *are suitable numbers* (*cf.* Figure 10). *Furthermore*

$$f(z) = k\ z^k \int_0^{+\infty} \phi_g\left(\frac{1}{t}\right) \exp\left\{-(tz)^k\right\}\ t^{k-1}\,dt \tag{A.4.0.6}$$

for $\Re(z^k) > \rho$ *for a sufficiently large positive number* ρ.

REMARK A.4.0.4.

(1) Theorem A.4.0.3 also implies that the maps A_{sp} are injective but not surjective.
(2) Under assumption (A.4.0.1) the domain $\{z\ ;\ \Re(z^k) > \rho\}$ is contained in the sectorial domain $\mathscr{D}(R_0, -\frac{\beta_0}{2}, \frac{\beta_0}{2})$ if ρ is sufficiently large.

(3) The right-hand side of (A.4.0.6) can be writen as $\ell_{+\infty,k}(\phi_g)$ if we use Definition (A.2.2.15) of the incomplete Leroy transforms of order k.

In the following definition, we assume that β_0 satisfies condition (A.4.0.1).

DEFINITION A.4.0.5 (J.-P. Ramis [124]).

I: If $f \in \mathscr{A}_s\left(R_0, \alpha - \frac{\beta_0}{2}, \alpha + \frac{\beta_0}{2}\right)$ and $g \in \mathbf{C}[[z^{-1}]]_s$ satisfy the condition

$$A_{sp}(f) = g,$$

the power series g is said to be **k-summable in the direction** α and the function f is called the sum of the power series g in the direction α, where $k = \frac{1}{s}$ (cf. (A.4.0.2)).

II: $g \in \mathbf{C}[[z^{-1}]]_s$ is said to be **k-summable** if g is k-summable in every direction α except for a finite number of directions $\alpha_1, \cdots, \alpha_p$. The finite set

$$\Sigma(g) = \{\alpha_1, \cdots, \alpha_p\} \tag{A.4.0.7}$$

is called the singular support of g.

REMARK A.4.0.6.

(a) If k is a positive integer and if a power series $g \in \mathbf{C}[[z^{-1}]]$ is k-summable in the directions $\alpha + \left(\frac{2\pi}{k}\right)h$ $(h = 0, 1, \cdots, k-1)$, then the sum of g in the direction α has an inverse factorial series expansion

$$\sum_{p=0}^{k-1} \frac{1}{z^p} \sum_{n=0}^{\infty} \frac{b_{p,n}}{(ze^{-i\alpha})^k[(ze^{-i\alpha})^k + \omega][(ze^{-i\alpha})^k + 2\omega]\cdots[(ze^{-i\alpha})^k + n\omega]},$$

where the $b_{p,n}$ are suitable numbers and ω is a sufficiently large positive number. This series expansion is uniformly convergent if $\Re((ze^{-i\alpha})^k)$ is sufficiently large (cf. F. Nevanlinna [111] and G. N. Watson [148]).

(b) If g is k-summable in every direction α belonging to an open interval $\alpha_1 < \alpha < \alpha_2$, the sum f_α of g in the directions α all together define a function $f \in \mathscr{A}_s\left(\rho(\theta), \alpha_1 - \frac{\varepsilon_1+\pi}{2k}, \alpha_2 + \frac{\varepsilon_2+\pi}{2k}\right)$, where $\rho(\theta)$ is a positive-valued continuous function on the open interval $\alpha_1 - \frac{\varepsilon_1+\pi}{2k} < \theta < \alpha_2 + \frac{\varepsilon_2+\pi}{2k}$ and ε_1 and ε_2 are nonnegative numbers.

(c) If g is k-summable and the singular support $\Sigma(g)$ is empty, then $g \in \mathbf{C}\{z^{-1}\}$.

(d) If we denote by $\mathbf{C}\{z^{-1}\}_s$ the set of all k-summable power series in $\mathbf{C}[[z^{-1}]]$, where k and s are related by (A.4.0.2), then $\mathbf{C}\{z^{-1}\}_s$ is a commutative differential algebra over **C**. Furthermore we have

$$\mathbf{C}\{z^{-1}\} \subset \mathbf{C}\{z^{-1}\}_s \subset \mathbf{C}[[z^{-1}]]_s. \tag{A.4.0.8}$$

(e) A formal power series $g \in \mathbf{C}[[z^{-1}]]$ is k-summable with the singular support $\Sigma(g) = \{\alpha_1, \cdots, \alpha_p\}$ if and only if

$$g = g_0 + \sum_{j=1}^{p} g_j \tag{A.4.0.9}$$

for some $g_0 \in \mathbf{C}\{z^{-1}\}$ and $g_j \in \mathbf{C}\{z^{-1}\}_s$ with $\Sigma(g_j) = \{\alpha_j\}$ $(j = 1, \cdots, p)$. The g_j $(j = 1, \cdots, p)$ are uniquely determined modulo $\mathbf{C}\{z^{-1}\}$. In fact suppose that $g_1 \in \mathbf{C}\{z^{-1}\}_s$ and $\tilde{g}_1 \in \mathbf{C}\{z^{-1}\}_s$ satisfy the condition

$$\begin{cases} \Sigma(g_1) = \Sigma(\tilde{g}_1) = \{\alpha_1\}, \\ \Sigma(g - g_1) = \Sigma(g - \tilde{g}_1) = \{\alpha_2, \cdots, \alpha_p\}. \end{cases}$$

Then $g_1 - \tilde{g}_1 \in \mathbf{C}\{z^{-1}\}_s$ and

$$\Sigma(g_1 - \tilde{g}_1) \subset \{\alpha_1\} \cap \{\alpha_2, \cdots, \alpha_p\} = \emptyset.$$

Hence $g_1 - \tilde{g}_1 \in \mathbf{C}\{z^{-1}\}$. Therefore we can use induction on p if we can find a $g_1 \in \mathbf{C}\{z^{-1}\}_s$ satisfying the condition

$$\Sigma(g_1) = \{\alpha_1\}, \qquad \Sigma(g - g_1) = \{\alpha_2, \cdots, \alpha_p\}. \tag{A.4.0.10}$$

We can construct such a g_1 by means of Theorem A.2.5.1 (or Theorem A.3.0.4).

(f) If a formal power series g is k_1-summable and k_2-summable for two distinct numbers k_1 and k_2, then g is convergent, i.e.,

$$\mathbf{C}\{z^{-1}\}_{s_1} \cap \mathbf{C}\{z^{-1}\}_{s_2} = \mathbf{C}\{z^{-1}\} \quad \text{if } s_1 \neq s_2. \tag{A.4.0.11}$$

As a matter of fact

$$\mathbf{C}\{z^{-1}\}_s \cap \mathbf{C}[[z^{-1}]]_\sigma = \mathbf{C}\{z^{-1}\} \quad \text{if } \sigma < s \tag{A.4.0.12}$$

(cf. J.-P. Ramis [126] and J.-P. Ramis–Y. Sibuya [129]).

§A.4.1. *k*-summability of formal solutions

Let us look at a differential equation:

$$\delta\vec{y} = z^q \left\{ \vec{g}_0(z) + \Phi(z)\,\vec{y} + \sum_{|\vec{p}|\geq 2} \vec{g}_{\vec{p}}(z)\,\vec{y}^{\vec{p}} \right\}, \tag{A.4.1.1}$$

where q is a positive integer and

$$\vec{g}_0 \in z^{-1}\left(\mathbf{C}\{z^{-1}\}_s\right)^n, \qquad \Phi \in gl(n, \mathbf{C}\{z^{-1}\}), \qquad \vec{g}_{\vec{p}} \in \mathbf{C}\{z^{-1}\}^n.$$

The series $\sum_{|\vec{p}|\geq 2} \vec{g}_{\vec{p}}(z)\, \vec{y}^{\vec{p}}$ is uniformly and absolutely convergent on a domain $|z| > R_0$, $|\vec{y}| < r_0$, where $|\vec{y}|$ is the maximum of absolute values of entries of the vector $\vec{y}$, and R_0 and r_0 are positive numbers.

OBSERVATION A.4.1.1. Assume that there exists a matrix $A_0 \in GL(n, \mathbf{C})$ such that $\Phi(z) - A_0 \in z^{-1} gl(n, \mathbf{C}\{z^{-1}\})$. By virtue of Theorem A.2.1.9 differential equation (A.4.1.1) admits a unique solution $\vec{g} \in z^{-1}\left(\mathbf{C}[[z^{-1}]]_\sigma\right)^n$, where

$$\sigma = \max\left\{s, \frac{1}{q}\right\}. \tag{A.4.1.2}$$

The matrix A_0 has nonzero eigenvalues $\lambda_1, \cdots, \lambda_n$. For each j let Σ_j be the set of all directions α such that

$$\Im(\lambda_j\, e^{iq\alpha}) = 0, \qquad \Re(\lambda_j\, e^{iq\alpha}) < 0. \tag{A.4.1.3}$$

THEOREM A.4.1.2 (J.-P. Ramis–Y. Sibuya [129]). *The unique solution* $\vec{g} \in z^{-1}\left(\mathbf{C}[[z^{-1}]]_\sigma\right)^n$ *is q-summable if*

$$s = \frac{1}{q}. \tag{A.4.1.4}$$

Furthermore

$$\Sigma(\vec{g}) \subset \Sigma(\vec{g}_0) \cup \left(\bigcup_{j=1}^{n} \Sigma_j\right). \tag{A.4.1.5}$$

EXPLANATION. This result is based on the following facts:

(i) If $\alpha \notin \Sigma(\vec{g}_0)$, there exist two positive numbers R and ε such that the sum $\vec{f}_0$ of $\vec{g}_0$ in the direction α is in
$$z^{-1}\left(\mathscr{A}_s\left(R, \alpha - \varepsilon - \frac{s\pi}{2}, \alpha + \varepsilon + \frac{s\pi}{2}\right)\right)^n.$$

(ii) If $s \geq \frac{1}{q}$ and $\alpha \notin \Sigma(\vec{g}_0) \cup \left(\bigcup_{j=1}^{n} \Sigma_j\right)$, there exist two positive numbers $\tilde{R}$, $\tilde{\varepsilon}$ and
$$\vec{f} \in z^{-1}\left(\mathscr{A}_s\left(\tilde{R}, \alpha - \tilde{\varepsilon} - \frac{\pi}{2q}, \alpha + \tilde{\varepsilon} + \frac{\pi}{2q}\right)\right)^n$$
such that
$$\begin{cases} \delta \vec{f}(z) = z^q \left\{ \vec{f}_0(z) + \Phi(z)\vec{f}(z) + \sum_{|\vec{p}|\geq 2} \vec{g}_{\vec{p}}(z)\, \vec{f}(z)^{\vec{p}} \right\}, \\ A_{sp}(\vec{f}) = \vec{g}. \end{cases}$$

(iii) If $0 < s < \frac{1}{q}$, then $\vec{g} \in z^{-1}\left(\mathbf{C}[[z^{-1}]]_{\frac{1}{q}}\right)^n$. Since $\mathscr{D}(R, \alpha - \varepsilon - \frac{s\pi}{2}, \alpha + \varepsilon + \frac{s\pi}{2})$ is too small, $\vec{g}$ is in general not q-summable in the direction α.

(iv) If $s > \frac{1}{q}$, then $\vec{g} \in z^{-1}\left(\mathbf{C}[[z^{-1}]]_s\right)^n$. Since $\mathscr{D}(\tilde{R}, \alpha-\tilde{\varepsilon}-\frac{\pi}{2q}, \alpha+\tilde{\varepsilon}+\frac{\pi}{2q})$ is too small, $\vec{g}$ is in general not $\frac{1}{s}$-summable.

REMARK A.4.1.3. To prove the existence of $\vec{f}$ of (ii), we must study the existence and uniqueness of Gevrey asymptotic solutions more carefully (cf. (1) of Remark A.2.3.4).

REMARK A.4.1.4. Since $\mathbf{C}\{z^{-1}\} \subset \mathbf{C}\{z^{-1}\}_s$ for every positive s, theorem A.4.1.2 applies to the case when $\vec{g}_0 \in \left(\mathbf{C}\{z^{-1}\}\right)^n$.

Next let us look at the differential equation

$$P[y] = z^{-M}\, G(z, y, \delta y, \cdots, \delta^{n-1} y), \tag{A.4.1.6}$$

where

$$\begin{cases} P = \displaystyle\sum_{h=0}^{n} a_h(z)\, \delta^h, \\ G(z, y_0, y_1, \cdots, y_{n-1}) = G_0(z) + \displaystyle\sum_{|\vec{p}|\geq 1} G_{\vec{p}}(z)\, \vec{y}^{\vec{p}}. \end{cases} \tag{A.4.1.7}$$

We assume that

(1) $a_h \in \mathbf{C}\{z^{-1}\}$, $G_0 \in \mathbf{C}\{z^{-1}\}_s$ and $G_{\vec{p}} \in \mathbf{C}\{z^{-1}\}$,
(2) $a_n \neq 0$,
(3) the series G is uniformly and absolutely convergent on a domain $|z| > R_0$, $|y_h| < r_0$ $(h = 0, 1, \cdots, n-1)$, where R_0 and r_0 are two positive numbers.

OBSERVATION A.4.1.5. Let

$$0 < k_1 < k_2 < \cdots < k_p < +\infty \tag{A.4.1.8}$$

be all of the distinct positive slopes of sides of the Newton polygon $\mathscr{N}(P)$ and let

$$\{(X, Y_j + k_j(X - X_j))\ ;\ X_j \leq X \leq X_{j+1}\}$$

be the side of $\mathscr{N}(P)$ whose slope is k_j, where $Y_{j+1} = Y_j + k_j(X_{j+1} - X_j)$. Then the $X_{j+1} - X_j$ are positive integers. Set

$$n_j = X_{j+1} - X_j \qquad (j = 1, 2, \cdots, p).$$

The linear homogeneous differential equation $P[y] = 0$ admits n_j linearly independent formal solutions of the form

$$y_{j,h}(z) = z^{\mu_{j,h}} \exp[\lambda_{j,h} z^{k_j}\{1 + o(1)\}]\, f_{j,h}(z) \qquad (h = 1, \cdots, n_j),$$

where the $\mu_{j,h}$ are constants, the $\lambda_{j,h}$ are nonzero constants and $f_{j,h} \in \mathbf{C}[[z^{-\frac{1}{\sigma}}]][z^{\frac{1}{\sigma}}, \log\ z]$ for some positive integer σ (cf. Remark A.1.4.2.).

For every (p,h) $(h=1,\cdots,n_p)$ we denote by Σ_h the set of all directions α such that

$$\Im(\lambda_{p,h}z^{k_p})=0, \qquad \Re(\lambda_{p,h}z^{k_p})<0.$$

Assuming that the positive integer M on the right-hand side of differential equation (A.4.1.6) is sufficiently large, we can prove the theorem below.

THEOREM A.4.1.6 (J.-P. Ramis–Y. Sibuya [129]). *If* $g\in z^{-1}\mathbf{C}[[z^{-1}]]_s$ *satisfies differential equation* (A.4.1.6) *and if* $0<s\le\frac{1}{k_p}$, *then* g *is* $\frac{1}{s}$-*summable and*

$$\Sigma(g)\subset\begin{cases}\Sigma(G_0)\cup\left(\bigcup_{h=1}^{n_p}\Sigma_h\right) & \text{if}\quad s=\dfrac{1}{k_p},\\ \Sigma(G_0) & \text{if}\quad 0<s<\dfrac{1}{k_p}.\end{cases}$$

The main ideas of the proof are similar to those of the proof of Theorem A.4.1.2.

EXAMPLE A.4.1.7. J.-P. Ramis showed that

$$g=\sum_{m=0}^{\infty}(-1)^m\,m!\,x^{m+1}+\sum_{m=0}^{\infty}(-1)^m\,m!\,x^{2m+2}$$

satisfies the differential equation

$$\begin{cases}P[y]=0,\\ P=D^5[x^5(2-x)D^2-x^2(2x^3-5x^2+4)D+2(x^2-x+2)],\end{cases}$$

where $D=\frac{d}{dx}$. He also showed that $g\left(\frac{1}{z}\right)$ is not k-summable for any $k>0$ (cf. J.-P. Ramis–Y. Sibuya [129]).

§A.4.2. The Turrittin problem

In this section, we shall consider a differential equation

$$\delta W+A(z)W=0\qquad\left(\delta=z\frac{d}{dz}\right), \tag{A.4.2.1}$$

where W is an $n\times n$ unknown matrix and $A(z)\in gl(n,\mathbf{C}\{z^{-1}\}[z])$. Let

$$\tilde\delta U+B(\zeta)U=0\qquad\left(\tilde\delta=\zeta\frac{d}{d\zeta}\right) \tag{A.4.2.2}$$

be a differential equation in the standard form given in the Hukuhara–Turrittin Theorem 6.8.1. By virtue of Theorem 6.8.1 differential equation (A.4.2.1) can be changed to the differential equation (A.4.2.2) by a transformation

$$z=\zeta^\sigma,\qquad W=P(\zeta)U, \tag{A.4.2.3}$$

where σ is a positive integer, $P \in gl(n, \mathbf{C}[[\zeta^{-1}]])$, and $\det(P) \neq 0$ as an element of $\mathbf{C}[[z^{-1}]]$. In §6.8 we gave a sketch of the proof of Theorem 6.8.1 in terms of three cases. The matrix P of transformation (A.4.2.3) can be written as a product

$$P(\zeta) = \hat{P}_1(\zeta)\hat{P}_2(\zeta)\cdots\hat{P}_N(\zeta)$$

of a finite number of matrices of transformations $W_j = \hat{P}_j(\zeta)U_j$ ($j = 1, \cdots, N$) each of which is of one of those three cases (cf. Remark 6.8.2). The main concern of this section is to show that each $\hat{P}_j$ is k_j-summable as a formal power series in ζ^{-1} for some positive integer k_j.

OBSERVATION A.4.2.1. We are not going to prove Theorem 6.8.1 again by another method. We want to find out more information concerning transformation (A.4.2.3). The two matrices P and B are related through the equation

$$\tilde{\delta}P(\zeta) + \sigma A(\zeta^{\sigma})P(\zeta) - P(\zeta)B(\zeta) = 0.$$

Theorem 6.8.1 guarantees the existence of P and B. We can write P in the form

$$P(\zeta) = P_0(\zeta)\left[I + \zeta^{-M} \sum_{m \geq 0} \zeta^{-m} P_m \right],$$

where $P_0 \in GL(n, \mathbf{C}[\zeta, \zeta^{-1}])$, I is the $n \times n$ identity matrix, M is a positive integer, and $P_m \in gl(n, \mathbf{C})$. The transformation

$$z = \zeta^{\sigma}, \qquad W = P_0(\zeta)V, \tag{A.4.2.4}$$

changes differential equation (A.4.2.1) to

$$\tilde{\delta}V + [\, B(\zeta) + \zeta^{-L}E(\zeta)\,]\, V = 0, \tag{A.4.2.5}$$

where $E \in gl(n, \mathbf{C}\{\zeta^{-1}\})$ and L is a positive integer. The integer L tends to $+\infty$ as M tends to $+\infty$. The transformation

$$V = \left[I + \zeta^{-M} \sum_{m \geq 0} \zeta^{-m} P_m \right] U \tag{A.4.2.6}$$

changes equation (A.4.2.5) to equation (A.4.2.2). Hence the matrix

$$X = I + \zeta^{-M} \sum_{m \geq 0} \zeta^{-m} P_m \tag{A.4.2.7}$$

satisfies the differential equation

$$\tilde{\delta}X + [\, B(\zeta) + \zeta^{-L}E(\zeta)\,]X - X\, B(\zeta) = 0. \tag{A.4.2.8}$$

OBSERVATION A.4.2.2. Let

$$0 < k_1 < k_2 < \cdots < k_\ell < +\infty \tag{A.4.2.9}$$

be all of the distinct positive slopes of the Newton polygon of the differential equation

$$\tilde{\delta} Y + \sigma\{A(\zeta^\sigma)\, Y \; - \; Y\, A(\zeta^\sigma)\} \; = \; 0. \tag{A.4.2.10}$$

The Newton polygons of differential equations (A.4.2.8) and (A.4.2.10) have the same shape.

THEOREM A.4.2.3. *Suppose that L is sufficiently large. Then matrix* (A.4.2.7) *can be written as a product*

$$X \; = \; P_\ell(\zeta)P_{\ell-1}(\zeta)\cdots P_1(\zeta), \tag{A.4.2.11}$$

where

$$P_j \in gl(n, \mathbf{C}\{\zeta^{-1}\}_{\frac{1}{k_j}}) \qquad (j = 1, \cdots, \ell). \tag{A.4.2.12}$$

OBSERVATION A.4.2.4. To prove Theorem A.4.2.3, it suffices to prove the following refinement of Lemma 5.5.2. Let us assume that the matrix $A(z)$ of differential equation (A.4.2.1) has the form

$$A(z) \; = \; z^q\{A_0 \; + z^{-1}\tilde{A}(z)\},$$

where q is a positive integer, $\tilde{A} \in gl(n, \mathbf{C}\{z^{-1}\})$, and $A_0 \in gl(n, \mathbf{C})$. We also assume that A_0 has the form

$$A_0 = \begin{bmatrix} E_1 & 0 & \cdots & 0 \\ 0 & E_2 & \cdots & 0 \\ \cdots & \cdots & \cdots & \cdots \\ 0 & 0 & \cdots & E_p \end{bmatrix}, \tag{A.4.2.13}$$

where

(i) for each j, E_j is an $n_j \times n_j$ matrix and $\sum_{p \geq j \geq 1} n_j \; = \; n$,
(ii) for each j, there exists a complex number λ_j such that $E_j - \lambda_j I_j$ is nilpotent; here I_j is the $n_j \times n_j$ identity matrix,
(iii) $p \geq 2$,
(iv) $\lambda_j \neq \lambda_k$ if j and k are distinct.

LEMMA A.4.2.5. *There exists an $n \times n$ matrix $T(z)$ such that*

(1) $T \; - \; I \in z^{-1}\; gl(n, \mathbf{C}\{z^{-1}\}_{\frac{1}{q}})$, *where I is the $n \times n$ identity matrix,*
(2) *the transformation*

$$W = T(z)V \tag{A.4.2.14}$$

changes differential equation (A.4.2.1) *to a differential equation*

$$\delta V \; + \; z^q B(z) V \; = \; 0 \tag{A.4.2.15}$$

with a matrix $B(z)$ *of the form*

$$B(z) = \begin{bmatrix} F_1(z) & 0 & \cdots & 0 \\ 0 & F_2(z) & \cdots & 0 \\ \cdots & \cdots & \cdots & \cdots \\ 0 & 0 & \cdots & F_p(z) \end{bmatrix}, \qquad \text{(A.4.2.16)}$$

where, for each j,

$$F_j(z) - E_j \in z^{-1} gl(n_j, \mathbf{C}\{z^{-1}\}). \qquad \text{(A.4.2.17)}$$

REMARK A.4.2.6.

(a) We can prove Lemma A.4.2.5 by means of the argument given in §5.5 together with Theorem A.2.1.9, the analytization Theorem A.3.3.1, and Theorem A.4.1.2.

(b) The uniqueness of factors $P_1, \cdots, P_\ell$ on the right-hand side of factorization (A.4.2.11) can be explained as follows. Suppose that another factorization of X is given as

$$X = Q_\ell(\zeta) Q_{\ell-1}(\zeta) \cdots Q_1(\zeta),$$

where

$$Q_j \in gl(n, \mathbf{C}\{\zeta^{-1}\}_{\frac{1}{k_j}}) \qquad (j = 1, \cdots, \ell).$$

Then

$$Q_1 P_1^{-1} = \{P_\ell \cdots P_2\}\{Q_\ell \cdots Q_2\}^{-1}. \qquad \text{(A.4.2.18)}$$

The left-hand side of (A.4.2.18) belongs to $GL(n, \mathbf{C}\{\zeta^{-1}\}_{\frac{1}{k_1}})$, whereas the right-hand side of (A.4.2.18) belongs to $GL(n, \mathbf{C}[[\zeta^{-1}]]_{\frac{1}{k_2}})$. Hence by virtue of (A.4.0.2) (cf. (f) of Remark A.4.0.6) we have $Q_1 P_1^{-1} \in GL(n, \mathbf{C}\{z^{-1}\})$. In this way we can prove that the factors P_j are unique modulo $GL(n, \mathbf{C}\{z^{-1}\})$, i.e.,

$$Q_j = P_j R_j, \qquad R_j \in GL(n, \mathbf{C}\{z^{-1}\}) \qquad (j = 1, \cdots, \ell).$$

Note that the R_j are not independent of each other.

(c) Each P_j has a convergent inverse factorial series expansion (cf. (a) of Remark A.4.0.6).

(d) Matrix (A.4.2.7) always has a formal inverse factorial series expansion. H. L. Turrittin [147] showed that such an expansion of the matrix X is convergent if $\ell = 1$. He left other cases open. B. Malgrange and J.-P. Ramis gave an answer to the Turrittin problem by proving Theorem A.4.2.3 (cf. J.-P. Ramis [124, 127]). J.-P. Ramis [126, 128] applied factorization (A.4.2.11) to the study of Galois groups of linear differential equations. Note that, if a direction α is not in any of the singular supports of P_j, the sums F_j of P_j define an analytic solution $F_\ell F_{\ell-1} \cdots F_1$ of differential equation (A.4.2.8). J.-P.Ramis used this fact to establish a canonical isomorphism between the formal Galois group and the analytic Galois group.

(e) For some other information concerning factorization (A.4.2.11) see also W. Jurkat [77], W. Balser [44, 45, 47] and Balser–Jurkat–Lutz [50].

APPENDIX 5

Differential Equations Containing Parameters

§A.5.0. Preliminaries

In the case when a given linear differential equation contains auxiliary parameters $\vec{t} = (t_1, \cdots, t_p)$, the structure of analytic continuation (or of monodromy) and the Stokes structure defined by this equation also depend on $\vec{t}$. Generally speaking, it takes much harder work than what is usually required to solve *good exercises* if we try to find the dependence of those structures on the parameters $\vec{t}$ (cf. for example Y. Sibuya [9, 29, 30, 31], W. Balser [46], D. G. Babbitt–V. S. Varadarajan [37, 38, 39, 40, 41] and R. Schäfke [131]). In this appendix we shall explain some aspects of

(1) isomonodromic deformations (§A.5.2),
(2) isoformal deformations (§A.5.3), and
(3) singular perturbations (§A.5.4).

REMARK A.5.0.1. Given a particular situation in which we study a problem of asymptotic analysis containing some auxiliary parameters, we must use a definition of asymptotic expansions in many variables that is most suitable for the situation. H. Majima generalized Theorem 6.4.1 to the case of several complex variables by utilizing a concept of *strong asymptotic expansions* in many variables. We shall explain the Majima asymptotics in §A.5.1. For details of the Majima asymptotics, see H. Majima [95, 96, 97].

§A.5.1. The Majima asymptotics

In any asymptotic analysis, it is very important to define *flatness* in a suitable way. Let us consider a sectorial domain

$$\mathscr{D}(\vec{R}, \vec{a}, \vec{b}) = \left\{ \prod_{m=1}^{n_0} \mathscr{D}(R_m, a_m, b_m) \right\} \times \left\{ \prod_{m=n_0+1}^{n} \mathscr{D}(R_m) \right\} \qquad \text{(A.5.1.1)}$$

in $\mathbf{C}^n$, where $\vec{R} = (R_1, \cdots, R_n)$, $\vec{a} = (a_1, \cdots, a_{n_0})$, $\vec{b} = (b_1, \cdots, b_{n_0})$, and

$$\begin{cases} \mathscr{D}(R_m, a_m, b_m) = \{z_m \; ; \; a_m < \arg z_m < b_m, \; |z_m| > R_m\}, \\ \mathscr{D}(R_m) = \{z_m \; ; \; |z_m| > R_m\}. \end{cases}$$

Here a point in $\mathbf{C}^n$ is denoted by $\vec{z} = (z_1, \cdots, z_n)$. H. Majima [95] defines a flat function $f(\vec{z})$ as follows. An $f \in \mathscr{O}(\mathscr{D}(\vec{R}, \vec{a}, \vec{b}))$ is said to be flat as $(z_1, \cdots, z_{n_0}) \to (\infty, \cdots, \infty)$ uniformly on $\prod_{m=n_0+1}^{n} \mathscr{D}(R_m)$ if, for every n_0-tuple $\vec{N} = (N_1, \cdots, N_{n_0})$ of nonnegative integers and every $(\vec{\rho}, \vec{\alpha}, \vec{\beta})$ satisfying the conditions

$$\begin{cases} \rho_m > R_m & (m = 1, \cdots, n), \\ a_m < \alpha_m < \beta_m < b_m & (m = 1, \cdots, n_0), \end{cases} \tag{A.5.1.2}$$

there exists a nonnegative constant $K[\vec{N}, \vec{\rho}, \vec{\alpha}, \vec{\beta}]$ such that

$$|f(\vec{z})| \leq K[\vec{N}, \vec{\rho}, \vec{\alpha}, \vec{\beta}] \prod_{m=1}^{n_0} |z_m|^{-N_m} \quad \text{for} \quad \vec{z} \in \mathscr{D}(\vec{\rho}, \vec{\alpha}, \vec{\beta}). \tag{A.5.1.3}$$

Note that, if f is flat in this sense, then its integral $\int_a^b f(\vec{z})\, dz_j$ with respect to an entry z_j of $\vec{z}$ is also flat as a function of other entries $(\cdots, z_{j-1}, z_{j+1}, \cdots)$.

Majima defines strongly asymptotically developable functions as follows: Denote by $\mathscr{G}$ the set of all nonempty subsets of $\{1, \cdots, n_0\}$ and by $\sigma(J)$ $(J \in \mathscr{G})$ the number of elements of J. Let us identify with $\mathbf{C}^{n-\sigma(J)}$ the subspace of $\mathbf{C}^n$ defined by $z_j = \infty$ $(j \in J)$, where $\mathbf{C}^0 = \{0\}$. Also we define maps

$$p_J : \mathbf{C}^n \to \mathbf{C}^{n-\sigma(J)}$$

by

$$p_J(\vec{z}) = \vec{z}\,\big|_{z_j = \infty \text{ for } j \in J} \tag{A5.1.4}$$

respectively. Set

$$p_J(\mathscr{D}(\vec{R}, \vec{a}, \vec{b})) = \left\{ \prod_{m \notin J} \mathscr{D}(R_m, a_m, b_m) \right\} \times \left\{ \prod_{m=n_0+1}^{n} \mathscr{D}(R_m) \right\}.$$

DEFINITION A.5.1.1. A function $f(\vec{z})$ is said to be **strongly asymptotically developable** on $\mathscr{D}(\vec{R}, \vec{a}, \vec{b})$ if

(i) $f \in \mathscr{O}(\mathscr{D}(\vec{R}, \vec{a}, \vec{b}))$ and bounded on $\mathscr{D}(\vec{R}, \vec{a}, \vec{b})$,

(ii) there exists a set of functions

$$\mathscr{F} = \{ f_J(p_J(\vec{z}) \; ; \; \vec{\ell}) \; ; \; J \in \mathscr{G}, \; \vec{\ell} \in N^{\sigma(J)} \} \tag{A.5.1.5}$$

such that

(1) $f_J(p_J(\vec{z})\,;\,\vec{\ell})$ is holomorphic and bounded on $p_J(\mathscr{D}(\vec{R},\vec{a},\vec{b}))$,

(2)

$$\lim_{t_j\to 0\ (j\in J)} \left[\prod_{j\in J}\left(\frac{\partial}{\partial t_j}\right)^{\ell_j}\right] f\left(\frac{1}{\vec{t}}\right) = \left(\prod_{j\in J}(\ell_j!)\right) f_J\left(p_J\left(\frac{1}{\vec{t}}\right)\,;\,\vec{\ell}\right)$$

uniformly on every sectorial domain $\mathscr{D}(\vec{\rho},\vec{\alpha},\vec{\beta})$ (of $\vec{z}=\frac{1}{\vec{t}}$) satisfying condition (A.5.1.2), where N is the set of all non-negative integers, the ℓ_j $(j\in J)$ are the entries of $\vec{\ell}$ and $\frac{1}{\vec{t}}=\left(\frac{1}{t_1},\cdots,\frac{1}{t_n}\right)$. The set $\mathscr{F}$ is called **the set of coefficients of asymptotic expansion** of f on $\mathscr{D}(\vec{R},\vec{a},\vec{b})$.

DEFINITION A.5.1.2. A set of functions

$$\mathscr{F} = \{f_J(p_J(\vec{z})\,;\,\vec{\ell})\,;\ J\in\mathscr{G},\ \vec{\ell}\in N^{\sigma(J)}\}$$

is said to be **consistent** on $\mathscr{D}(\vec{R},\vec{a},\vec{b})$ if

(i) $f_J(p_J(\vec{z})\,;\,\vec{\ell})\in\mathscr{O}(p_J(\mathscr{D}(\vec{R},\vec{a},\vec{b})))$ and bounded on $p_J(\mathscr{D}(\vec{R},\vec{a},\vec{b}))$,

(ii)

$$\lim_{t_j\to 0\ (j\in J-K)} \left[\prod_{j\in J-K}\left(\frac{\partial}{\partial t_j}\right)^{\ell_j}\right] f_K\left(p_K\left(\frac{1}{\vec{t}}\right)\,;\,\vec{k}\right) = \left(\prod_{j\in J-K}(\ell_j!)\right) f_J\left(p_J\left(\frac{1}{\vec{t}}\right)\,;\,(\vec{\ell},\vec{k})\right),$$

whenever $K\subset J$, uniformly on every sectorial domain $\mathscr{D}(\vec{\rho},\vec{\alpha},\vec{\beta})$ (of $\vec{z}=\frac{1}{\vec{t}}$) satisfying condition (A.5.1.2), where $\vec{\ell}\in N^{\sigma(J-K)}$, $\vec{k}\in N^{\sigma(K)}$ and $(\vec{\ell},\vec{k})\in N^{\sigma(J)}$.

OBSERVATION A.5.1.3. Let us denote by $\mathscr{A}(\vec{R},\vec{a},\vec{b})$ the set of all strongly asymptotically developable functions on $\mathscr{D}(\vec{R},\vec{a},\vec{b})$. Then $\mathscr{A}(\vec{R},\vec{a},\vec{b})$ is a commutative differential algebra over $\mathbf{C}$. Furthermore, for every consistent set $\mathscr{F}$ on $\mathscr{D}(\vec{R},\vec{a},\vec{b})$, there exists an $f\in\mathscr{A}(\vec{R},\vec{a},\vec{b})$ such that $\mathscr{F}$ is the set of coefficients of asymptotic expansion of f on $\mathscr{D}(\vec{R},\vec{a},\vec{b})$. Note also that, if $f\in\mathscr{A}(\vec{R},\vec{a},\vec{b})$, then the set $\mathscr{F}$ of coefficients of asymptotic expansion of f is consistent on $\mathscr{D}(\vec{R},\vec{a},\vec{b})$.

REMARK A.5.1.4. If $\sigma(J) = n_0$ (*i.e.*, $J = \{1, \cdots, n_0\}$), then $f_J(p_J(\vec{z})\,;\,\vec{\ell}) \in \mathscr{O}(\prod_{m=n_0+1}^{n} \mathscr{D}(R_m))$. Furthermore, if $n_0 = n$, then f_J is a constant. Conversely, for any power series

$$\sum_{\vec{\ell}\in N^{n_0}} a_{\vec{\ell}}(z_{n_0+1}, \cdots, z_n)\, z_1^{-\ell_1} \cdots z_{n_0}^{-\ell_{n_0}} \in \mathbf{C}\{z_{n_0+1}^{-1}, \cdots, z_n^{-1}\}[[z_1^{-1}, \cdots, z_{n_0}^{-1}]]$$

and any sectorial domain $\mathscr{D}(\vec{R}, \vec{a}, \vec{b})$, there exists a set $\mathscr{F}$ of functions

$$\mathscr{F} = \{f_J(p_J(\vec{z})\,;\,\vec{\ell})\,;\, J \in \mathscr{G},\ \vec{\ell} \in N^{\sigma(J)}\}$$

such that

(1) $\mathscr{F}$ is consistent on $\mathscr{D}(\vec{R}, \vec{a}, \vec{b})$,

(2) if $J = \{1, \cdots, n_0\}$, we have $f_J(p_J(\vec{z})\,;\,\vec{\ell}) = a_{\vec{\ell}}(z_{n_0+1}, \cdots, z_n)$ for all $\vec{\ell} \in N^{n_0}$.

OBSERVATION A.5.1.5. A function $f \in \mathscr{A}(\vec{R}, \vec{a}, \vec{b})$ is flat in the sense of Majima if all coefficients $f_J(p_J(\vec{z})\,;\,\vec{\ell})$ of asymptotic expansion of f are identically equal to zero on $\mathscr{D}(\vec{R}, \vec{a}, \vec{b})$.

OBSERVATION A.5.1.6. For $f \in \mathscr{A}(\vec{R}, \vec{a}, \vec{b})$ we can derive various expansions by computing the integrals of

$$\left[\prod_{j\in J}\left(\frac{\partial}{\partial t_j}\right)^{\ell_j}\right] f\left(\frac{1}{\vec{t}}\right)$$

with respect to t_j $(j \in J)$. For example, let us consider the case when $n = n_0 = 2$. In this case the set $\mathscr{F}$ of coefficients of asymptotic expansion of $f \in \mathscr{A}(\vec{R}, \vec{a}, \vec{b})$ is given by

$$\mathscr{F} = \{f_\ell(z_1),\ g_k(z_2),\ a_{k\ell}\,;\, \ell,\ k \in N\},$$

where $f_\ell \in \mathscr{A}(R_1, a_1, b_1)$, $g_k \in \mathscr{A}(R_2, a_2, b_2)$, $a_{\ell k} \in \mathbf{C}$. We can show that

$$\begin{cases} \lim_{t_2\to 0} \dfrac{\partial^\ell}{\partial t_2^\ell} f\left(\dfrac{1}{t_1}, \dfrac{1}{t_2}\right) = (\ell!)\, f_\ell\left(\dfrac{1}{t_1}\right), \\[2ex] \lim_{t_1\to 0} \dfrac{\partial^k}{\partial t_1^k} f\left(\dfrac{1}{t_1}, \dfrac{1}{t_2}\right) = (k!)\, g_k\left(\dfrac{1}{t_2}\right), \\[2ex] \lim_{(t_1,t_2)\to (0,0)} \dfrac{\partial^{k+\ell}}{\partial t_1^k \partial t_2^\ell} f\left(\dfrac{1}{t_1}, \dfrac{1}{t_2}\right) = (k!)(\ell!)\, a_{k\ell}, \\[2ex] A_{sp}(f_\ell) = \sum_{k\ge 0} a_{k\ell}\, z_1^{-k}, \qquad A_{sp}(g_k) = \sum_{\ell\ge 0} a_{k\ell}\, z_2^{-\ell}. \end{cases}$$

Furthermore, for example, we can derive

$$\begin{cases} |\, f(z_1, z_2) - f_0(z_1) \,| = O(|z_2|^{-1}), \\ |\, f(z_1, z_2) - g_0(z_2) \,| = O(|z_1|^{-1}), \\ |\, f(z_1, z_2) - f_0(z_1) - g_0(z_2) + a_{00} \,| = O(|z_1|^{-1}|z_2|^{-1}), \\ |\, f(z_1, z_2) - f_0(z_1) - z_2^{-1} f_1(z_1) - g_0(z_2) - z_1^{-1} g_1(z_2) \\ + a_{00} + a_{10} z_1^{-1} + a_{01} z_2^{-1} + a_{11} z_1^{-1} z_2^{-1} | = O(|z_1|^{-2}|z_2|^{-2}), \text{ etc.} \end{cases}$$

Utilizing the concept of strong asymptotic developability, Majima generalized Theorem 6.4.1 to the case of several complex variables.

THEOREM A.5.1.7. *Assume that N sectorial domains*

$$\mathscr{S}_\ell = \mathscr{D}(\vec{R}, \vec{a}_\ell, \vec{b}_\ell) \qquad (\ell = 1, \cdots, N)$$

form a covering of the domain

$$\mathscr{D}(\vec{R}) = \prod_{m=1}^{n} \mathscr{D}(R_m).$$

and that $n \times n$ matrices $F_{k\ell}(\vec{z}) \in gl(n, \mathscr{A}(\mathscr{S}_k \cap \mathscr{S}_\ell))$ are defined for all (k, ℓ) satisfying the condition

$$\mathscr{S}_k \cap \mathscr{S}_\ell \neq \varnothing. \tag{A.5.1.6}$$

Also assume that, if

$$\mathscr{S}_h \cap \mathscr{S}_k \cap \mathscr{S}_\ell \neq \varnothing,$$

the relation

$$F_{hk}(\vec{z}) F_{k\ell}(\vec{z}) = F_{h\ell}(\vec{z}) \qquad (\vec{z} \in \mathscr{S}_h \cap \mathscr{S}_k \cap \mathscr{S}_\ell)$$

holds. Furthermore assume that the matrices

$$F_{k\ell} - I$$

are flat on $\mathscr{S}_k \cap \mathscr{S}_\ell$, where I is the $n \times n$ identity matrix. Then there exist positive numbers R'_m greater than R_m $(m = 1, \cdots, n)$ respectively and $n \times n$ matrices $Q_\ell(\vec{z})$ $(\ell = 1, \cdots, N)$ such that

(i) *$Q_\ell(\vec{z})$ and $Q_\ell(\vec{z})^{-1}$ belong to $gl(n, \mathscr{A}(\vec{R}', \vec{a}_\ell, \vec{b}_\ell))$,*
(ii) *$Q_\ell(\vec{z}) \to I$ as $(z_1, \cdots, z_{n_0}) \to (\infty, \cdots, \infty)$,*
(iii) *the relation*

$$F_{k\ell}(\vec{z}) = Q_k(\vec{z})^{-1} Q_\ell(\vec{z}) \qquad (\vec{z} \in \mathscr{S}'_k \cap \mathscr{S}'_\ell)$$

holds when condition (A.5.1.6) *is satisfied, where $\mathscr{S}'_\ell = \mathscr{D}(\vec{R}', \vec{a}_\ell, \vec{b}_\ell)$.*

REMARK A.5.1.8.

(i) For details of what we explained above, see H. Majima [95].

(ii) Majima derived existence and uniqueness of strongly asymptotically developable solutions of completely integrable Pfaffian systems in various cases. Also the Gevrey asymptotics were developed within the Majima asymptotics. For that information, see H. Majima [96, 97], Y. Haraoka [62] and T. Yagami [149].

§A.5.2. Isomonodromic deformations

Let $\mathscr{D}$ be a connected open set in the z-plane and let $A(z,\vec{t})$ be a matrix in $gl(n,\mathscr{O}(\mathscr{D}\times\mathscr{D}(\vec{r},\vec{\tau})))$, where $\vec{r}=(\rho_1,\cdots,\rho_p)$ is a p-tuple of positive numbers, $\vec{t}=(t_1,\cdots,t_p)\in\mathbf{C}^p$, and

$$\mathscr{D}(\vec{r},\vec{\tau}) = \prod_{h=1}^{p}\{t_h\ ;\ |t_h-\tau_h|<\rho_h\}.$$

Assume that $\{\Delta(z_j,r_j), W_j(z,\vec{t})\}$ is a system of fundamental solutions of the differential equation (cf. Definition 1.9.2)

$$\frac{dW}{dz} = A(z,\vec{t}) \tag{A.5.2.1}$$

for each fixed $\vec{t}\in\mathscr{D}(\vec{r},\vec{\tau})$, where $W_j(z,\vec{t})\in GL(n,\mathscr{O}(\Delta(z_j,r_j)\times\mathscr{D}(\vec{r},\vec{\tau})))$. Set

$$C_{jk}(\vec{t}) = W_j(z,\vec{t})^{-1}W_k(z,\vec{t}) \tag{A.5.2.2}$$

whenever $\Delta(z_j,r_j)\cap\Delta(z_k,r_k)\neq\emptyset$. Then $C_{jk}(\vec{t})\in GL(n,\mathscr{O}(\mathscr{D}(\vec{r},\vec{\tau})))$. Moreover if $\Delta(z_j,r_j)\cap\Delta(z_k,r_k)\cap\Delta(z_h,r_h)\neq\emptyset$, we have

$$C_{jk}(\vec{t})C_{kh}(\vec{t}) = C_{jh}(\vec{t}).$$

Hence $\{\Delta(z_j,r_j), C_{jk}(\vec{t})\}$ is a system of connection matrices on $\mathscr{D}$ for each fixed $\vec{t}\in\mathscr{D}(\vec{r},\vec{\tau})$ defined by the system $\{\Delta(z_j,r_j), W_j(z,\vec{t})\}$ of fundamental solutions of (A.5.2.1) (cf. Definition 1.10.1).

DEFINITION A.5.2.1. Differential equation (A.5.2.1) is said to be **isomonodromic** on $\mathscr{D}\times\mathscr{D}(\vec{r},\vec{\tau})$ if there exists a system $\{\Delta(z_j,r_j), W_j(z,\vec{t})\}$ of fundamental solutions of differential equation (A.5.2.1) such that

(i) $W_j(z,\vec{t})\in GL(n,\mathscr{O}(\Delta(z_j,r_j)\times\mathscr{D}(\vec{r},\vec{\tau})))$,

(ii) the connection matrices $C_{jk}=W_j(z,\vec{t})^{-1}W_k(z,\vec{t})$ defined on $\mathscr{D}$ by $\{\Delta(z_j,r_j), W_j(z,\vec{t})\}$ are independent of $\vec{t}\in\mathscr{D}(\vec{r},\vec{\tau})$.

REMARK A.5.2.2. Condition (ii) of Definition A.5.2.1 implies that the structure of analytic continuation (or of monodromy) defined on $\mathscr{D}$ by differential equation (A.5.2.1) is independent of $\vec{t}$. The author does not know whether the converse is also true. Note that the analyticity of $W_j(z,\vec{t})$ with respect to $\vec{t}$ is included in condition (i) of Definition A.5.2.1.

THEOREM A.5.2.3. *Differential equation* (A.5.2.1) *is isomonodromic on* $\mathscr{D} \times \mathscr{D}(\vec{r}, \vec{\tau})$ *if and only if there exist p matrices* $B_h(z, \vec{t}) \in gl(n, \mathscr{O}(\mathscr{D} \times \mathscr{D}(\vec{r}, \vec{\tau})))$ $(h = 1, \cdots, p)$ *such that the Pfaffian system*

$$\begin{cases} \dfrac{\partial W}{\partial z} = A(z, \vec{t})W, \\ \dfrac{\partial W}{\partial t_h} = B_h(z, \vec{t})W \qquad (h = 1, \cdots, p) \end{cases} \tag{A.5.2.3}$$

is completely integrable.

In fact, if $\{\Delta(z_j, r_j), W_j(z, \vec{t})\}$ satisfies all of the requirements of Definition A.5.2.1, we have

$$W_j(z, \vec{t})C_{jk} = W_k(z, \vec{t}), \qquad \frac{\partial W_j}{\partial t_h}(z, \vec{t})C_{jk} = \frac{\partial W_k}{\partial t_h}(z, \vec{t})$$

on $(\Delta(z_j, r_j) \cap \Delta(z_k, r_k)) \times \mathscr{D}(\vec{r}, \vec{\tau})$. Hence

$$\frac{\partial W_j}{\partial t_h}(z, \vec{t})W_j(z, \vec{t})^{-1} = \frac{\partial W_k}{\partial t_h}(z, \vec{t})W_k(z, \vec{t})^{-1},$$

on $(\Delta(z_j, r_j) \cap \Delta(z_k, r_k)) \times \mathscr{D}(\vec{r}, \vec{\tau})$. Therefore we can define $B_h(z, \vec{t})$ by

$$B_h(z, \vec{t}) = \frac{\partial W_j}{\partial t_h}(z, \vec{t})W_j(z, \vec{t})^{-1} \quad \text{on} \quad \Delta(z_j, r_j) \times \mathscr{D}(\vec{r}, \vec{\tau}). \tag{A.5.2.4}$$

If Pfaffian system (A.5.2.3) is completely integrable, then we can define $\{\Delta(z_j, r_j), W_j(z, \vec{t})\}$ by solving system (A.5.2.3) on $\Delta(z_j, r_j) \times \mathscr{D}(\vec{r}, \vec{\tau})$.

OBSERVATION A.5.2.4. Pfaffian system (A.5.2.3) is completely integrable if and only if

$$\begin{cases} \dfrac{\partial B_h}{\partial z} - \dfrac{\partial A}{\partial t_h} = [A, B_h] \qquad (h = 1, \cdots, p), \\ \\ \dfrac{\partial B_h}{\partial t_k} - \dfrac{\partial B_k}{\partial t_h} = [B_k, B_h] \qquad (h, k = 1, \cdots, p), \end{cases} \tag{A.5.2.5}$$

where

$$[X, Y] = XY - YX.$$

REMARK A.5.2.5. To determine whether differential equation (A.5.2.1) is isomonodromic on $\mathscr{D} \times \mathscr{D}(\vec{r}, \vec{\tau})$, we must try to find p matrices $B_h(z, \vec{t}) \in gl(n, \mathscr{O}(\mathscr{D} \times \mathscr{D}(\vec{r}, \vec{\tau})))$ $(h = 1, \cdots, p)$ by solving equation (A.5.2.5). In some particular cases where more detailed information of the matrix $A(z, \vec{t})$ is available, we can find some necessary properties of $B_h(z, \vec{t})$ through (A.5.2.4). For example, if $A(z, \vec{t})$ is rational in z and if differential equation (A.5.2.1) is of Fuchsian type (cf. Definition 4.3.5), the matrices $B_h(z,\vec{t})$ are also rational in z. In such cases we might be able to derive necessary and sufficient conditions on the matrix $A(z,\vec{t})$ such that differential equation (A.5.2.1) is isomonodromic.

EXAMPLE A.5.2.6 (L. Schlesinger [132]). Let us consider differential equation (A.5.2.1) in the case when

$$A(z,\vec{t}) = \sum_{h=1}^{p} \frac{A_h(\vec{t})}{z - t_h}, \tag{A.5.2.6}$$

where $A_h \in gl(n,\mathscr{O}(\mathscr{D}(\vec{r},\vec{\tau})))$. Set

$$\mathscr{D} = \{z;\ |z-\tau_h| > \rho_h \qquad (h = 1,\cdots,p)\}.$$

We assume that the positive numbers ρ_h are sufficiently small. Assume also that

(1) any two eigenvalues of $A_h(\vec{t})$ do not differ by integer,
(2) any two eigenvalues of $\sum_{1\le h\le p} A_h(\vec{t})$ do not differ by integer,
(3) the matrix $\sum_{1\le h\le p} A_h(\vec{t})$ is independent of $\vec{t}$.

Then, if differential equation (A.5.2.1) is isomonodromic, the matrices $B_h(z,\vec{t})$ defined by (A.5.2.4) must have the form

$$B_h(z,\vec{t}) = -\frac{A_h(\vec{t})}{z - t_h} + C_h(\vec{t}) \qquad (h = 1,\cdots,p), \tag{A.5.2.7}$$

where $C_h(\vec{t}) \in gl(n,\mathscr{O}(\mathscr{D}(\vec{r},\vec{\tau})))$. Inserting (A.5.2.7) into equation (A.5.2.5), we derive

$$\frac{\partial C_h}{\partial t_k} - \frac{\partial C_k}{\partial t_h} = [C_k, C_h].$$

This means that the Pfaffian system

$$\frac{\partial R}{\partial t_h} = C_h(\vec{t})R \qquad (h = 1,\cdots,p)$$

is completely integrable on $\mathscr{D}(\vec{r},\vec{\tau})$. Therefore there exists a matrix $R \in GL(n,\mathscr{O}(\mathscr{D}(\vec{r},\vec{\tau})))$ such that

$$C_h(\vec{t}) = \frac{\partial R}{\partial t_h} R(\vec{t})^{-1} \qquad (h = 1,\cdots,p).$$

From this fact we can easily conclude that it will suffice to consider the case when all of the matrices $C_h(\vec{t})$ are identically equal to zero. Now we look at the Pfaffian system

$$\begin{cases} \dfrac{\partial W}{\partial z} = \displaystyle\sum_{h=1}^{p} \frac{A_h(\vec{t})}{z - t_h} W, \\[2ex] \dfrac{\partial W}{\partial t_h} = -\dfrac{A_h(\vec{t})}{z - t_h} W \qquad (h = 1,\cdots,p). \end{cases} \tag{A.5.2.8}$$

THEOREM A.5.2.7 (L. Schlesinger [132]). *Pfaffian system* (A.5.2.8) *is completely integrable if and only if*

$$\begin{cases} \dfrac{\partial A_h}{\partial t_k} = \dfrac{[A_h, A_k]}{t_h - t_k} \quad (h \neq k), \\ \\ \dfrac{\partial A_h}{\partial t_h} = -\displaystyle\sum_{k \neq h} \dfrac{[A_h, A_k]}{t_h - t_k}. \end{cases} \tag{A.5.2.9}$$

REMARK A.5.2.8.

(I) To prove Theorem A.5.2.7 it is not necessary to assume condition (3) given above. As a matter of fact, the Schlesinger equation (A.5.2.9) implies that $\sum_{h=1}^{p} A_h(\vec{t})$ is independent of $\vec{t}$.

(II) Pfaffian system (A.5.2.9) is completely integrable. Hence for any initial condition at an ordinary point, there exists a unique solution of (A.5.2.9) in a neighborhood of the initial point.

OBSERVATION A.5.2.9. In the case when differential equation (A.5.2.1) admits irregular singular points, the matrices $B_h(z, \vec{t})$ $(h = 1, \cdots, p)$ defined by (A.5.2.4) are not necessarily rational in z even if (A.5.2.1) is isomonodromic. For example if the entries of the matrix $A(z, \vec{t})$ are polynomials in z, fundamental solutions $W_j(z, \vec{t})$ are entire in z no matter what. Hence (A.5.2.1) is isomonodromic. However $B_h(z, \vec{t})$ $(h = 1, \cdots, p)$ defined by (A.5.2.4) are not necessarily polynomials in z. In order that the B_h have at worst poles at an irregular singular point ζ, it is necessary and sufficient that

$$\frac{\partial W_j}{\partial t_h} W_j^{-1} = O(|z - \zeta|^{-M})$$

for some positive number M in the neighborhood of ζ. To explain how to construct such a system $\{\Delta(z_j, r_j), W_j(z, \vec{t})\}$ of fundamental solutions of differential equation (A.5.2.1), let us make the following observation. Assuming that $z = \infty$ is an irregular singular point of (A.5.2.1), we shall try to define a Stokes phenomenon according to Definitions 6.10.2, 6.10.3 and 6.10.5.

Step I: We fix another differential equation

$$\frac{dU}{dz} = A_0(z, \vec{t})U \tag{A.5.2.10}$$

so that $A_0 \in gl(n, \mathcal{O}(\mathcal{D}(R) \times \mathcal{D}(\vec{r}, \vec{t})))$, where $\mathcal{D}(R) = \{z;\ |z| > R\}$ and that there exists $\hat{P}(z, \vec{t}) \in GL(n, \mathcal{O}(\mathcal{D}(\vec{r}, \vec{t}))[[z^{-1}]][z])$ such that the transfomation

$$W = \hat{P}(z, \vec{t})U \tag{A.5.2.11}$$

changes (A.5.2.1) to (A.5.2.10).

Step II: We fix a covering $\{\mathscr{S}_1, \cdots, \mathscr{S}_N\}$ at $z = \infty$ and fundamental solutions $\Phi_j(z, \vec{t})$ of (A.5.2.10) on $\mathscr{S}_j$ $(j = 1, \cdots, N)$ in such a way that

(a) $\Phi_j \in GL(n, \mathscr{O}(\mathscr{S}_j \times \mathscr{D}(\vec{r}, \vec{t})))$,

(b)

$$\frac{\partial \Phi_j}{\partial t_h}(z, \vec{t})\Phi_j(z, \vec{t})^{-1} = O(|z|^M) \qquad as \quad z \to \infty \text{ in } \mathscr{S}_j \quad (j = 1, \cdots, N), \tag{A.5.2.12}$$

where M is a positive number.

Step III: We fix a system $\{\mathscr{S}_j, P_j(z, \vec{t})\Phi_j(z, \vec{t}); j = 1, \cdots, N\}$ of fundamental solutions of differential equation (A.5.2.1) in the neighborhood of $z = \infty$ (cf. Definition 6.10.2), where $P_j(z, \vec{t})$ and $P_j(z, \vec{t})^{-1}$ are in $gl(n, \mathscr{A}(\mathscr{S}_j \times \mathscr{D}(\vec{r}, \vec{\tau})))$. [The definition of $\mathscr{A}(\mathscr{S}_j \times \mathscr{D}(\vec{r}, \vec{\tau}))$ is analogous to the definition of $\mathscr{A}(\vec{R}, \vec{a}, \vec{b})$ in §A.5.1 (cf. Observation A.5.1.3).]

Now let us set

$$W_j(z, \vec{t}) = P_j(z, \vec{t})\Phi_j(z, \vec{t}) \qquad (j = 1, \cdots, N).$$

Then we have

$$\frac{\partial W_j}{\partial t_h}(z, \vec{t})W_j(z, \vec{t})^{-1} = O(|z|^{M'})$$

for some positive number M' as $z \to \infty$ in $\mathscr{S}_j$. Note that if we set

$$W_j(z, \vec{t})^{-1}W_k(z, \vec{t}) = C_{jk}(\vec{t}) \tag{A.5.2.13}$$

whenever $\mathscr{S}_j \cap \mathscr{S}_k \neq \varnothing$, then $\{A_0(z, \vec{t}), \{\mathscr{S}_j, \Phi_j\}, C_{jk}(\vec{t})\}$ is the Stokes phenomenon defined at $z = \infty$ by the system $\{\mathscr{S}_j, W_j(z, \vec{t});\ j = 1, \cdots, N\}$ of fundamental solutions of differential equation (A.5.2.1). If we choose the system $\{\Delta(z_j, r_j), W_j(z, \vec{t})\}$ of fundamental solutions of Definition A.5.2.1 in this way and if the connection matrices C_{jk} are independent of $\vec{t}$, then the matrices B_h will be rational in z even if (A.5.2.1) admits irregular singular points.

REMARK A.5.2.10.

(α) Isomonodromic differential equations of form (A.5.2.1) are extensively studied by T. Miwa [110], Jimbo–Miwa [73, 74], Jimbo–Miwa–Ueno [76], K. Ueno [46] and B. Malgrange [102]. [See also Lin-Sibuya [88, 89]].

(β) Steps I and II of Observation A.5.2.9 can be worked out under certain assumptions by utilizing the Hukuhara–Turrittin Theorem 6.8.1 together with suitable asymptotic analysis (cf. K. Ueno [146] and Y. Sibuya [29]).

OBSERVATION A.5.2.11. Let us look at the second-order linear differential equation

$$\frac{d^2y}{dz^2} + p_1(z,t)\frac{dy}{dz} + p_2(z,t)y = 0 \tag{A.5.2.14}$$

where $(z,t) \in \mathbf{C}^2$ and $p_j \in \mathscr{O}(\mathscr{D}\times\mathscr{D}(\rho,\tau))$ $(j = 1,2)$. Differential equation (A.5.2.14) is equivalent to differential equation (A.5.2.1) if we set

$$A(z,t) = \begin{bmatrix} 0 & 1 \\ -p_2(z,t) & -p_1(z,t) \end{bmatrix}. \tag{A.5.2.15}$$

Pfaffian system (A.5.2.3) and integrability condition (A.5.2.5) become

$$\begin{cases} \dfrac{\partial W}{\partial z} = A(z,t)W, \\ \dfrac{\partial W}{\partial t} = B(z,t)W \end{cases} \tag{A.5.2.16}$$

and

$$\frac{\partial B}{\partial z} - \frac{\partial A}{\partial t} = [A,B], \tag{A.5.2.17}$$

respectively. If we set

$$B = \begin{bmatrix} B_{11} & B_{12} \\ B_{21} & B_{22} \end{bmatrix},$$

condition (A.5.2.17) is equivalent to

$$\begin{cases} \dfrac{1}{2}\dfrac{\partial^3 B_{12}}{\partial z^3} - 2p\dfrac{\partial B_{12}}{\partial z} - \dfrac{\partial p}{\partial z}B_{12} + \dfrac{\partial p}{\partial t} = 0, \\[2ex] \dfrac{\partial B_{11}}{\partial z} = \dfrac{1}{2}\left\{\dfrac{\partial(p_1B_{12})}{\partial z} - \dfrac{\partial^2 B_{12}}{\partial z^2} - \dfrac{\partial p_1}{\partial t}\right\}, \\[2ex] B_{21} = \dfrac{\partial B_{11}}{\partial z} - p_2B_{12}, \\ B_{22} = \dfrac{\partial B_{12}}{\partial z} + B_{11} - p_1B_{12}, \end{cases} \tag{A.5.2.18}$$

where $p = \frac{1}{2}\frac{\partial p_1}{\partial z} + \frac{1}{4}p_1^2 - p_2$.

OBSERVATION A.5.2.12. Assume that

$$\begin{cases} p_1(z,t) = \dfrac{1-\theta_0}{z} + \dfrac{1-\theta_1}{z-1} + \dfrac{1-\theta}{z-t} - \dfrac{1}{z-\lambda}, \\ p_2(z,t) = \dfrac{\kappa}{z(z-1)} + \dfrac{\lambda(\lambda-1)\mu}{z(z-1)(z-\lambda)} - \dfrac{t(t-1)H}{z(z-1)(z-t)}, \end{cases} \tag{A.5.2.19}$$

where θ_0, θ_1, θ, and κ are independent of t, while λ, μ, and H are functions of t. The points at $z = 0, 1, t$, and ∞ are regular singular points. We also assume that the point at $z = \lambda$ is an apparent singular point. This requirement fixes the function H as follows:

$$H = \frac{1}{t(t-1)}[\lambda(\lambda-1)(\lambda-t)\mu^2 - \{\theta_0(\lambda-1)(\lambda-t) + \theta_1\lambda(\lambda-t) + (\theta-1)\lambda(\lambda-1)\}\mu + \kappa(\lambda-t)]. \tag{A.5.2.20}$$

THEOREM A.5.2.13 (K. Okamoto [117]). *Under the assumption given above, differential equation* (A.5.2.14) *is isomonodromic if and only if* (λ, μ) *satisfies the Hamiltonian system*

$$\frac{d\lambda}{dt} = \frac{\partial H}{\partial \mu}, \qquad \frac{d\mu}{dt} = -\frac{\partial H}{\partial \lambda}. \tag{A.5.2.21}$$

Furthermore system (A.5.2.21) *is equivalent to the Painlevé equation* VI:

$$\frac{d^2\lambda}{dt^2} = \frac{1}{2}\left(\frac{1}{\lambda} + \frac{1}{\lambda-1} + \frac{1}{\lambda-t}\right)\left(\frac{d\lambda}{dt}\right)^2 - \left(\frac{1}{t} + \frac{1}{t-1} + \frac{1}{\lambda-t}\right)\frac{d\lambda}{dt}$$

$$+ \frac{\lambda(\lambda-1)(\lambda-t)}{2t^2(t-1)^2}\left[\theta_\infty^2 - \theta_0^2\frac{t}{\lambda^2} + \theta_1^2\frac{t-1}{(\lambda-1)^2} + (1-\theta^2)\frac{t(t-1)}{(\lambda-t)^2}\right],$$

where $\theta_\infty^2 = (\theta_0 + \theta_1 + \theta - 1)^2 - 4\kappa$.

REMARK A.5.2.14.

(1) By coalescing singular points in suitable ways, we can derive similar results for other Painlevé equations (cf. K. Okamoto [115, 117]).
(2) General solutions of Painlevé equations do not have *movable* branch points. Solutions to Pfaffian system (A.5.2.9) also do not have movable branch points. In more general situations, the Painlevé properties of isomonodromic differential equations were extensively investigated by T. Miwa [109] and B. Malgrange [102].
(3) In order to derive the Painlevé properties of isomonodromic differential equations, we must choose parameters in suitable ways. Generally speaking, as we mentioned at the end of Chapter 4 (§4.5), it becomes necessary to introduce some apparent singular points if we consider a Riemann problem over the Riemann sphere. In the case of a single nth-order linear differential equation

$$\sum_{0 \leq h \leq n} a_h(z)y^{(h)} = 0$$

of Fuchsian type, the number of apparent singular points that should be introduced to solve a given Riemann problem is known in terms of the number of prescribed singular points and the order n of the differential equation provided that the prescribed monodromy data be irreducible (cf. M. Ohtsuki [119]). Let $\vec{t} = (t_1, \cdots, t_p)$ be the prescribed singular points and let $\vec{\lambda} = (\lambda_1, \cdots, \lambda_k)$ be the introduced apparent singular points. Then, in certain cases of second-order linear differential equations, conditions that the given differential equation be isomonodromic are given by means of a system of differential equations on $\vec{\lambda}$ as a function of $\vec{t}$. Furthermore, such a system on $\vec{\lambda}$ are equivalent to a Hamiltonian system which is in turn equivalent to a 2×2 Schlesinger system (A.5.2.9) (cf. K. Okamoto [116, 117] and H. Kimura–K. Okamoto [79, 80]). There are many interesting open questions in this territory (cf. T. Kimura [78]).

(4) W. Dekkers [54] showed that, in the case of 2×2 matrices, it is not necessary to introduce any apparent singular points to solve the Riemann problem on the Riemann sphere.

(5) K. Okamoto [112, 113, 118] studied isomonodromic differential equations also on a torus. Recently, an intensive study on isomonodromic differential equations on a closed Riemann surface has been done by K. Iwasaki [71]. [See also D. A. Hejhal [19]].

(6) Although the Painlevé equations I $\sim$ VI and their relations with isomonodromic differential equations had been known for a long time, B. M. McCoy–C. A. Tracy–T. T. Wu [108] and Jimbo–Miwa–Sato–Môri [75] revived strong interest in that territory.

§A.5.3. Isoformal deformations

In §A.5.2 we explained isomonodromic differential equations with irregular singular points (cf. Observation A.5.2.9). There we suggested that, in Steps I and II, we may use the Hukuhara–Turrittin transformation (6.8.2) (cf. (β) of remark A.5.2.10). However, to do this, we must show that such a transformation is analytic with respect to the auxiliary variable $\vec{t}$. Generally speaking this is not at all a trivial enterprise. As a matter of fact, we can not expect such a property of the Hukuhara–Turrittin transformation except for very restricted cases. The main concern is to find a decent sufficient condition such that the Hukuhara–Turrittin transformation is analytic in $\vec{t}$. So far the sufficient condition due to D. G. Babbitt and V. S. Varadarajan [37, 38, 41] has been the best. Their condition applies to the case when the standard form (6.8.3) of **Hukuhara–Turrittin** is independent of $\vec{t}$, even though the matrix A of the given differential equation depends on $\vec{t}$ (i.e., the case when the given differential equation is *isoformal*). By utilizing such a result, Babbitt–Varadarajan [39, 40] investigated the analytic structure of the moduli space of analytic linear differential equations with respect to the meromorphic classification (cf. the end of §A.3.2). [Their method is based on the theory of Lie algebra. Recently, R. Schäfke [131] indicated that the Babbitt–Varadarajan transformation could be constructed by means of a more traditional method.]

§A.5.4. Singular perturbations and the Gevrey asymptotics

As we indicated in §A.2.5 some results concerning the Gevrey asymptotics were obtained through studies of singular perturbations. However, so far there have been only a limited number of significant applications of the Gevrey asymptotics to singular perturbations. We must realize that many works of singular perturbations do not utilize complex analysis other than results concerning turning points. Nevertheless, we believe that the Gevrey asymptotics are useful whenever we need sharp results.

In this section, we shall explain how to derive the Gevrey property of formal solutions of a singularly perturbed differential equation:

$$\frac{d\vec{y}}{dx} = \lambda^{\sigma}\left\{\vec{f}(x,\lambda) + A(x,\lambda)\vec{y} + \sum_{|\vec{p}|\geq 2} \vec{f}_{\vec{p}}(x,\lambda)\,\vec{y}^{\vec{p}}\right\}, \qquad \text{(A.5.4.1)}$$

where

(i) σ is a positive integer; $\vec{y}$, $\vec{f}$, and $\vec{f}_{\vec{p}}$ are n-dimensional column vectors; A is an $n \times n$ matrix; $\vec{p} = (p_1, \cdots, p_n)$ is an n-tuple of nonnegative integers; $|\vec{p}| = \sum_{1\leq h\leq n} p_h$; $\vec{y}^{\vec{p}} = y_1^{p_1}\cdots y_n^{p_n}$, and y_j is the jth entry of the column vector $\vec{y}$,

(ii) the entries of $\vec{f}$, $\vec{f}_{\vec{p}}$ and A belong to $\mathcal{O}(\mathcal{D}(r,R))$, where r and R are positive numbers and

$$\mathcal{D}(r,R) = \{(x,\lambda)\ ;\ |x| < r,\ \ |\lambda| > R\};$$

these functions are also assumed to be bounded on $\mathcal{D}(r,R)$,

(iii) the series $\sum_{|\vec{p}|\geq 2} \vec{f}_{\vec{p}}(x,\lambda)\,\vec{y}^{\vec{p}}$ is convergent uniformly on the domain

$$\mathcal{D}(r,R) \times \{\vec{y}\ ;\ |y_j| < \rho \qquad j = 1, \cdots, n\},$$

where ρ is a positive number,

(iv) $\vec{f}(0,\infty) = 0$,

(v) $A(,0,\infty)$ is invertible.

Under the assumptions given above, we shall explain how to prove the following theorem.

THEOREM A.5.4.1 (Y. Sibuya [139]). *Differential equation* (A.5.4.1) *admits a unique formal solution*

$$\vec{y} = \sum_{m \geq 0} \vec{a}_m(x)\lambda^{-m}, \tag{A.5.4.2}$$

such that $\vec{a}_0(0) = 0$ *and the entries of* $\vec{a}_m$ *belong to* $\mathcal{O}(\mathcal{D}(r_1))$, *where* r_1 *is a suitable positive number and* $\mathcal{D}(r_1) = \{x;\ |x| < r_1\}$. *Furthermore, the* $\vec{a}_m$ *satisfy the inequalities*:

$$|\vec{a}_m(x)| \leq C_0(m!)^{\frac{1}{\sigma}}\, C_1^m \qquad (m = 0, 1, \cdots) \tag{A.5.4.3}$$

on $\mathcal{D}(r_1)$, *where* C_0 *and* C_1 *are suitable positive numbers.*

OBSERVATION A.5.4.2. To illustrate the situation, let us look at the scalar equation

$$\frac{dy}{dx} - \lambda\{y - f_0(x)\} = 0.$$

The unique formal solution of this equation is given by

$$y = \sum_{m \geq 0} \frac{d^m f_0}{dx^m}(x)\, \lambda^{-m}.$$

Suppose that $f_0 \in \mathcal{O}(\mathcal{D}(r_0))$, $|f_0(x)| \leq M$ on $\mathcal{D}(r_0)$ and $0 < r_1 < r_0$. Then

$$\left|\frac{d^m f_0}{dx^m}(x)\right| \leq \left(\frac{r_0 M}{(r_0 - r_1)^{m+1}}\right) m!$$

on $\mathcal{D}(r_1)$. This example also shows that the assertion of Theorem A.5.4.1 is generally false if we do not assume analyticity of the entries of $\vec{f}_0$, $\vec{f}_{\vec{p}}$, and A with respect to x.

OBSERVATION A.5.4.3. The existence and uniqueness of formal solution (A.5.4.2) can be verified easily. The main problem is to verify inequalities (A.5.4.3). We can derive *recurrence formulas* to compute the $\vec{a}_m$ successively. However, since such formulas involve not only the $\vec{a}_m$ but also their derivatives with respect to x, it is extremely difficult to verify (A.5.4.3) by means of *recurrence formulas*.

OBSERVATION A.5.4.4. For every direction θ in the λ-plane, there exist a sectorial domain $\mathcal{S}(\theta) = \{\lambda\ ;\ |\arg\lambda - \theta| < \alpha(\theta),\ |\lambda| > R(\theta)\}$ and a disc $\Delta(\theta) = \{x\ ;\ |x| < r(\theta)\}$ such that differential equation (A.5.4.1) has a solution $\vec{y} = \vec{\phi}(x, \lambda; \theta)$ satisfying the inequalities:

$$\left|\vec{\phi}(x, \lambda; \theta) - \sum_{m=0}^{N} \vec{a}_m(x)\lambda^{-m}\right| \leq K_N\, |\lambda|^{-N-1} \tag{A.5.4.4}$$

on $\Delta(\theta) \times \mathscr{S}(\theta)$ for every nonnegative integer N, where $\alpha(\theta)$, $R(\theta)$, $r(\theta)$ and the K_N are suitable positive numbers (cf. Y. Sibuya [133]).

OBSERVATION A.5.4.5. We can choose a finite number of $\theta_1, \cdots, \theta_p$ in such a way that p sectorial domains $\mathscr{S}_j = \{\lambda\, ;\, |\arg\lambda - \theta_j| < \alpha(\theta_j),\ |\lambda| > R_0\}$ $(j = 1, \cdots, p)$ form a good covering at $\lambda = \infty$ (cf §A.2.0), where R_0 is a positive number greater than $R(\theta_1), \cdots, R(\theta_p)$. Set $\mathscr{D} = \cap_{1\le j\le p}\Delta(\theta_j)$. Now by utilizing differential equation (A.5.4.1), we can derive the following estimates:

$$|\vec{\phi}(x,\lambda;\theta_j) - \vec{\phi}(x,\lambda;\theta_{j+1})| < \gamma \exp(-\mu|\lambda|^\sigma) \quad \text{on} \quad \mathscr{D}\times(\mathscr{S}_j \cap \mathscr{S}_{j+1}), \tag{A.5.4.5}$$

where γ and λ are suitable positive numbers. In fact by setting

$$\vec{\delta}_j(x,\lambda) = \vec{\phi}(x,\lambda;\theta_j) - \vec{\phi}(x,\lambda;\theta_{j+1}),$$

we derive

$$|\delta_j(x,\lambda)| \le 2K_N\,|\lambda|^{-N-1} \quad \text{on} \quad \mathscr{D}\times(\mathscr{S}_j \cap \mathscr{S}_{j+1}) \tag{A.5.4.6}$$

from (A.5.4.4) for every nonnegative integer N, and

$$\begin{aligned}\frac{d\vec{\delta}_j(x,\lambda)}{dx} &\\ = \lambda^\sigma \Big\{&A(x,\lambda)\vec{\delta}_j(x,\lambda) + \vec{F}(x,\lambda,\vec{\phi}(x,\lambda;\theta_{j+1}) + \vec{\delta}_j(x,\lambda)) \\ &-\vec{F}(x,\lambda,\vec{\phi}(x,\lambda;\theta_{j+1}))\Big\} \quad \text{on} \quad \mathscr{D}\times(\mathscr{S}_j \cap \mathscr{S}_{j+1})\end{aligned} \tag{A.5.4.7}$$

from (A.5.4.1), where $\vec{F} = \sum_{|\vec{p}|\ge 2} \vec{f}_{\vec{p}}(x,\lambda)\,\vec{y}^{\vec{p}}$. As we did in §A.2.4, we can write (A.5.4.7) in the form

$$\frac{d\vec{\delta}_j(x,\lambda)}{dx} = \lambda^\sigma\,\{A(x,\lambda) + B(x,\lambda)\}\vec{\delta}_j(x,\lambda), \tag{A.5.4.8}$$

where

$$B(x,\lambda) = \int_0^1 \vec{F}_{\vec{y}}(x,\lambda,\vec{\phi}(x,\lambda;\theta_{j+1}) + t\vec{\delta}_j(x,\lambda))dt.$$

Here $\vec{F}_{\vec{y}}$ denotes the Jacobian matrix of $\vec{F}$ with respect to $\vec{y}$. Note that $B(0,\infty) = 0$. We can derive (A.5.4.5) from (A.5.4.6) and (A.5.4.8).

OBSERVATION A.5.4.6. To complete the proof of Theorem A.5.4.1, it suffices to generalize Lemma A.2.0.5 to the case when $\phi_1, \cdots, \phi_N$ depend on auxiliary parameters. In doing this, we can also improve estimates (A.5.4.4) as

$$|\vec{\phi}(x,\lambda;\theta) - \sum_{m=0}^{N} \vec{a}_m(x)\lambda^{-m}| \le K(\theta)\,((N+1)!)^{\frac{1}{\sigma}}L(\theta)^{N+1}|\lambda|^{-N-1} \tag{A.5.4.9}$$

for some positive numbers $K(\theta)$ and $L(\theta)$.

References

I. Books and lecture notes:

1. P. Deligne, *Equations différentielles à points singuliers réguliers*, Lecture Notes in Math., vol. 163, Springer-Verlag, Berlin and New York, 1970.

2. R. Gunning and H. Rossi, *Analytic functions of several complex variables*, Prentice-Hall, 1965.

3. R. Gunning, *Lectures on Riemann surfaces*, Princeton, 1966.

4. M. Hukuhara, *Ordinary differential equations* (in Japanese), Iwanami-Zensho **116**, Iwanami-Shoten, 1950.

5. Hurwitz–Courant, *Funktionentheorie* (*Vierte Auflage*), Springer-Verlag, 1964.

6. N. Iwahori *Theory of Lie Groups* I (in Japanese), Iwanami-Koza Gendai-Oyosugaku **A.3.I.**, Iwanami-Shoten, 1957.

7. T. Kimura *Analytic theory of linear ordinary differential equations on a Riemann surface—Riemann's Problem*, Lecture Notes, University of Minnesota, 1973.

8. E. R. Kolchin, *Differential algebra and algebraic groups*, Academic Press, 1973.

9. Y. Sibuya, *Global theory of a second order linear ordinary differential equation with a polynomial coefficient*, Math. Studies **18**, North-Holland, 1975.

10. N. Steenrod, *The topology of fibre bundles*, Princeton, 1951.

11. W. Wasow, *Asymptotic expansions for ordinary differential equations*, Dover, 1976.

II. Papers:

12. G. D. Birkhoff, *Singular points of ordinary linear differential equations*, Trans. Amer. Math. Soc. **10** (1909), 436–470.

13. ——, *A theorem on matrices of analytic functions*, Math. Ann. **74** (1913), 122–133.

14. ——, *The generalized Riemann problem for linear differential equations and the allied problems for linear difference and q-difference equations*, Proc. Amer. Acad. Arts and Sci. **49** (1913), 521–568.

15. D. Drasin, *The inverse problem of the Nevanlinna theory*, Acta Math. **138** (1977), 83–151.

16. R. Gérard and A. H. M. Levelt, *Invariants mesurant l'irrégularité en un point singulier des systèmes d'équations différentielles linéaires*, Ann. Inst. Fourier **23** (1973), 157–195.

17. H. Grauert, *Approximationssätze für holomorphe Funktionen mit Werten in komplexen Räumen*, Math. Ann. **133** (1957), 139–159.

18. ——, *Holomorphe Funktionen mit Werten in komplexen Lieschen Gruppen*, Math. Ann. **133** (1957), 450–472.

19. D. A. Hejhal, *Monodromy groups and linear polymorphic functions*, Acta Math., **135** (1975), 1–55.

20. T. Kimura, *On the Riemann problem on Riemann surfaces*, Japan–U.S. Seminar on Ordinary Differential and Functional Equations, Lecture Notes in Math., vol. 243, Springer-Verlag, Berlin and New York, 1971, pp. 219–228.

21. W. B. Jurkat and D. A. Lutz, *On the order of solutions of analytic linear differential equations*, Proc. London Math. Soc. **22** (1971), 465–482.

22. D.A. Lutz, *Some characterizations of systems of linear differential equations having regular singular solutions*, Trans. Amer. Math. Soc. **126** (1967), 427–441.

23. ——, *Asymptotic behavior of solutions of linear systems of ordinary differential equations near an irregular singular point*, Amer. J. Math. **91** (1969), 95–105.

24. ——, *Perturbations of matrix differential equations in the neighborhood of a singular point*, Funkcial. Ekvac. **13** (1970), 97–107.

25. L. Markus, *Exponentials in algebraic matrix groups*, Adv. in Math. **11** (1973), 351–367.

26. J. Moser, *The order of a singularity in Fuch's Theory*, Math. Z. **72** (1960), 379–398.

27. H. Röhrl, *Das Riemann–Hilbertsche Problem der Theorie der Linearen Differentialgleichungen*, Math. Ann. **133** (1957), 1–25.

28. Y. Sibuya, *A theorem of G. D. Birkhoff concerning matrices and its applications* (I) *and* (II) (in Japanese), Kansuhoteisiki **16** (1963), no. 1, 3–19; no.2, 3–18.

29. ——, *Simplification of a system of linear ordinary differential equations about a singular point*, Funkcial. Ekvac. **4** (1962), 29–56.

30. ——, *Uniform simplification in a full neighborhood of a transition point*, Mem. Amer. Math. Soc. **149**, 1974.

31. ——, *Subdominant solutions admitting a prescribed Stokes phenomenon*, International Conf. on Diff. Equations, Academic Press, 1975, pp. 709–738.

32. K. Takahasi, *Analytic reduction of a system of differential equations to a normal form*, Comment. Math. Univ. St. Paul **22** (1973/74), 57–94.

33. W. J. Trjitzinsky, *Analytic theory of linear differential equations*, Acta Math. **62** (1933), 167–226.

34. H. L. Turrittin, *Reduction of ordinary differential equations to the Birkhoff canonical form*, Trans. Amer. Math. Soc. **107** (1963), 485–507.

III. References added to the translation:

35. R. C. Ackerberg and R. E. O'Malley, *Boundary layer problems exhibiting resonance*, Stud. Appl. Math. **49** (1970), 277–295.

36. M. F. Atiyah and I. G. Macdonald, *Introduction to commutative algebra*, Addison-Wesley, 1969.

37. D. G. Babbitt and V. S. Varadarajan, *Formal reduction of meromorphic differential equations: a group theoretic view*, Pacific J. Math. **108** (1983), 1–80.

38. ——, *Deformations of nilpotent matrices over rings and reduction of analytic families of meromorphic differential equations*, Mem. Amer. Math. Soc. **55** (325) (1985).

39. ——, *Local moduli for meromorphic differential equations*, Bull. Amer. Math. Soc. **12** (1985), 95–98.

40. ——, *Local moduli for meromorphic differential equations*, Astérisque, **169/170** (1989), 1–217.

41. ——, *Some remarks on the asymptotic existence theorem for meromorphic differential equations*, J. Fac. Sci. Univ. Tokyo, Sect. IA Math. **36** (1989), 247–262.

42. F. Baldassarri, *Differential modules and singular points of p-adic differential equations*, Adv. in Math. **44** (1982), 155–179.

43. W. Balser, *Zum Einzigkeitssatz in der Invariantentheorie meromorpher Differentialgleichungen*, J. für Math. **318** (1980), 51–82.

44. ——, *Solutions of first level of meromorphic differential equations*, Proc. Edinburgh Math. Soc. **25** (1982), 183–207.

45. ——, *Growth estimates for the coefficients of generalized formal solutions, and representation of solutions using Laplace integrals and factorial series*, Hiroshima Math. J. **12** (1982), 11–42.

46. ——, *Convergent power series expansions for the Birkhoff invariants of meromorphic differential equations: Part* I. *Definition of the coefficient functions*, Yokohama Math. J. **32** (1984), 15–29; *Part* II. *A closer study of the coefficients* **33** (1985), 5–19.

47. ——, *A constructive existence proof for first level formal solutions of meromorphic differential equations*, Hiroshima Math. J. **15** (1985), 411–427.

48. W. Balser, W. Jurkat and D. A. Lutz, *A general theory of invariants for meromorphic differential equations: Part* I. *Formal invariants*, Funkcial. Ekvac. **22** (1979), 197–221; *Part* II. *Proper invariants*, **22** (1979), 257–283; *Part* III. *Applications*, Houston J. of Math. **6** (1980), 149–189.

49. ——, *Birkhoff invariants and Stokes' multipliers for meromorphic linear differential equations*, J. Math. Anal. Appl. **71** (1979), 48–94.

50. ——, *Characterization of first level formal solutions by means of the growth of their coefficients*, J. Differential Equations **51** (1984) 48–77.

51. D. Bertrand, *Travaux récents sur les points singuliers des équations différentielles linéaires*, Sém. Bourbaki 1978/79, Exp. 525–542, Lecture Notes in Math., vol. 770, Springer-Verlag, Berlin and New York, 1980, pp. 228–243.

52. H. Charrière, *Triangulation formelle de certains systèmes de Pfaff complètement intégrables et application à l'étude C^∞ des systèmes linéaires*, Ann. Scuola Norm. Sup. Pisa Cl. Sci. (4) **7** (1980), 625–714.

53. H. Charrière and R. Gérard, *Formal reduction of integrable linear connections having a certain kind of irregualr singularities*, Analysis **1** (1981), 85–115.

54. W. Dekkers, *The matrix of a connection having reglar singularities on a vector bundle of rank* 2 *on* $\mathbf{P}^1(\mathbf{C})$, Equations Différentielles et Systèmes de Pfaff dans le Champ Complexe, Lecture Notes in Math., vol. 712, Springer-Verlag, Berlin and New York, 1979, pp. 33–43.

55. P. Deligne, Letters to Malgrange, 1977–1978.

56. A. R. P. van den Essen and A. H. M. Levelt, *Irregular singularities in several variables*, Mem. Amer. Math. Soc. **270**, 1982.

57. L. Fuchs, *Infinite abelian groups*, I, Academic Press, 1970.

58. S. Gacsályi, *On algebraically closed abelian groups*, Publ. Math. Debrecen **2** (1952), 292-296.

59. M. Gevrey, *Sur la nature analytique des solutions des équations aux dérivées partielles*, Ann. Sci. École Norm. Sup. (3) **25** (1918), 129–190.

60. H. Gingold, *A necessary condition for a power series to be a formal solution of a singular linear differential equation of order k*, J. Math. Anal. Appl. **52** (1975) 546-552.

61. R. Goldement, *Topologie algébrique et théorie des faisceaux*, Hermann, 1964.

62. Y. Haraoka, *Theorems of Sibuya–Malgrange type for Gevrey functions of several variables*, Funkcial. Ekvac. **32** (1989), 365-388.

63. W. A. Harris, Y. Sibuya and L. Weinberg, *Holomorphic solutions of linear differential systems at singular points*, Arch. Rational Mech. Anal. **35** (1969), 245–248.

64. W. A. Harris and Y. Sibuya, *The reciprocals of solutions of linear ordinary differential equations*, Adv. in Math. **58** (1985), 119–132.

65. ——, *The* n*th roots of solutions of linear ordinary differential equations*, Proc. Amer. Math. Soc. **97** (1986), 207–211.

66. P. Hartman, *Ordinary differential equations*, John Wiley, 1964.

67. M. Hukuhara, *Sur les points singuliers des équations différentielles* linéaires, II, J. Fac. Sci. Hokkaido Univ. **5** (1937), 123–166; III, Mem. Fac. Sci. Kyusyu Univ., Ser. A, **2** (1942), 125–137.

68. M. Hukuhara and M. Iwano, *Étude de la convergence des solutions formelles d'un système différentielles ordinaires linéaires*, Funkcial. Ekvac. **2** (1959), 1–18.

69. M. Iwano, *Sur les points singuliers d'une équation différentielle ordinaire linéaire du* $n^{ième}$ *ordre*, J. Fac. Sci. Univ. Tokyo, Sect. IA Math. **7** (1956), 343–351.

70. ——, *Bounded solutions and stable domains of nonlinear ordinary differential equations*, Analytic Theory of Differential Equations, Lecture Notes in Math., vol. 183, Springer-Verlag, Berlin and New York, 1971, pp. 59–127.

71. K. Iwasaki, *Structure of the moduli space of SL-operators on a Riemann surface and the monodromy preserving deformation*, Ordinary Differential Equations in the Complex Domain and Special Functions, Math. Forschungsinstitut Oberwolfach, 1989, p. 6.

72. N. Jacobson, *Lectures in Abstract Algebra*, vol. I, van Nostrand, 1951.

73. M. Jimbo and T. Miwa, *Deformation of linear ordinary differential equations*: I, Proc. Japan Acad. Ser. A Math. Sci. **56A** (1980), 143–148; II, **56A** (1980), 149–153; III, **56A** (1980), 269–274.

74. ——, *Monodromy preserving deformation of linear ordinary differential equations with rational coefficients* II, Phys. **2D** (1981), 407–448.

75. M. Jimbo, T. Miwa, M. Sato and Y. Môri, *Holonomic quantum fields–The unanticipated link between deformation theory of differential equations and quantum fields*, Mathematical Problem in Theoretical Physics, Lectures Notes in Phys., vol. 116, Springer–Verlag, Berlin and New York, 1980, pp. 123–154.

76. M. Jimbo, T. Miwa and K. Ueno, *Monodromy preserving deformation of linear ordinary differential equations with rational coefficients:* I *general theory and* τ*-function*, Phys. **2D** (1981), 306–352.

77. W. Jurkat, *Meromorphe Differentialgleichungen*, Lecture Notes in Math., vol. 637, Springer–Verlag, Berlin and New York, 1978.

78. T. Kimura, *On the isomonodromic deformation for linear ordinary differential equations of the second order*: I, Proc. Japan Acad. **57A** (1981), 285-290; II, **58A** (1982), 294–297.

79. H. Kimura and K. Okamoto, *On the isomonodromic deformation of linear ordinary differential equations of higher order*, Funkcial. Ekvac. **26** (1983), 37–50.

80. ——, *On the polynomial Hamiltonian structure of the Garnier systems*, J. Math. Pures Appl. (9) **63** (1984), 129–146.

81. H. Komatsu, *On the index of ordinary differential operators*, J. Fac. Sci. Univ. Tokyo Sect. IA Math. **18** (1971), 379–398.

82. *Ultradistributions*, I. *Structure theorems and a characterization*, J. Fac. Sci. Univ. Tokyo, Sect. IA Math. **20** (1973), 25–105.

83. ——, *Ultradistribution, hyperfunctions and linear differential equations*, Astérisque **2/3** (1973), 252–271.

84. N. Kopell, *A geometric approach to boundary layer problems exhibiting resonance*, SIAM J. Appl. Math. **37** (1979), 436–458.

85. ——, *The singularly perturbed turning-point problem: a geometric approach*, Singular Perturbations and Asymptotics, Academic Press, 1980, pp. 173–190.

86. C.-H. Lin, *Phragmén–Lindelöf theorem in a cohomological form*, Proc. Amer. Math. Soc. **89** (1983), 589-597.

87. ——, *The sufficiency of the Matkowsky condition in the problem of resonance*, Trans. Amer. Math. Soc. **278** (1983), 647–670.

88. C.-H. Lin and Y. Sibuya, *Some applications of isomonodromic deformations to the study of Stokes multipliers (a summary)*, Differential Equations and Applications, vol. II, Ohio Univ. Press, 1988, pp. 146–154.

89. ——, *Some applications of isomonodromic deformations to the study of Stokes multipliers*, J. Fac. Sci. Univ. Tokyo Sect. IA Math. **36** (1989), 649–663.

90. F. Lettenmeyer, *Über die an einer Unbestimmtheitsstelle regulären Lösungen eines Systemes homogener linearen Differentialgleichungen*, S.-B. Bayer. Akad. Wiss. München Math.-nat. Abt.(1926), 287–307.

91. A. H. M. Levelt, *Jordan decomposition for a class of singular differentiual operators*, Ark. Mat. **13** (1975), 1–27.

92. K. Mahler, *On formal power series as integrals of algebraic differential equations*, Lincei Rend. Sc. fistran. mat. e nat. **50** (1971), 70–89.

93. ——, *Lectures on Transcendental Numbers*, Lecture Notes in Math., vol. 546, Springer–Verlag, Berlin and New York, 1980.

94. E. Maillet, *Sur les séries divergentes et les équations différentielles*, Ann. École. Norm. Sup. (4) **20** (1903), 487–517.

95. H. Majima, *Analogues of Cartan's decomposition theorems in asymptotic analysis*, Funkcial. Ekvac. **26** (1983), 131–154.

96. ——, *Asymptotic Analysis for Integrable Connections with Irregular Singular Points*, Lecture Notes in Math., vol. 1075, Springer–Verlag, Berlin and New York, 1984.

97. ——, *Vanishing theorems in asymptotic analysis and their applications to differential equations* (in Japanese), Sûgaku **37** (1985), 225–244.

98. B. Malgrange, *Sur les points singuliers des équations différentielles*, Enseign. Math. (2) **20** (1974), 147–176.

99. ——, *Remarques sur les équations différentielles à points singuliers irréguliers*, Équations Différentielles et Systèmes de Pfaff dans le Champ Complexe, Lecture Notes in Math., vol. 712, Springer-Verlag, Berlin and New York, 1979, pp. 77–86.

100. ——, *Sur la réduction formelle des équations différentielles à singularités irrégulières*, Grenoble, 1979.

101. ——, *La classification des connexions irrégulières à une variable*, Sém. Ec. Norm. Sup. 1979–1982, Progr. Math., vol. 37, Birkhäuser, 1983, pp. 381–400.

102. ——, *Sur les déformations isomonodromiques:* I. *Singularités regulières, and* II. *Singularités irréguilères*, Progr. Math., vol. 37, Birkhäuser, 1983, pp. 401–438.

103. ——, *Sur le théorème de Maillet*, Asymptotic Anal. **2** (1989), 1–4.

104. I. J. Manin, *Moduli fuchsiani*, Ann. Scuola Norm. Sup. Pisa Cl. Sci. (4) **19** (1965), 113–126.

105. J. Martinet and J.-P. Ramis, *Problèmes de modules pour des équations différentielles nonlinéaires du premier ordre*, Publ. I.H.E.S. **55** (1982), 63-164.

106. ——, *Classification analytique des équations différentielles nonlinéaires résonnantes du premier order*, Ann. Sci. École Norm. Sup. (4) **16** (1983), 571–621.

107. B. J. Matkowsky, *On boundary layer problems exhibiting resonance*, SIAM Rev. **17** (1975), 82–100.

108. B. M. McCoy, C. A. Tracy and T. T. Wu, *Painlevé functions of the third kind*, J. Math. Phys. **18** (1977), 1058–1092.

109. T. Miwa, *Painlevé property of monodromy preserving deformation equations and the analyticity of* τ *functions*, Publ. RIMS, Kyoto Univ. **17** (1981), 703–721.

110. ——, *Deformation of linear ordinary differential equations*: IV, Proc. Japan Acad. Ser. A Math. Sci. **56A** (1980), 301–305.

111. F. Nevanlinna, *Zur theorie der asymptotischen potenzreihen*, Ann. Acad. Sci. Fenn. Ser. A I Math. Dissertationes **12** (1919).

112. K. Okamoto, *On Fuchs's problem on a torus*, I, Funkcial. Ekvac. **14** (1971), 137–152.

113. ——, *Sur le problème de Fuchs sur un tore*, II, J. Fac. Sci. Univ. Tokyo Sect. IA Math. **24** (1977), 357–372.

114. ——, *Déformation d'une équation différentielle linéaire avec une singularité irrégulière sur un tore*, J. Fac. Sci. Univ. Tokyo Sect. IA Math. **26** (1979), 501–518.

115. ——, *On the τ-function of the Painlevé equations*, Phys. **2D** (1981), 525-535.

116. ——, *Introduction to the Painlevé equations* (in Japanese), Sophia Kokyuroku in Math. **19**, Sophia Univ., 1985.

117. ——, *Isomonodromic deformation and Painlvé equations, and the Garnier system*, J. Fac. Sci. Univ. Tokyo Sect. IA Math. **33** (1986), 575–618.

118. ——, *The Hamiltonian structure derived from the holonomic deformation of the linear ordinary differential equations on an elliptic curve*, Sci. Papers College Arts and Sci. Univ. Tokyo, vol. 37, (1987), pp. 1–11.

119. M. Ohtsuki, *On the number of apparent singularities of a linear differential equation*, Tokyo J. Math. **5** (1982), 23–29.

120. F. Pham, *Singularités des systèmes différentiels de Gauss–Manin*, Progr. Math., vol. 2, Birkhäuser, 1979.

121. O. Perron, *Über diejenigen Integrale linearer Differentialgleichungen, welche sich an einer Unbestimmtheitsstelle bestimmt verhalten*, Math. Ann. **70** (1911), 1–32.

122. H. Poincaré, *Les méthodes nouvelles de la mécanique céleste*, vol. II, Gauthier-Villars, Paris, 1893.

123. J.-P. Ramis, *Dévissage Gevrey*, Astérisque **59/60** (1978), 173–204.

124. ——, *Les séries k-sommables et leurs applications*, Complex Analysis, Microlocal Calculus and Relativistic Quantum Theory, Lecture Notes in Phys., vol. 126, Springer–Verlag, Berlin and New York, 1980, pp. 178–199.

125. ——, *Théorèmes d'indices Gevrey pour les équations différentielles ordinaires*, Mem. Amer. Math. Soc. **296** (1984).

126. ——, *Filtration Gevrey sur le groupe de Picard Vessiot d'un équation différentielle irrégulière*, Instituto de Matemática Pura e Aplicada, Informes de Mathemática, Série A-045/85, Rio de Janeiro, 1985.

127. ——, *Phénomène de Stokes et resommation*, C. R. Acad. Sci. Paris **301** (1985), 99–102.

128. ——, *Phénomène de Stokes et filtration Gevrey sur le groupe de Picard–Vessiot*, C. R. Acad. Sci. Paris **301** (1985), 165–167.

129. J.-P. Ramis and Y. Sibuya, *Hukuhara domains and fundamental existence and uniqueness theorems for asymptotic solutions of Gevrey type*, Asymptotic. Anal. **2** (1989), 39–94.

130. P. Robba, *Lemmes de Hensel pour les opérateurs différentiels*, Application à la réduction formelle des équations différentielles, Paris, 1979.

131. R. Schäfke, *On formal fundamental solutions of irregular singular differential equations depending upon parameters*, Ordinary Differential Equations in the Complex Domain and Special Functions, Math. Forschungsinstitut Oberwolfach, 1989, pp. 9–10.

132. L. Schlesinger, *Über eine Klasse von Differentialsystemen beliebiger Ordnung mit festen kritischen Punkten*, J. für Math. **141** (1912), 96–145.

133. Y. Sibuya, *Sur réduction analytique d'un système d'équations différentielles ordinaries linéaires contenant un paramètre*, J. Fac. Sci. Univ. Tokyo Sect. IA Math. **7** (1958), 527–540.

134. ——, *Stokes phenomena*, Bull. Amer. Math. Soc. **83** (1977), 1075–1977.

135. ——, *Differential modules and theorem of Hukuhara–Turrittin*, Proc. of International Conf. on Nonlinear Phenomena in Math. Sci., Academic Press, 1982, pp. 927–937.

136. ——, *A theorem concerning uniform simplification at a transition point and the problem of resonance*, SIAM J. Math. Anal. **12** (1981), 653–668.

137. ——, *Some results concerning injectivity of D-modules*, Funkcial. Ekvac. **29** (1986), 91–97; II, Funkcial. Ekvac. **33** (1990), 269–275.

138. ——, *Analytizations of D(A)-modules*, Analysis **7** (1987), 33–45.

139. ——, *Gevrey property of formal solutions in a parameter*, Proc. of International Symposium on Asymp. and Comput. Anal., Winnipeg, Canada, 1989, (to appear).

140. ——, *A remark on the Gevrey property of formal solutions*, Preliminary Report, Special Session on Diff. Equations, Amer. Math. Soc. 852d Meeting, Muncie, Indiana, 1989.

141. Y. Sibuya and S. Sperber, *Some new results on power series solutions of algebraic differential equations*, Proc. of Advanced Sem. on Singular Perturbations and Asymptotics, Academic Press, 1980, pp. 379–404.

142. ——, *Power Series Solutions of Algebraic Differential Equations*, Lecture Notes, IRMA, Univ. de Strasbourg, 1982.

143. M. Singer, *Algebraic relations among solutions of linear differential equations*, Trans. Amer. Math. Soc. **295** (1986), 753–763.

144. S. Sperber, *On solutions of differential equations which satisfy certain algebraic relations*, Pacific J. Math. **124** (1986), 249–256.

145. W. Sternberg, *Über die asymptotische Integration von Differentialgleichungen*, Math. Ann. **81** (1920), 119–186.

146. K. Ueno, *Monodromy preserving deformation of linear differential equations with irregular singlar points*, Proc. Japan Acad. Ser. A Math. Sci. **56A** (1980), 97–102.

147. H. L. Turrittin, *Convergent solutions of ordinary linear homogeneous differential equations in the neighborhood of an irregular singular point*, Acta Math. **93** (1955), 27–66.

148. G. N. Watson, *A theory of asymptotic series*, Philos. Trans. Roy. Soc. London Ser. A **211** (1911), 279–313.

149. T. Yagami, *On linear Pfaffian systems with Gevrey coefficients*, Thesis for Master of Science, Univ. of Tokyo, 1983.

Index